Lecture Notes in Computer Science 7736

Commenced Publication in 1973
Founding and Former Series Editors:
Gerhard Goos, Juris Hartmanis, and Jan van Leeuwer

James F. Peters   Andrzej Skowron
Sheela Ramanna   Zbigniew Suraj
Xin Wang (Eds.)

Transactions on
Rough Sets XVI

 Springer

Editors-in-Chief

James F. Peters
University of Manitoba, Winnipeg, MB, Canada
E-mail: jfpeters@ee.umanitoba.ca

Andrzej Skowron
University of Warsaw, Poland
E-mail: skowron@mimuw.edu.pl

Guest Editors

Sheela Ramanna
University of Winnipeg, MB, Canada
E-mail: s.ramanna@uwinnipeg.ca

Zbigniew Suraj
University of Rzeszów, Poland
E-mail: z.suraj@univ.rzeszow.pl

Xin Wang
University of Calgary, AB, Canada
E-mail: xcwang@ucalgary.ca

ISSN 0302-9743 (LNCS)   e-ISSN 1611-3349 (LNCS)
ISSN 1861-2059 (TRS)   e-ISSN 1861-2067 (TRS)
ISBN 978-3-642-36504-1   e-ISBN 978-3-642-36505-8
DOI 10.1007/978-3-642-36505-8
Springer Heidelberg Dordrecht London New York

CR Subject Classification (1998): I.5.1-3, I.2.3-4, I.2.6, F.4.1, G.1.2, I.4.1

Typesetting: Camera-ready by author, data conversion by Scientific Publishing Services, Chennai, India

Printed on acid-free paper

Springer is part of Springer Science+Business Media (www.springer.com)

Preface

Volume XVI of the *Transactions on Rough Sets* (TRS) includes extensions of papers from a very successful conference, Rough Sets and Knowledge Technology (RSKT 2011), held in Banff, Canada, in October 2011 and published in the *Lecture Notes in Artificial Intelligence* series as volume 6954. The papers in this volume were accepted after two rounds of reviewing by at least two reviewers for each paper. This volume also includes a long paper based on a Ph.D. thesis. This volume offers a continuation of a number of research streams that have grown out of the seminal work by Zdzisław Pawlak[1] during the first decade of the 21st century.

The research streams represented in the papers cover both theory and applications of rough, fuzzy, and near sets. The following topics are discussed: generalized probabilistic approximations defined using both rough set theory and probability theory, an approximation framework based on partial covering of the universe, theoretical basis for higher order granulations using rough and near sets, unified mathematical definition of different forms of reducts in rough sets, investigation of topological structures of rough fuzzy sets, rough set based c-means clustering using boundary elements, application of an anisotropic wavelet-based nearness measure in classifying arthritic hand-finger movement images, semantic clustering of scientific articles related to rough set theory, and application of maximal clique enumeration in computing near neighbourhoods.

The editors of this special issue would like to express their gratitude to the authors of all submitted papers. Special thanks are due to the following reviewers: Jan Bazan, Jerzy Grzymała-Busse, Davide Ciucci, Christopher Henry, Andrzej Janusz, Pawan Lingras, Mikhail Moshkov, Leszek Puzio, Dominik Ślęzak, Marcin Szczuka, Surabhi Tiwari, Jarosław Stepaniuk, Wojciech Rząsa, Marcin Wolski, Wei-Zhi Wu, and Wojciech Ziarko.

The editors and authors of this volume extend their gratitude to Alfred Hofmann and the LNCS staff at Springer for their support in making this volume of the TRS possible.

The Editors-in-Chief were supported by the research grant 2011/01/D/ST6/06981 from the Polish National Science Centre; grant SP/I/1/77065/10 from the National Centre for Research and Development (NCBiR) in frame of the Strategic Scientific Research and Experimental Development Program:

---

[1] See, *e.g.*, Pawlak, Z., A Treatise on Rough Sets, *Transactions on Rough Sets* IV, (2006), 1–17. See, also, Pawlak, Z., Skowron, A.: Rudiments of rough sets, *Information Sciences* 177 (2007) 3–27; Pawlak, Z., Skowron, A.: Rough sets: Some extensions, *Information Sciences* 177 (2007) 28–40; Pawlak, Z., Skowron, A.: Rough sets and Boolean reasoning, *Information Sciences* 177 (2007) 41–73.

"Interdisciplinary System for Interactive Scientific and Scientific-Technical Information"; the individual research grant by the program Homing Plus, edition 3/2011, from the Foundation for Polish Science; the Natural Sciences and Engineering Research Council of Canada (NSERC); research grant 185986; the Canadian Network of Excellence (NCE); and the Canadian Arthritis Network (CAN) grant SRI-BIO-05.

December 2012

Sheela Ramanna
Zbigniew Suraj
Xin Wang
James F. Peters
Andrzej Skowron

LNCS Transactions on Rough Sets

The *Transactions on Rough Sets* series has as its principal aim the fostering of professional exchanges between scientists and practitioners who are interested in the foundations and applications of rough sets. Topics include foundations and applications of rough sets as well as foundations and applications of hybrid methods combining rough sets with other approaches important for the development of intelligent systems. The journal includes high-quality research articles accepted for publication on the basis of thorough peer reviews. Dissertations and monographs up to 250 pages that include new research results can also be considered as regular papers. Extended and revised versions of selected papers from conferences can also be included in regular or special issues of the journal.

Editors-in-Chief: James F. Peters, Andrzej Skowron
Managing Editor: Sheela Ramanna
Technical Editor: Marcin Szczuka

Editorial Board

Table of Contents

Table of Contents

Generalized Probabilistic Approximations

Jerzy W. Grzymala-Busse[1,2]

[1] Department of Electrical Engineering and Computer Science,
University of Kansas, Lawrence, KS 66045, USA
[2] Institute of Computer Science, Polish Academy of Sciences,
01–237 Warsaw, Poland
jerzy@ku.edu

Abstract. We study generalized probabilistic approximations, defined using both rough set theory and probability theory. The main objective is to study, for a given subset of the universe U, all such probabilistic approximations, i.e., for all parameter values. For an approximation space (U, R), where R is an equivalence relation, there is only one type of such probabilistic approximations. For an approximation space (U, R), where R is an arbitrary binary relation, three types of probabilistic approximations are introduced in this paper: singleton, subset and concept. We show that for a given concept the number of probabilistic approximations of given type is not greater than the cardinality of U. Additionally, we show that singleton probabilistic approximations are not useful for data mining, since such approximations, in general, are not even locally definable.

Keywords: Probabilistic approximations, parameterized approximations, generalization of probabilistic approximations, singleton, subset and concept probabilistic approximations.

1 Introduction

The entire rough set theory is based on ideas of the lower and upper approximations. Complete data sets, presented as decision tables, are well described by an indiscernibility relation, yet another fundamental idea of rough set theory. The indiscernibility relation is an equivalence relation. Standard lower and upper approximations were extended, using probability theory, to probabilistic (parameterized) approximations. Such approximations were studied, among others, in [1,2,3,4,5,6,7]. The parameter, called a threshold and associated with the probabilistic approximation, may be interpreted as a probability. The threshold is, in general, a real number.

So far probabilistic approximations were usually defined as lower and upper approximations. As it was observed in [8], the only difference between so called lower and upper probabilistic approximations is in the choice of the value of the threshold.

Due to the fact that we explore the set of all probabilistic approximations of a given type, the distinction between lower and upper approximations is blurred.

J.F. Peters et al. (Eds.): Transactions on Rough Sets XVI, LNCS 7736, pp. 1–16, 2013.

Therefore, we will define only one kind of probabilistic approximations for an approximation space (U, R), where U is a finite set and R is an equivalence relation on U.

This paper, for a given decision table and a subset of the universe explores the set of all probabilistic approximations. It is shown that the number of all distinct probabilistic approximations is quite limited.

Additionally, this paper generalizes the usual three types of approximations: singleton, subset and concept, used for approximation spaces (U, R), where R is an arbitrary binary relation. Similarly as for singleton standard approximations, a singleton probabilistic approximation of a subset X of the universe U is, in general, not definable. There are two types of definability, local and global. If the set X is globally definable, it is locally definable, the converse is, in general, not true. Sets that is the singleton probabilistic approximation of X are, in general, not even locally definable. The idea of probabilistic approximations is applied to incomplete data sets. It is well known [9,10] that incomplete data sets, i.e., data sets with missing attribute values, are described by characteristic relations, which are reflexive but, in general, neither symmetric nor transitive.

A preliminary version of this paper was prepared for the 6-th International Conference on Rough Sets and Knowledge Technology, Banff, Canada, October 9–12, 2011 [11].

2   Equivalence Relations

In this section we will discuss data sets without missing attribute values, i.e., complete. Complete data sets are describable by equivalence relations. Then we will discuss all probabilistic partitions defined over a space approximation (U, R), where U is a finite set and R is an equivalence relation.

2.1   Complete Data

Many real-life data sets have conflicting cases, characterized by identical values for all attributes but belonging to different concepts (classes). Data sets with conflicting cases are called inconsistent. An example of the inconsistent data set is presented in Table 1. The data set presented in Table 1 is inconsistent since it contains conflicting cases: the cases 2 and 4 are in conflict with the case 3 and the case 6 is in conflict with case 8.

In Table 1, the set A of all attributes consists of three variables *Temperature*, *Headache* and *Cough*. A *concept* is a set of all cases with the same decision value. There are two concepts in Table 1, the first one contains cases 1, 2, 4 and 6 and is characterized by the decision value *no* of decision *Flu*. The other concept contains cases 3, 5, 7 and 8 and is characterized by the decision value *yes*.

The fact that an attribute a has the value v for the case x will be denoted by $a(x) = v$. The set of all cases will be denoted by U. In Table 1, $U = \{1, 2, 3, 4, 5, 6, 7, 8\}$.

Table 1. An inconsistent data set

	Attributes			Decision
Case	Temperature	Headache	Cough	Flu
1	normal	no	yes	no
2	normal	no	no	no
3	normal	no	no	yes
4	normal	no	no	no
5	high	yes	no	yes
6	high	yes	yes	no
7	high	no	yes	yes
8	high	yes	yes	yes

For an attribute-value pair $(a, v) = t$, a *block* of t, denoted by $[t]$, is a set of all cases from U such that for attribute a have value v. An *indiscernibility relation* R on U is defined for all $x, y \in U$ by

$$xRy \text{ if and only if } a(x) = a(y) \text{ for all } a \in A.$$

Equivalence classes of R are called *elementary sets* of R. An equivalence class of R containing x is denoted $[x]$. Any finite union of elementary sets is called a *definable set* [12]. Let X be a concept. In general, X is not a definable set. However, set X may be approximated by two definable sets, the first one is called a *lower approximation* of X, denoted by $appr(X)$ and defined as follows

$$\cup \{[x] \mid x \in U, \ [x] \subseteq X\},$$

The second set is called an *upper approximation* of X, denoted by $\overline{appr}(X)$ and defined as follows

$$\cup \{[x] \mid x \in U, \ [x] \cap X \neq \emptyset\}.$$

For example, for the concept $[(Flu, no)] = \{1, 2, 4, 6\}$,

$$appr(\{1, 2, 4, 6\}) = \{1\},$$

and

$$\overline{appr}(\{1, 2, 4, 6\}) = \{1, 2, 3, 4, 6, 8\}.$$

2.2 Probabilistic Approximations

Let (U, R) be an approximation space, where R is an equivalence relation on U. A probabilistic approximation of the set X with the threshold α, $0 < \alpha \leq 1$, is denoted by $appr_\alpha(X)$ and defined as follows

$$\cup \{[x] \mid x \in U, \ Pr(X|[x]) \geq \alpha\},$$

where $[x]$ is an elementary set of R and $Pr(X|[x]) = \frac{|X \cap [x]|}{|[x]|}$ is the conditional probability of X given $[x]$.

Obviously, the equivalence relation R uniquely defines a partition on U defined as the family of all elementary sets of R. Such a partition will be denoted by R^*. For Table 1, $R^* = \{\{1\}, \{2, 3, 4\}, \{5\}, \{6, 8\}, \{7\}\}$.

Table 2. Conditional probabilities

$[x]$	$\{1\}$	$\{2, 3, 4\}$	$\{5\}$	$\{6, 8\}$	$\{7\}$
$Pr(\{1, 2, 4, 6\} \mid [x])$	1	0.667	0	0.5	0

The number of distinct probabilistic approximations of the concept X is smaller than or equal to the number n of distinct thresholds α from the definition of a probabilistic approximation. The number n is equal to the number of distinct positive conditional probabilities $Pr(X|[x])$, where $x \in U$. Additionally, the number n is smaller than or equal to the number m of elementary sets $[x]$ of R. Finally, $m \leq |U|$. Thus the number of distinct probabilistic approximations of the given concept is smaller than or equal to the cardinality of U.

Table 2 shows conditional probabilities for all members of R^*. In Table 2 there are three positive conditional probabilities: 0.5, 0.667 and 1. Therefore there are only three probabilistic approximations:

$appr_{0.5}(\{1, 2, 4, 6\}) = \{1, 2, 3, 4, 6, 8\}$,

$appr_{0.667}(\{1, 2, 4, 6\}) = \{1, 2, 3, 4\}$,

and

$appr_1(\{1, 2, 4, 6\}) = \{1\}$.

Obviously, for the concept X, the probabilistic approximation of X computed for the threshold equal to the smallest positive conditional probability $Pr(X \mid [x])$ is equal to the upper approximation of X. Additionally, the probabilistic approximation of X computed for the threshold equal to 1 is equal to the lower approximation of X.

2.3   Rule Induction

In this subsection we assume that R is an equivalence relation. We will discuss how the existing rough set based data mining systems, such as LERS (Learning from Examples based on Rough Sets), may be used to induce rules using probabilistic approximations. As we will show, all what is necessary is, for every concept, to modify the input data set, run LERS, and then edit the induced rule set. We will illustrate this procedure by inducing a rule set for Table 1 and the concept $[(Flu, no)] = \{1, 2, 4, 6\}$ using the probabilistic approximation $appr_{0.667}(\{1, 2, 4, 6\}) = \{1, 2, 3, 4\}$. First, a new data set should be created in which for all cases that are members of the set $appr_{0.667}(\{1, 2, 4, 6\})$ the decision values are copied from the original data set (Table 1) and for all remaining cases, not being in the set $appr_{0.667}(\{1, 2, 4, 6\})$, a new decision value is introduced, say SPECIAL. Thus a new data set is created, see Table 3.

Table 3. A new data set

	Attributes			Decision
Case	Temperature	Headache	Cough	Flu
1	normal	no	yes	no
2	normal	no	no	no
3	normal	no	no	yes
4	normal	no	no	no
5	high	yes	no	SPECIAL
6	high	yes	yes	SPECIAL
7	high	no	yes	SPECIAL
8	high	yes	yes	SPECIAL

The LERS data mining system may be used to induce certain rules (from lower approximations) or possible rules (from upper approximations) [13]. Any rule r is characterized by the conditional probability $Pr(X|Y)$, where X is the concept and Y is the domain of r (the set of all cases described by the rule conditions). For a certain rule r, by definition, $Pr(X|Y) = 1$. We are interested in inducing probabilistic rules, with $Pr(X|Y > 0$, so we need to induce possible rules. For example, for Table 3,

$$\underline{appr}([(Flu, no)]) = \underline{appr}(\{1, 2, 4\}) = \{1\},$$

and

$$\overline{appr}([(Flu, no)]) = \overline{appr}(\{1, 2, 4\}) = \{1, 2, 3, 4\}.$$

Therefore, certain rules for [(*Flu, no*)] will describe only the set {1}, while possible rules for the same concept will describe all cases from the set {1, 2, 3, 4}, so the obvious choice is to use possible rules.

The data set presented in Table 3 should be inputted to the LERS system, where first the ordinary upper approximations of all concepts, [(*Flu, no*)], [(*Flu, yes*)] and [(*Flu, SPECIAL*)] are computed and then the MLEM2 algorithm [14] is applied. For Table 3, the MLEM2 algorithm will return the following rule set

1, 3, 4
(Temperature, normal) -> (Flu, no),
2, 1, 3
(Temperature, normal) & (Cough, no) -> (Flu, yes),
1, 4, 4
(Temperature, high) -> (Flu, SPECIAL).

Rules are presented in the LERS format, every rule is associated with three numbers: the total number of attribute-value pairs on the left-hand side of the rule, the total number of cases correctly classified by the rule during training, and the total number of training cases matching the left-hand side of the rule, i.e., the rule domain size. These numbers are computed by comparing induced rules with Table 3. In this rule set only rules describing the concept [(*Flu, no*)] are useful, the remaining rules should be deleted. Hence, only one rule is useful

1, 3, 4

(Temperature, normal) -> (Flu, no).

This rule describes the set {1, 2, 3, 4}, three cases (1, 2, and 4) truly belong to the concept.

For the second concept from Table 1, [(*Flu, yes*)] = {3, 5, 7, 8}, and for the following probabilistic approximation

$$appr_{0.667}(\{3, 5, 7, 8\}) = \{5, 7\},$$

the corresponding rule set may be induced from the data set presented in Table 4.

Table 4. A new data set

	Attributes			Decision
Case	Temperature	Headache	Cough	Flu
1	normal	no	yes	SPECIAL
2	normal	no	no	SPECIAL
3	normal	no	no	SPECIAL
4	normal	no	no	SPECIAL
5	high	yes	no	yes
6	high	yes	yes	SPECIAL
7	high	no	yes	yes
8	high	yes	yes	SPECIAL

The MLEM2 algorithm returns the following rule set

1, 4, 4

(Temperature, normal) -> (Flu, SPECIAL),

2, 2, 2

(Headache, yes) & (Cough, yes) -> (Flu, SPECIAL),

2, 1, 1

(Headache, yes) & (Cough, no) -> (Flu, yes),

2, 1, 1

(Temperature, high) & (Headache, no) -> (Flu, yes).

Among these four rules only the following two rules

2, 1, 1

(Headache, yes) & (Cough, no) -> (Flu, yes),

2, 1, 1

(Temperature, high) & (Headache, no) -> (Flu, yes).

describe the concept [(*Flu, yes*)]. Finally, the following rule set

1, 3, 4

(Temperature, normal) -> (Flu, no),

2, 1, 1

(Headache, yes) & (Cough, no) -> (Flu, yes),

2, 1, 1

(Temperature, high) & (Headache, no) -> (Flu, yes).

describes both concepts of the probabilistic approximations associated with the parameter $\alpha = 0.667$.

3   Arbitrary Binary Relations

In this section, first we will study approximations defined on the approximations space $A = (U, R)$ where U is a finite nonempty set and R is an arbitrary binary relation. Then we will extend corresponding definitions to generalized probabilistic approximations.

3.1   Non-parameterized Approximations

First we will quote some definitions from [15]. Let x be a member of U. The R-*successor* set of x, denoted by $R_s(x)$, is defined as follows

$$R_s(x) = \{y \mid xRy\}.$$

The R-*predecessor* set of x, denoted by $R_p(x)$, is defined as follows

$$R_p(x) = \{y \mid yRx\}.$$

For the rest of the paper we will discuss only R-successor sets and corresponding approximations.

Let X be a subset of U. The R-*singleton lower approximation* of X, denoted by $\underline{appr}^{singleton}(X)$, is defined as follows

$$\{x \mid x \in U, R_s(x) \subseteq X\}.$$

The singleton lower approximations were studied in many papers, see, e.g., [9,10,16,17,18,19,20,21,22,23].

The R-*singleton upper approximation* of X, denoted by $\overline{appr}^{singleton}(X)$, is defined as follows

$$\{x \mid x \in U, R_s(x) \cap X \neq \emptyset\}.$$

The singleton upper approximations, like singleton lower approximations, were also studied in many papers, e.g., [9,10,16,17,20,21,22,23].

The R-*subset lower approximation* of X, denoted by $\underline{appr}^{subset}(X)$, is defined as follows

$$\cup \{R_s(x) \mid x \in U, R_s(x) \subseteq X\}.$$

The subset lower approximations were introduced in [9,10].

The R-*subset upper approximation* of X, denoted by $\overline{appr}^{subset}(X)$, is defined as follows

$$\cup \{R_s(x) \mid x \in U, R_s(x) \cap X \neq \emptyset\}.$$

The subset upper approximations were introduced in [9,10].

The R-*concept lower approximation* of X, denoted by $\underline{appr}^{concept}(X)$, is defined as follows

$$\cup \{R_s(x) \mid x \in X, R_s(x) \subseteq X\}.$$

The concept lower approximations were introduced in [9,10].

The R-*concept successor upper approximation* of X, denoted by $\overline{appr}^{concept}(X)$, is defined as follows

$$\cup \{R_s(x) \mid x \in X, R_s(x) \cap X \neq \emptyset\} = \cup \{R_s(x) \mid x \in X\}.$$

The concept upper approximations were studied in [9,10,19].

3.2   Probabilistic Approximations

By analogy with standard approximations defined for arbitrary binary relations, we will introduce three kinds of probabilistic approximations for such relations: singleton, subset and concept.

The singleton probabilistic approximation of X with the threshold α, $0 < \alpha \leq 1$, denoted by $appr_\alpha^{singleton}(X)$, is defined as follows

$$\{x \mid x \in U, \ Pr(X|R_s(x)) \geq \alpha\},$$

where $Pr(X|R_s(x)) = \frac{|X \cap R_s(x)|}{|R_s(x)|}$ is the conditional probability of X given $R_s(x)$.

A subset probabilistic approximation of the set X with the threshold α, $0 < \alpha \leq 1$, denoted by $appr_\alpha^{subset}(X)$, is defined as follows

$$\cup\{R_s(x) \mid x \in U, \ Pr(X|R_s(x)) \geq \alpha\},$$

where $Pr(X|R_s(x)) = \frac{|X \cap R_s(x)|}{|R_s(x)|}$ is the conditional probability of X given $R_s(x)$.

A concept probabilistic approximation of the set X with the threshold α, $0 < \alpha \leq 1$, denoted by $appr_\alpha^{concept}(X)$, is defined as follows

$$\cup\{R_s(x) \mid x \in X, \ Pr(X|R_s(x)) \geq \alpha\},$$

where $Pr(X|R_s(x)) = \frac{|X \cap R_s(x)|}{|R_s(x)|}$ is the conditional probability of X given $R_s(x)$.

It is not difficult to see that the number of different probabilistic approximations of a given type (singleton, subset or concept) is not greater than the cardinality of U.

Obviously, for the concept X, the probabilistic approximation of a given type (singleton, subset or concept) of X computed for the threshold equal to the smallest positive conditional probability $Pr(X \mid [x])$ is equal to the standard upper approximation of X of the same type. Additionally, the probabilistic approximation of a given type of X computed for the threshold equal to 1 is equal to the standard lower approximation of X of the same type.

3.3   Incomplete Data Sets

It is well-known that any incomplete data set is described by a *characteristic relation* R, a generalization of the indiscernibility relation. The characteristic relation is reflexive but, in general, is neither symmetric nor transitive. For incomplete data sets R-definable sets are called *characteristic sets*, a generalization of elementary sets.

We distinguish between two types of missing attribute values: *lost* (e.g., the value was erased) and *"do not care" conditions* (such a value may be any value of the attribute), see [9,10].

An example of incomplete data set is presented in Table 5.

For incomplete decision tables the definition of a block of an attribute-value pair must be modified in the following way:

Table 5. An incomplete data set

	Attributes			Decision
Case	Temperature	Headache	Cough	Flu
1	normal	no	*	no
2	?	no	no	no
3	normal	*	no	yes
4	normal	no	?	no
5	high	yes	*	yes
6	high	yes	yes	no
7	high	?	yes	yes
8	high	yes	yes	yes

- If for an attribute a there exists a case x such that $a(x) =?$, i.e., the corresponding value is lost, then the case x should not be included in any blocks $[(a, v)]$ for all values v of attribute a,
- If for an attribute a there exists a case x such that the corresponding value is a "do not care" condition, i.e., $a(x) = *$, then the case x should be included in blocks $[(a, v)]$ for all specified values v of attribute a.

For a case $x \in U$ the *characteristic set* $K_B(x)$ is defined as the intersection of the sets $K(x, a)$, for all $a \in B$, where the set $K(x, a)$ is defined in the following way:

- If $a(x)$ is specified, then $K(x, a)$ is the block $[(a, a(x))]$ of attribute a and its value $a(x)$,
- If $a(x) =?$ or $a(x) = *$ then the set $K(x, a) = U$.

The characteristic set $K_B(x)$ may be interpreted as the set of cases that are indistinguishable from x using all attributes from B and using a given interpretation of missing attribute values.

For the data set from Table 5, the set of blocks of attribute-value pairs is
$[(Temperature, normal)] = \{1, 3, 4\}$,
$[(Temperature, high)] = \{5, 6, 7, 8\}$,
$[(Headache, no)] = \{1, 2, 3, 4\}$,
$[(Headache, yes)] = \{3, 5, 6, 8\}$,
$[(Cough, no)] = \{1, 2, 3, 5\}$,
$[(Cough, yes)] = \{1, 5, 6, 7, 8\}$.

The corresponding characteristic sets are

$K_A(1) = K_A(4) = \{1, 3, 4\}$,
$K_A(2) = \{1, 2, 3\}$,
$K_A(3) = \{1, 3\}$,
$K_A(5) = K_A(6) = K_A(8) = \{5, 6, 8\}$,
$K_A(7) = \{5, 6, 7, 8\}$.

Conditional probabilities of the concept $\{1, 2, 4, 6\}$ given a characteristic set $K_A(x)$ are presented in Table 6.

For Table 5, all probabilistic approximations (singleton, subset and concept) are

Table 6. Conditional probabilities

$K_A(x)$	$\{1, 3, 4\}$	$\{1, 2, 3\}$	$\{1, 3\}$	$\{5, 6, 8\}$	$\{5, 6, 7, 8\}$
$Pr(\{1, 2, 4, 6\} \mid K_A(x))$	0.667	0.667	0.5	0.333	0.25

$$appr_{0.25}^{singleton}(\{1,2,4,6\}) = U,$$

$$appr_{0.333}^{singleton}(\{1,2,4,6\}) = \{1,2,3,4,5,6,8\},$$

$$appr_{0.5}^{singleton}(\{1,2,4,6\}) = \{1,2,3,4\},$$

$$appr_{0.667}^{singleton}(\{1,2,4,6\}) = \{1,2,4\},$$

$$appr_1^{singleton}(\{1,2,4,6\}) = \emptyset,$$

$$appr_{0.25}^{subset}(\{1,2,4,6\}) = U,$$

$$appr_{0.333}^{subset}(\{1,2,4,6\}) = \{1,2,3,4,5,6,8\},$$

$$appr_{0.5}^{subset}(\{1,2,4,6\}) = \{1,2,3,4\},$$

$$appr_{0.667}^{subset}(\{1,2,4,6\}) = \{1,2,3,4\},$$

$$appr_1^{subset}(\{1,2,4,6\}) = \emptyset,$$

$$appr_{0.25}^{concept}(\{1,2,4,6\}) = \{1,2,3,4,5,6,8\},$$

$$appr_{0.333}^{concept}(\{1,2,4,6\}) = \{1,2,3,4,5,6,8\},$$

$$appr_{0.5}^{concept}(\{1,2,4,6\}) = \{1,2,3,4\},$$

$$appr_{0.667}^{concept}(\{1,2,4,6\}) = \{1,2,3,4\},$$

$$appr_1^{concept}(\{1,2,4,6\}) = \emptyset.$$

3.4   Definability

Definability for completely specified decision tables should be modified to fit into incomplete decision tables. For incomplete decision tables, a union of some intersections of attribute-value pair blocks, where such attributes are members of B and are distinct, will be called B-*locally definable* sets. A union of characteristic sets $K_B(x)$, where $x \in X \subseteq U$ will be called a B-*globally definable* set. Any set X that is B-globally definable is B-locally definable, the converse is not true. For example, the set $\{1\}$ is A-locally definable since $\{1\} = [(Temperature, normal)] \cap [(Cough, yes)]$. However, the set $\{1\}$ is not A-globally definable. On the other hand, the set $\{1, 2, 4\} = appr_{0.667}^{singleton}(\{1,2,4,6\})$ is not

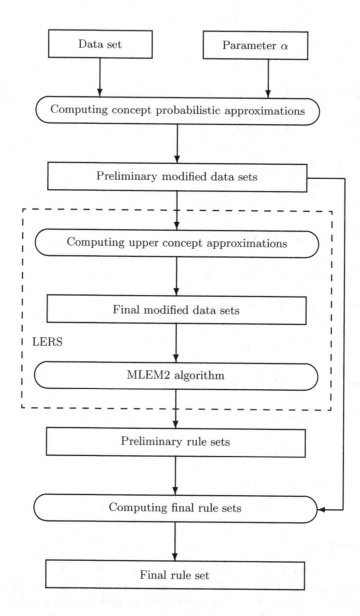

Fig. 1. Rule induction from probabilistic approximations using the LERS data mining system

even locally definable since in all blocks of attribute-value pairs containing the case 4 contain also the case 3 as well. Obviously, if a set is not B-locally definable then it cannot be expressed by rule sets using attributes from B. This is why it is so important to distinguish between B-locally definable sets and those that are not B-locally definable. In general, subset and concept probabilistic approximations are globally definable while singleton probabilistic approximations are not even locally definable.

3.5   Rule Induction

We will study how to adapt the LERS data mining system for rule induction from probabilistic approximations of the given concept. We will use a similar technique as in Subsection 3.3, i.e., for a concept and the probabilistic approximation of the concept we will create a new decision table. However, we have more choices since we may use a few different types of approximations.

Let us say that we want to induce rules for the concept $[(Flu, no)]$ and the concept probabilistic approximation with the parameter $\alpha = 0.5$. The preliminary modified data set, constructed in the same way as described in Subsection 2.3, is presented in Table 7.

Table 7. A preliminary modified data set

	Attributes			Decision
Case	Temperature	Headache	Cough	Flu
1	normal	no	*	no
2	?	no	no	no
3	normal	*	no	yes
4	normal	no	?	no
5	high	yes	*	SPECIAL
6	high	yes	yes	SPECIAL
7	high	?	yes	SPECIAL
8	high	yes	yes	SPECIAL

This data set is inputted to the LERS data mining system, see Figure 1. The LERS system computes the upper concept approximation of the set $\{1, 2, 4, 6\}$, in our example it is $\{1, 2, 3, 4\}$, and the corresponding final modified data set. The MLEM2 algorithm induces the following preliminary rule set from the final modified data sets

1, 3, 4
(Headache, no) -> (Flu, no),
2, 1, 2
(Temperature, normal) & (Cough, no) -> (Flu, yes),
1, 4, 4
(Temperature, high) -> (Flu, SPECIAL).

where the three numbers that precede every rule are computed from Table 7. Obviously, only the first rule

1, 3, 4
(Headache, no) -> (Flu, no),

should be saved and the remaining two rules should be deleted in computing the final rule set.

Note that, in general, the result of computing the upper concept approximation by LERS results in the set

$$\cup \{K_A(y) \mid y \in \cup \{K_A(x) \mid x \in X, \ Pr(X|K_A(x)) \geq \alpha\}\}$$

which is a superset of the concept probabilistic approximation of X. For some data sets, for example for incomplete data sets with only lost values, both sets are identical. Nevertheless, in the preliminary rule set the three numbers that precede every rule are adjusted taking into account the preliminary modified data set. Thus during classification of unseen cases by the LERS classification system rules describe the original concept probabilistic approximation of the concept X.

4   Conclusions

In this paper we study a set of all probabilistic approximations, first for the approximation space (U, R), where U is a nonempty finite set and R is an equivalence relation, and then for the approximation space (U, R), where R is an arbitrary binary relation. For an arbitrary binary relation R standard definitions of singleton, subset and concept approximations are generalized to probabilistic approximations. It is shown that the set of such probabilistic approximations, even if R is an arbitrary binary relation, is finite and quite limited. Moreover, singleton probabilistic approximations of a subset X of the universe U is, in general, not even locally definable, so X is not expressible by a rule set. Therefore, singleton probabilistic approximations should not be used for data mining.

Acknowledgement. The author would like to thank the anonymous referees for all their valuable suggestions.

References

1. Pawlak, Z., Wong, S.K.M., Ziarko, W.: Rough sets: probabilistic versus deterministic approach. International Journal of Man-Machine Studies 29, 81–95 (1988)
2. Tsumoto, S., Tanaka, H.: PRIMEROSE: probabilistic rule induction method based on rough sets and resampling methods. Computational Intelligence 11, 389–405 (1995)
3. Yao, Y.: Decision-Theoretic Rough Set Models. In: Yao, J., Lingras, P., Wu, W.-Z., Szczuka, M.S., Cercone, N.J., Ślęzak, D. (eds.) RSKT 2007. LNCS (LNAI), vol. 4481, pp. 1–12. Springer, Heidelberg (2007)
4. Yao, Y.Y., Wong, S.K.M.: A decision theoretic framework for approximate concepts. International Journal of Man-Machine Studies 37, 793–809 (1992)
5. Yao, Y.Y., Wong, S.K.M., Lingras, P.: A decision-theoretic rough set model. In: Proceedings of the 5th International Symposium on Methodologies for Intelligent Systems, pp. 388–395 (1990)
6. Ziarko, W.: Variable precision rough set model. Journal of Computer and System Sciences 46(1), 39–59 (1993)
7. Ziarko, W.: Probabilistic approach to rough sets. International Journal of Approximate Reasoning 49, 272–284 (2008)
8. Grzymala-Busse, J.W., Marepally, S.R., Yao, Y.: An Empirical Comparison of Rule Sets Induced by LERS and Probabilistic Rough Classification. In: Szczuka, M., Kryszkiewicz, M., Ramanna, S., Jensen, R., Hu, Q. (eds.) RSCTC 2010. LNCS, vol. 6086, pp. 590–599. Springer, Heidelberg (2010)
9. Grzymala-Busse, J.W.: Rough set strategies to data with missing attribute values. In: Workshop Notes, Foundations and New Directions of Data Mining, in Conjunction with the 3rd International Conference on Data Mining, pp. 56–63 (2003)
10. Grzymała-Busse, J.W.: Data with Missing Attribute Values: Generalization of Indiscernibility Relation and Rule Induction. In: Peters, J.F., Skowron, A., Grzymała-Busse, J.W., Kostek, B.z., Swiniarski, R., Szczuka, M.S. (eds.) Transactions on Rough Sets I. LNCS, vol. 3100, pp. 78–95. Springer, Heidelberg (2004)
11. Grzymała-Busse, J.W.: Generalized Parameterized Approximations. In: Yao, J., Ramanna, S., Wang, G., Suraj, Z. (eds.) RSKT 2011. LNCS, vol. 6954, pp. 136–145. Springer, Heidelberg (2011)
12. Pawlak, Z.: Rough Sets. Theoretical Aspects of Reasoning about Data. Kluwer Academic Publishers, Dordrecht (1991)
13. Grzymala-Busse, J.W.: A new version of the rule induction system LERS. Fundamenta Informaticae 31, 27–39 (1997)
14. Grzymala-Busse, J.W.: MLEM2: A new algorithm for rule induction from imperfect data. In: Proceedings of the 9th International Conference on Information Processing and Management of Uncertainty in Knowledge-Based Systems, pp. 243–250 (2002)
15. Grzymala-Busse, J.W., Rząsa, W.: Definability and Other Properties of Approximations for Generalized Indiscernibility Relations. In: Peters, J.F., Skowron, A. (eds.) Transactions on Rough Sets XI. LNCS, vol. 5946, pp. 14–39. Springer, Heidelberg (2010)
16. Kryszkiewicz, M.: Rough set approach to incomplete information systems. In: Proceedings of the Second Annual Joint Conference on Information Sciences, pp. 194–197 (1995)
17. Kryszkiewicz, M.: Rules in incomplete information systems. Information Sciences 113, 271–292 (1999)

18. Lin, T.Y.: Neighborhood systems and approximation in database and knowledge base systems. In: Proceedings of the ISMIS 1989, the Fourth International Symposium on Methodologies of Intelligent Systems, pp. 75–86 (1989)
19. Lin, T.Y.: Topological and fuzzy rough sets. In: Slowinski, R. (ed.) Intelligent Decision Support. Handbook of Applications and Advances of the Rough Sets Theory, pp. 287–304. Kluwer Academic Publishers, Dordrecht (1992)
20. Slowinski, R., Vanderpooten, D.: A generalized definition of rough approximations based on similarity. IEEE Transactions on Knowledge and Data Engineering 12, 331–336 (2000)
21. Stefanowski, J., Tsoukiàs, A.: On the Extension of Rough Sets under Incomplete Information. In: Zhong, N., Skowron, A., Ohsuga, S. (eds.) RSFDGrC 1999. LNCS (LNAI), vol. 1711, pp. 73–82. Springer, Heidelberg (1999)
22. Stefanowski, J., Tsoukias, A.: Incomplete information tables and rough classification. Computational Intelligence 17(3), 545–566 (2001)
23. Yao, Y.Y.: Relational interpretations of neighborhood operators and rough set approximation operators. Information Sciences 111, 239–259 (1998)

An Extension to Rough c-Means Clustering Algorithm Based on Boundary Area Elements Discrimination

Fan Li and Qihe Liu

School of Computer Science and Engineering
University of Electronic Science and Technology of China
Chengdu, 610051, P.R. China
{lifan,qiheliu}@uestc.edu.cn

Abstract. Rough c-means algorithm has gained increasing attention in recent years. However, the original Rough c-means algorithm does not distinguish data points in the boundary area while computing the new centroid of each cluster. In this paper, we consider the distinction between data points in the boundary area and present an extended Rough c-means algorithm which benefits from this information. The distinction is reflected by the degree of the data point in the boundary area being close to its corresponding lower approximation. This information is utilized in the step of calculating the new centroid of each cluster. The algorithm is tested on four UCI machine learning repository data sets. Experimental results indicate that the proposed algorithm yields more desirable clustering results than the original Rough c-means algorithm.

Keywords: Rough c-means algorithm, Rough Sets, Lower approximation, Upper approximation, Boundary area.

1   Introduction

Clustering plays an important role in various areas [1, 2]. It can be viewed as the problem of dividing a potentially large data (patterns) set X of n data points in p-dimensional space, i.e. $X = \{\mathbf{x}_1, \mathbf{x}_2..., \mathbf{x}_n\} \subset \mathfrak{R}^p$, into a few $c < n$ compact subsets $C_1, C_2, \cdots, C_c$. Those data points within each cluster are more closely related to one another than data points assigned to different clusters. In the past years, a large number of different clustering algorithms were proposed to handle this problem, such as c-means [3, 4], fuzzy c-means (FCM) [5, 6], spectral clustering [7, 8] hierarchical clustering [9, 10], and model based clustering [11, 12].

Recently, Rough Sets theory [13, 14] has been incorporated in the famous c-means framework to develop Rough c-means (RCM) algorithm [15] (For more aspects of Rough Sets theory and its applications, please refer to [16–29]). Since its introduction, RCM has gained increasing attention. Further researches include parameter selection [30], method extension and modification [31–36], and applications for real-life problems [15, 37–39].

J.F. Peters et al. (Eds.): Transactions on Rough Sets XVI, LNCS 7736, pp. 17–33, 2013.
© Springer-Verlag Berlin Heidelberg 2013

Roughly speaking, clustering algorithms can be divided into two types: partitional and hierarchical. Like c-means, RCM can be classified into the partitional clustering methods. In c-means, c clusters construct a partition of X, i.e. $C_i \cap C_j = \phi$ if $i \neq j$. This might not be appropriate in many real-life scenarios. FCM uses the membership value to make it possible to assign a data point to potentially all clusters. RCM can also assign a data point to more than one cluster, but uses more restrictions. In RCM, each cluster is represented by its lower and upper approximations. A data point may be assigned to the lower approximation of a certain cluster (and hence in the upper approximation of this cluster) or the boundary areas of two or more clusters (and hence in the upper approximations of these clusters) according to its distances to cluster centroids. RCM iteratively updates cluster centroids and the assignment of all data points until a certain termination criterion has been satisfied [15, 33].

In this paper, we present an extended RCM algorithm, which intends to capture the distinction between data points in the boundary area and uses this information in the clustering procedure (the preliminary results of this paper were presented in [36]). According to the definition in Rough Sets theory, data points in the boundary area of a cluster possibly (but not definitely) belong to this cluster. So distinguishing these data points according to the degree of possibility may be useful for clustering. On the other hand, by the definition of the lower approximation, data points in the lower approximation of a cluster all belong to this cluster with certainty. Thus it is not necessary to distinguish these data points. In this way, the presented algorithm is more descriptive than the original RCM, but less descriptive than FCM.

This idea is motivated by several works (e.g. [31, 39]). But there are important distinctions between approaches discussed in these works and ours. Firstly, in [31, 39], when computing the new centroid of a cluster, the membership value is used to reflect the difference between all data points in the lower approximation and boundary area. This membership value is calculated by using the distances of a data point to all cluster centroids. But these distances may be close, which means some data points in the boundary area may not be distinguished effectively. In our approach, the distances of a data point in the boundary area to its closet neighbors in the corresponding lower approximations are used. Secondly, in [31, 39] when computing the new centroid of each cluster, the membership value of each data point in the lower approximation is also used. In our approach, data points in the lower approximation are not distinguished. (see section 3.1 for more details).

Four UCI data sets are employed to compare the performance of RCM and the presented algorithm. We use the modified DB index [31] to evaluate the performance of the both algorithms. Experimental results indicate that the presented algorithm provides superior clustering results contrasted with the original RCM, both for the overall performance and the optimal performance.

The rest of this paper is organized as follows. Section 2 briefly reviews the basic notions. Section 3 presents the motivations and the algorithm framework.

Section 4 presents the experiment process and lists the results. Finally, Section 5 presents the conclusion and some future research perspectives.

2   Basic Concepts

In this section, we briefly review the basic notions of Rough Sets and RCM.

2.1   Rough Sets

Let U be a nonempty finite set of objects and R an equivalence relation on U. R constitutes a partition of U, denoted by U/R. Let $[x]_R$ denotes the equivalence class of x, then $U/R = \{[x]_R : x \in U\}$.

The R-lower and R-upper approximations of a set $X \subseteq U$ are defined as follows [13, 14, 40]:

$$\underline{R}(X) = \{x \in U : [x]_R \subseteq X\}, \tag{1}$$

$$\overline{R}(X) = \{x \in U : [x]_R \cap X \neq \emptyset\}. \tag{2}$$

$\underline{R}(X)$ is the set of objects that belong to X with certainty, whereas $\overline{R}(X)$ is the set of objects that possibly belong to X.

The boundary area of the set X is defined as:

$$bn_R(X) = \overline{R}(X) - \underline{R}(X). \tag{3}$$

The properties of Rough Sets can be found in [13, 14, 40]. RCM does not verify all of these properties but only uses the following rules to assign each data point.

1. $\underline{R}(X_i) \subseteq \overline{R}(X_i) \subseteq U, \forall X_i \subseteq U.$
2. $\underline{R}(X_i) \cap \underline{R}(X_j) = \phi, \forall X_i, X_j \subseteq U, i \neq j.$
3. $\underline{R}(X_i) \cap \overline{R}(X_j) = \phi, \forall X_i, X_j \subseteq U, i \neq j.$
4. If an object $x \in U$ is not a part of any lower approximations, then it must belong to the boundary areas of two or more clusters, and hence belongs to the upper approximations of these clusters.

2.2   Rough c-Means Algorithm

Let $X = \{\mathbf{x}_1, \mathbf{x}_2, \cdots, \mathbf{x}_n\} \subset \mathfrak{R}^p$ be a data set, which is to be partitioned to c clusters $C_1, C_2, \cdots, C_c$. The centroid (mean) of the i-th cluster C_i is denoted by $\mathbf{v}_i, i = 1, 2, \cdots, c$. The distance between $\mathbf{x}_i$ and $\mathbf{v}_k$ is denoted by $d(\mathbf{x}_i, \mathbf{v}_k)$, $d(\mathbf{x}_i, \mathbf{v}_k) = \| \mathbf{x}_i - \mathbf{v}_k \|$. In the rest of this paper, the cardinality of a set C is denoted by $|C|$.

In RCM, the equivalence relation R is not constructed explicitly. So in the rest of this paper, $\underline{R}(C_i)$, $\overline{R}(C_i)$ and $bn_R(C_i)$ are denoted by $\underline{C}_i$, $\overline{C}_i$ and $bn(C_i)$, respectively.

A cluster C_i is denoted by a pair $(\underline{C_i}, \overline{C_i})$ in RCM. The assignment criteria to determine whether a data point belongs to the upper approximation or the lower approximation of a certain cluster are as follows.

$\forall \mathbf{x}_i \in X$, let $d(\mathbf{x}_i, \mathbf{v}_h) = min_{k=1,\cdots,c} d(\mathbf{x}_i, \mathbf{v}_k)$. $\mathbf{v}_h$ is the closest centroid to $\mathbf{x}_i$. Then, the index set of all centroids that are also close to $\mathbf{x}_i$ is found:

$$J = \{j : d(\mathbf{x}_i, \mathbf{v}_j) - d(\mathbf{x}_i, \mathbf{v}_h) \le \varepsilon \wedge j \ne h\}, \tag{4}$$

where ε is a predefined threshold. If $J = \emptyset$, which means $\mathbf{v}_h$ is the only centroid similar to $\mathbf{x}_i$. So $\mathbf{x}_i$ is assigned to $\underline{C_h}$, and hence belongs to $\overline{C_h}$. On the other hand, if $J \ne \emptyset$, which means $\mathbf{x}_i$ is similar to two or more centroids. According to the properties described in section 2.1, $\mathbf{x}_i$ cannot be assigned to any lower approximations, but assigned to the upper approximations of all clusters determined by set J.

In [33], the formula which decides the centroids close to $\mathbf{x}_i$ (Eq. (4)) is replaced by:

$$J = \left\{j : \frac{d(\mathbf{x}_i, \mathbf{v}_j)}{d(\mathbf{x}_i, \mathbf{v}_h)} \le 1 + \varepsilon \wedge j \ne h\right\}. \tag{5}$$

By using Eq. (5), RCM shows a superior performance (see [33] for more details). So in the rest of this paper we use this equation to decide all centroids close to a data point (both in RCM and our extended method).

According to the result of data points assignment, RCM uses the following equation to determine the new centroid of each cluster.

$$\mathbf{v}_i = \begin{cases} w_l \dfrac{\sum_{\mathbf{x}_k \in \underline{C_i}} \mathbf{x}_k}{|\underline{C_i}|} + w_b \dfrac{\sum_{\mathbf{x}_k \in bn(C_i)} \mathbf{x}_k}{|bn(C_i)|}, & \text{if } \underline{C_i} \ne \emptyset \wedge bn(C_i) \ne \emptyset \\[3mm] \dfrac{\sum_{\mathbf{x}_k \in bn(C_i)} \mathbf{x}_k}{|bn(C_i)|}, & \text{if } \underline{C_i} = \emptyset \wedge bn(C_i) \ne \emptyset \\[3mm] \dfrac{\sum_{\mathbf{x}_k \in \underline{C_i}} \mathbf{x}_k}{|\underline{C_i}|}, & \text{otherwise.} \end{cases} \tag{6}$$

RCM iteratively updates the assignment of all data points and cluster centroids until a certain termination criterion has been met.

The original form of RCM algorithm [15, 33] is outlined in Algorithm 1.

In Eq. (6), the parameters w_l and w_b define the importance of the lower approximation and the boundary area during clustering, respectively, satisfying $w_l + w_b = 1$ and $0.5 < w_l < 1$.

In line 1 of RCM (and our extended method in Section 3.3), the initial centroids are assigned by randomly choosing c data points in X. The benefit of this choice is that every cluster has at least one point in its lower approximation, which avoids invalid initial assignment.

3   The Extended Rough c-Means Algorithm

In this section, we propose the extended algorithm based on boundary area elements discrimination. First, we show the basic ideas and motivations. Then we describe the technical details of the proposed approach. Finally we present all steps of the proposed algorithm.

Algorithm 1: RCM algorithm

> **input** : The given data set $X = \{\mathbf{x}_1, \mathbf{x}_2, \cdots, \mathbf{x}_n\}$, c, w_l, ε
> **output**: Clustering result: $(\underline{C_1}, \overline{C_1}), \cdots, (\underline{C_c}, \overline{C_c})$

1 randomly assign the initial centroid $\mathbf{v}_i$ for C_i, $i = 1, 2, \cdots, c$;
2 **repeat**
3   **for** $i \leftarrow 1$ **to** n **do**
4   for a data point $\mathbf{x}_i$, determine its closest centroid $\mathbf{v}_h$: $d(\mathbf{x}_i, \mathbf{v}_h) = min_{k=1,\ldots,c} d(\mathbf{x}_i, \mathbf{v}_k)$;
5   assign $\mathbf{x}_i$ to the upper approximation of the cluster h, i.e. $\mathbf{x}_i \in \overline{C_h}$;
6   using Eq. (5) to determine the set J;
7   **if** $J = \emptyset$ **then**
8   assign $\mathbf{x}_i$ to the lower approximation of the cluster h, i.e. $\mathbf{x}_i \in \underline{C_h}$;
9   **else**
10   assign $\mathbf{x}_i$ to the upper approximations of the clusters determined by J, i.e. $\mathbf{x}_i \in \overline{C_j}, \forall j \in J$;
11   **end**
12   **end**
13   calculate the new centroid for each cluster using Eq. (6);
14 **until** *the termination criterion is met*;

Table 1. An illustrative data set

	$\mathbf{x}_1$	$\mathbf{x}_2$	$\mathbf{x}_3$	$\mathbf{x}_4$	$\mathbf{x}_5$	$\mathbf{x}_6$	$\mathbf{x}_7$	$\mathbf{x}_8$	$\mathbf{x}_9$	$\mathbf{x}_{10}$	$\mathbf{x}_{11}$
x	0.1	0.0	0.0	0.0	0.15	0.1	0.7	0.8	1.0	0.24	0.75
y	0.0	0.0	0.2	0.4	0.5	0.2	1.0	0.75	0.8	0.68	0.1

3.1 Discriminating Boundary Area Elements

Here, we employ an example to illustrate the motivations involved in the proposed method.

Example 1: A two-dimensional data set $X = \{\mathbf{x}_1, \mathbf{x}_2..., \mathbf{x}_{11}\}$ is listed in Table 1. With the following parameter setting: initial mean$= \left(\begin{smallmatrix} 0.0 & 0.8 \\ 0.0 & 0.8 \end{smallmatrix}\right)$, $c = 2$, $w_l = 0.9$, $w_b = 0.1$ and $\varepsilon = 0.25$, the result of RCM is shown in Fig.1. For cluster C_1, $\underline{C_1} = \{\mathbf{x}_1, \mathbf{x}_2, \mathbf{x}_3, \mathbf{x}_4, \mathbf{x}_5, \mathbf{x}_6\}$, $bn(C_1) = \{\mathbf{x}_{10}, \mathbf{x}_{11}\}$. For cluster C_2, $\underline{C_2} = \{\mathbf{x}_7, \mathbf{x}_8, \mathbf{x}_9\}$, $bn(C_2) = \{\mathbf{x}_{10}, \mathbf{x}_{11}\}$.

Although in RCM, the equivalence relation is not constructed explicitly, two data points can be regarded as similar according to their distances to cluster centroids. For example, $\mathbf{x}_{10}$ and $\mathbf{x}_{11}$ are similar since $\frac{d(\mathbf{x}_{10}, \mathbf{v}_2)}{d(\mathbf{x}_{10}, \mathbf{v}_1)} \leq 1 + \varepsilon$ and $\frac{d(\mathbf{x}_{11}, \mathbf{v}_2)}{d(\mathbf{x}_{11}, \mathbf{v}_1)} \leq 1 + \varepsilon$ hold. So in RCM, $\mathbf{x}_{10}$ and $\mathbf{x}_{11}$ are assigned to $bn(C_1)$ and $bn(C_2)$ (and hence in $\overline{C_1}$ and $\overline{C_2}$) without any difference. And in Eq. (6), when calculating new centroids for C_1 and C_2, $\mathbf{x}_{10}$ and $\mathbf{x}_{11}$ are also used without any difference. In other words, they are assigned the same weight in calculating new centroids for C_1 and C_2.

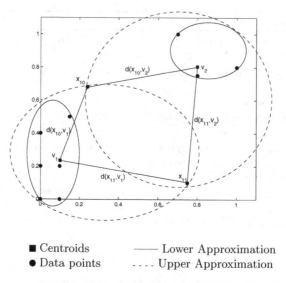

■ Centroids   —— Lower Approximation
● Data points   - - - - Upper Approximation

Fig. 1. Clustering result of RCM

On the other hand, there exists obvious difference between x_{10} and x_{11}. One can see that x_{10} is much closer to $\underline{C_1}$ than $\underline{C_2}$. Whereas the distance x_{11} to $\underline{C_1}$ is almost the same as the distance x_{11} to $\underline{C_2}$. Thus by using Eq. (6), the information about the distinction between x_{10} and x_{11} (in the sense of the degree of the two data points being close to their respective lower approximations) cannot be captured.

In general, a good clustering result should have the property that the cluster label of each data point shows consistency with its neighboring data points. For example, in the local learning approach [41, 42], for a given data point, a model is constructed by using its neighboring data points as training data, and then the label of the given data point is predicted by this model. It has been reported that local learning algorithms often show good performance. So it is useful to consider the neighboring information.

From this point of view, the neighboring information indicates that although x_{10} and x_{11} all possibly belong to C_1 and C_2, the degrees of x_{10} and x_{11} belong to C_1 and C_2 are different. Specifically, x_{10} is more likely to be in C_1 than in C_2, and the degrees of x_{11} to be in C_1 and C_2 are almost equal. This information should be used in clustering procedure. More specifically, in Eq. (6), when calculating new centroids for C_1 and C_2, x_{10} and x_{11} are all weighted by w_l to reflect their similarity. And at the same time, x_{10} and x_{11} should also be assigned different coefficients to reflect neighboring information. Thus intuitively, there is a need for a new approach to reflect the distinction between data points in the boundary area.

On the other hand, it is not necessary to distinguish data points in the lower approximation. Firstly, according to the definition in Rough Sets theory, these data points all belong to the corresponding cluster with certainty. Secondly, if these points should be distinguished, the distances of theses points to cluster prototypes (centroids) would be used. This means the difference between these points is relatively small. At last, from the practical perspective, a data point in the lower approximation shows consistency with its neighboring data points on the cluster label, apart from some special cases (e.g. outliers).

3.2 Calculating New Centroids

The typical case is shown in Fig. 2. $\mathbf{x}_k$ is in both $bn(C_1), bn(C_2), \cdots, bn(C_m)$. According to property 4 described in Section 2.1, $\mathbf{x}_k$ belongs to $\overline{C_1}, \overline{C_2}, \cdots, \overline{C_m}$.

Let $d_{1k}, d_{2k}, \cdots, d_{mk}$ denote the distances of $\mathbf{x}_k$ to its closest data point in $\underline{C_1}, \underline{C_2}, \cdots, \underline{C_m}$, respectively. We use $d_{1k}, d_{2k}, \cdots, d_{mk}$ to calculate a measure to evaluate the degree of $\mathbf{x}_k$ being close to $\underline{C_1}, \underline{C_2}, \cdots, \underline{C_m}$.

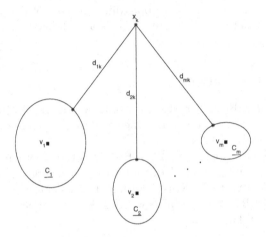

Fig. 2. Calculating new centroids

Formally, let F_k be the index set such that $F_k = \{i : \mathbf{x}_k \in \overline{C_i}\}$. With the properties described in Section 2.1 we know that if $|F_k| > 1$, $\forall j \in F_k$, $\mathbf{x}_k \in bn(C_j)$. The degree of $\mathbf{x}_k$ being close to the lower approximation of C_i is denoted as u_{ik} and calculated by:

$$u_{ik} = \frac{1}{\sum_{j \in F_k} \left(\frac{d_{ik}}{d_{jk}}\right)^{\frac{2}{m-1}}}, \tag{7}$$

where the exponent $m > 1$ is called a fuzzifier, and $\forall j \in F_k$,

$$d_{jk} = \begin{cases} min_{\mathbf{x}_l \in \underline{C_j}} \|\mathbf{x}_k - \mathbf{x}_l\|, & \text{if } \underline{C_j} \neq \emptyset, \\ \|\mathbf{x}_k - \frac{\sum_{\mathbf{x}_p \in \overline{C_j}} \mathbf{x}_p}{|\overline{C_j}|}\|, & \text{if } \underline{C_j} = \emptyset. \end{cases} \tag{8}$$

We note that Eq. (7) is a variation of the equation used in FCM to calculate the membership value of a data point belonging to a certain cluster.

This idea is motivated by [31, 39]. But in these researches the membership value is calculated by using the distances of a data point to all cluster centroids.

For example, in the case shown in Fig. 2, the membership values are determined by $d(\mathbf{x}_k, \mathbf{v}_1), d(\mathbf{x}_k, \mathbf{v}_2), \cdots, d(\mathbf{x}_k, \mathbf{v}_m)$. Since these distances are close, it is not effective to distinguish data points in the boundary area.

The new centroid for each cluster is calculated as follows:

$$
\mathbf{v}_i = \begin{cases} w_l \dfrac{\sum_{\mathbf{x}_k \in C_i} \mathbf{x}_k}{|C_i|} + w_b \dfrac{\sum_{\mathbf{x}_k \in bn(C_i)} u_{ik}^m \mathbf{x}_k}{\sum_{\mathbf{x}_k \in bn(C_i)} u_{ik}^m}, & \text{if } \underline{C_i} \neq \emptyset \wedge bn(C_i) \neq \emptyset \\[2ex] \dfrac{\sum_{\mathbf{x}_k \in bn(C_i)} u_{ik}^m \mathbf{x}_k}{\sum_{\mathbf{x}_k \in bn(C_i)} u_{ik}^m}, & \text{if } \underline{C_i} = \emptyset \wedge bn(C_i) \neq \emptyset \quad\quad (9) \\[2ex] \dfrac{\sum_{\mathbf{x}_k \in C_i} \mathbf{x}_k}{|C_i|}, & \text{otherwise.} \end{cases}
$$

We note that from Eq. (9) one can see that according to the motivation in section 3.1, data points in the lower approximation are not distinguished when calculating the centroid of each cluster. This is also different with the previous works in [31, 39].

3.3 Algorithm Descriptions

The extended RCM (ERCM hereafter) is shown in Algorithm 2. Lines 13 to 19 contain the key code of this framework. One can design different degree measures to obtain respective algorithms.

The termination criterion in line 21 of ERCM is that there are no more new assignments of data points, or the algorithm does not converge in S_{max} iterations.

The time complexity of ERCM is determined by calculating u_{il}. The overall time complexity of ERCM is $O(n^2 pT)$, where T is the number of iterations and p is the dimension of data points.

Example 2: For the data set given by Table 1, the clustering result of ERCM is shown in Fig. 3. For cluster C_1: $\underline{C_1} = \{\mathbf{x}_1, \mathbf{x}_2, \mathbf{x}_3, \mathbf{x}_4, \mathbf{x}_5, \mathbf{x}_6, \mathbf{x}_{10}\}$, $bn(C_1) = \{\mathbf{x}_{11}\}$. For cluster C_2, $\underline{C_2} = \{\mathbf{x}_7, \mathbf{x}_8, \mathbf{x}_9\}$, $bn(C_2) = \{\mathbf{x}_{11}\}$. In this result, $\mathbf{x}_{10}$ is assigned to the lower approximation of C_1 rather than being assigned to the upper approximations of both clusters. One can see that this result is more appropriate than the result depicted in Fig. 1.

4 Experimental Results

In this section, we conduct two sets of experiments. The first one aims at comparing the performance of ERCM and RCM. The second one is the sensitivity studies of the parameter m. In all experiments, we use four data sets in the UCI Machine Learning repository [43]: Iris, Wine, Wisconsin breast cancer (Wdbc) and Haberman. The characteristic of the four data sets is summarized in Table 2. We set c equal to the class number of each data set for ERCM and RCM. All data in the four data sets are normalized to the interval $[0, 1]$.

In all experiments, we use the following parameter setting: $w_l = 0.94$, $S_{max} = 100$. And we use the Davies-Bouldin index [44] (DB index hereafter) to evaluate the performance of ERCM and RCM. DB index is a widely used cluster

---

Algorithm 2: ERCM algorithm

---

    input : The given data set $X = \{\mathbf{x}_1, \mathbf{x}_2, \cdots, \mathbf{x}_n\}$, c, w_l, ε, m

    output: Clustering result: $(\underline{C_1}, \overline{C_1}), \cdots, (\underline{C_c}, \overline{C_c})$

1 randomly assign the initial centroid $\mathbf{v}_i$ for C_i, $i = 1, 2, \cdots, c$;

2 **repeat**

3   **for** $i \leftarrow 1$ **to** n **do**

4   for a data point $\mathbf{x}_i$, determine its closest centroid $\mathbf{v}_h$: $d(\mathbf{x}_i, \mathbf{v}_h) = min_{k=1,\ldots,c} d(\mathbf{x}_i, \mathbf{v}_k)$;

5   assign $\mathbf{x}_i$ to the upper approximation of the cluster h, i.e. $\mathbf{x}_i \in \overline{C_h}$;

6   using Eq. (5) to determine the set J;

7   **if** $J = \emptyset$ **then**

8   assign $\mathbf{x}_i$ to the lower approximation of the cluster h, i.e. $\mathbf{x}_i \in \underline{C_h}$;

9   **else**

10   assign $\mathbf{x}_i$ to the upper approximations of the clusters determined by J, i.e. $\mathbf{x}_i \in \overline{C_j}, \forall j \in J$;

11   **end**

12   **end**

13   determine the boundary area $L = X - (\underline{C_1} \cup \underline{C_2} \cup \cdots \cup \underline{C_c})$;

14   **for** $l \leftarrow 1$ **to** $|L|$ **do**

15   For $\mathbf{x}_l \in L$, determine the set F_l;

16   **foreach** $i \in F_l$ **do**

17   calculate u_{il} using Eq. (7)

18   **end**

19   **end**

20   calculate the new centroid for each cluster using Eq. (9);

21 **until** *the termination criterion is met*;

---

validity index, but the original form of DB index is designed for crisp clustering algorithms. So we use the extended form of DB index [31] for performance evaluation. The extended DB index is defined as:

$$DB = \frac{1}{c} \sum_{i=1}^{c} max_{j \neq i} \left(\frac{S_r(C_i) + S_r(C_j)}{d(\mathbf{v}_i, \mathbf{v}_j)} \right), \tag{10}$$

where $S_r(C_i)$ is the intra-cluster distance, and $d(\mathbf{v}_i, \mathbf{v}_j) = \| \mathbf{v}_i - \mathbf{v}_j \|$ denotes the inter-cluster separation.

$S_r(C_i)$ is defined as follows:

$$S_r(C_i) = \begin{cases} w_l \frac{\sum_{\mathbf{x}_k \in \underline{C_i}} \|\mathbf{x}_k - \mathbf{v}_i\|^2}{|\underline{C_i}|} + w_b \frac{\sum_{\mathbf{x}_k \in bn(C_i)} \|\mathbf{x}_k - \mathbf{v}_i\|^2}{|bn(C_i)|}, & \text{if } \underline{C_i} \neq \emptyset \wedge bn(C_i) \neq \emptyset \\ \frac{\sum_{\mathbf{x}_k \in bn(C_i)} \|\mathbf{x}_k - \mathbf{v}_i\|^2}{|bn(C_i)|}, & \text{if } \underline{C_i} = \emptyset \wedge bn(C_i) \neq \emptyset \\ \frac{\sum_{\mathbf{x}_k \in \underline{C_i}} \|\mathbf{x}_k - \mathbf{v}_i\|^2}{|\underline{C_i}|}. & \text{otherwise.} \end{cases} \tag{11}$$

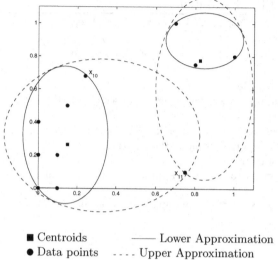

■ Centroids   ——— Lower Approximation
● Data points   - - - - Upper Approximation

Fig. 3. Clustering result of ERCM

Table 2. Characteristics of the data sets used in the experiments

Dataset	♯Dimension	♯Sample	♯Class
Iris	4	150	3
Wine	13	178	3
Wdbc	30	569	2
Haberman	3	306	2

A small value of DB index means a good clustering result, in which clusters are compact and well separated.

4.1   Performance Comparison

In this section, we apply ERCM and RCM to the four data sets and compare the performance for different values of ε, $\varepsilon = 0.005, 0.01, \cdots, 0.1$. In this experiment, m is set to 2.0 for ERCM. The following criteria are used in the comparison:

1. For a value of ε, RCM and ERCM start from the same randomly chosen center initialization, and DB index is calculated and compared when ERCM and RCM stop. This experiment is repeated 100 times. According to the comparison of DB index, the number that ERCM outperforms RCM or RCM outperforms ERCM in the 100 runs is recorded and denoted as N_b. The larger this value, the better the performance. This criterion aims at the overall performance comparison.

2. For a value of ε, the best DB index of RCM and ERCM in the 100 runs is recorded and denoted as DB_{best}. The smaller this value, the better the performance. This criterion aims at the optimal performance comparison.

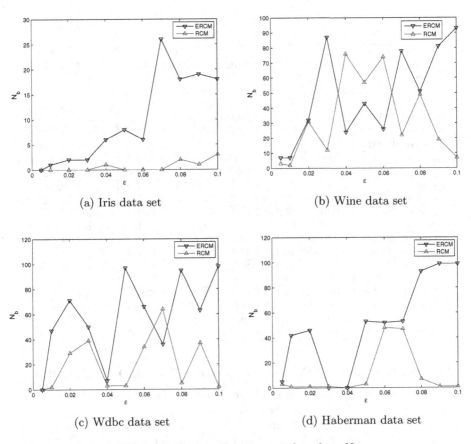

(a) Iris data set   (b) Wine data set

(c) Wdbc data set   (d) Haberman data set

Fig. 4. Performance comparison based on N_b

The results of criterion 1 is depicted in Fig. 4. One can see that:

1. For Iris dataset, ERCM outperforms RCM in 90.9% cases ($\varepsilon = 0.01 - 0.10$). RCM shows performance equivalent to ERCM in $\varepsilon = 0.005$.
2. For Wine dataset, ERCM outperforms RCM in 72.7% cases ($\varepsilon = 0.005, 0.01$ $- 0.03, 0.07 - 0.10$). RCM outperforms ERCM in 27.3% cases ($\varepsilon = 0.04, 0.05, 0.06$).
3. For Wdbc dataset, ERCM outperforms RCM in 81.8% cases ($\varepsilon = 0.01 - 0.06, 0.08 - 0.10$). RCM outperforms ERCM in 9.1% cases ($\varepsilon = 0.07$). RCM shows performance equivalent to ERCM in $\varepsilon = 0.005$.
4. For Haberman dataset, ERCM outperforms RCM in 81.8% cases ($\varepsilon = 0.005, 0.01, 0.02, 0.05 - 0.10$). RCM outperforms ERCM in 9.1% cases ($\varepsilon = 0.03$). RCM shows performance equivalent to ERCM in $\varepsilon = 0.04$.

Obviously, although RCM outperforms ERCM on few values of ε, ERCM shows better performance than RCM for the four tested data sets.

Table 3. Statistical results of DB index

ε	m&v	Iris		Wine		Wdbc		Haberman	
		RCM	ERCM	RCM	ERCM	RCM	ERCM	RCM	ERCM
0.005	mean	1.912E-1	1.912E-1	7.444E-1	7.399E-1	8.266E-1	8.266E-1	3.295E-1	3.293E-1
	variance	1.792E-4	1.792E-4	9.663E-4	5.165E-5	2.440E-30	2.440E-30	9.388E-5	9.288E-5
0.01	mean	1.926E-1	1.923E-1	7.693E-1	7.669E-1	8.981E-1	8.796E-1	3.356E-1	3.223E-1
	variance	2.133E-4	2.052E-4	2.126E-3	1.300E-3	6.047E-3	4.086E-3	1.942E-4	3.282E-13
0.02	mean	1.973E-1	1.971E-1	7.834E-1	7.650E-1	9.563E-1	9.543E-1	3.354E-1	3.366E-1
	variance	2.199E-4	2.195E-4	1.130E-3	1.520E-5	1.475E-4	3.187E-8	1.819E-4	2.240E-4
0.03	mean	1.923E-1	1.913E-1	7.907E-1	7.881E-1	9.578E-1	9.552E-1	3.220E-1	3.220E-1
	variance	2.052E-4	1.929E-4	2.213E-3	2.059E-3	1.555E-4	2.193E-6	4.482E-31	2.530E-13
0.04	mean	1.959E-1	1.939E-1	7.739E-1	7.770E-1	9.077E-1	9.067E-1	3.220E-1	3.220E-1
	variance	2.745E-4	2.994E-4	3.874E-4	7.195E-4	5.617E-5	3.765E-5	4.482E-31	4.482E-31
0.05	mean	1.912E-1	1.866E-1	7.587E-1	7.599E-1	8.934E-1	8.933E-1	3.361E-1	3.329E-1
	variance	1.792E-4	1.528E-4	1.189E-3	1.259E-3	2.604E-5	2.484E-5	1.833E-4	1.422E-4
0.06	mean	1.997E-1	1.909E-1	7.652E-1	7.737E-1	8.850E-1	8.838E-1	3.550E-1	3.541E-1
	variance	2.309E-4	2.958E-4	2.838E-4	7.189E-4	1.161E-5	3.234E-6	1.083E-4	3.823E-5
0.07	mean	2.088E-1	2.007E-1	7.551E-1	7.506E-1	8.755E-1	8.779E-1	3.346E-1	3.336E-1
	variance	1.666E-4	2.318E-4	5.286E-3	1.480E-3	4.528E-6	1.171E-6	2.641E-6	1.224E-5
0.08	mean	2.088E-1	2.008E-1	7.432E-1	7.407E-1	9.281E-1	9.236E-1	3.354E-1	3.335E-1
	variance	2.083E-4	2.919E-4	1.392E-3	3.929E-4	5.228E-5	3.733E-5	2.798E-6	6.816E-6
0.09	mean	2.004E-1	1.955E-1	7.458E-1	7.422E-1	9.307E-1	9.250E-1	3.354E-1	3.296E-1
	variance	1.698E-4	3.491E-4	1.785E-3	1.340E-3	4.703E-5	4.938E-7	1.525E-5	3.382E-6
0.10	mean	2.020E-1	1.968E-1	7.458E-1	7.392E-1	9.252E-1	9.224E-1	3.315E-1	3.281E-1
	variance	4.613E-4	3.362E-4	2.904E-3	1.325E-3	1.407E-6	2.104E-7	1.001E-5	4.224E-6

The statistical results of DB index are shown in Table 3. One can see that:

1. If the mean of DB index is used for overall performance comparison, the result is almost the same as the result shown in Fig.4.
2. For the Iris dataset, there are 45.5% cases that $V_{ERCM} < V_{RCM}$ and $V_{RCM} < V_{ERCM}$, where V_{ERCM} and V_{RCM} denote the variance of DB index for ERCM and RCM, respectively.
3. For the Wine dataset, there are 72.7% cases that $V_{ERCM} < V_{RCM}$ and 27.3% cases that $V_{RCM} < V_{ERCM}$.
4. For the Wdbc dataset, there are 90.9% cases that $V_{ERCM} < V_{RCM}$. $V_{RCM} < V_{ERCM}$ does not happen.
5. For the Haberman dataset, there are 54.5% cases that $V_{ERCM} < V_{RCM}$ and 36.4% cases that $V_{RCM} < V_{ERCM}$.

Thus, the statistical results also show that ERCM yields superior clustering results in comparison with RCM.

The results of criterion 2 is depicted in Fig. 5. One can see that:

1. For Iris dataset, ERCM outperforms RCM in 63.6% cases ($\varepsilon = 0.04 - 0.10$). RCM does not outperforms ERCM. RCM shows performance equivalent to ERCM in $\varepsilon = 0.005, 0.01 - 0.03$.

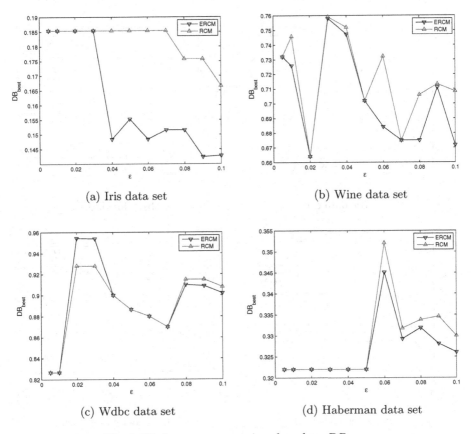

Fig. 5. Performance comparison based on DB_{best}.

2. For Wine dataset, ERCM outperforms RCM in 63.6% cases ($\varepsilon = 0.01, 0.03,$ 0.04, 0.06, 0.08 – 0.10). RCM outperforms ERCM in 9.1% cases ($\varepsilon = 0.07$). RCM shows performance equivalent to ERCM in $\varepsilon = 0.005, 0.02, 0.05$.
3. For Wdbc dataset, ERCM outperforms RCM in 27.3% cases ($\varepsilon = 0.08 -$ 0.10). RCM outperforms ERCM in 18.2% cases($\varepsilon = 0.02, 0.03$). RCM shows performance equivalent to ERCM in $\varepsilon = 0.005, 0.01, 0.04 - 0.07$.
4. For Haberman dataset, ERCM outperforms RCM in 45.5% cases ($\varepsilon = 0.06$ – 0.10). RCM does ont outperforms ERCM. RCM shows performance equivalent to ERCM in $\varepsilon = 0.005, 0.01 - 0.05$.

Thus, the optimal clustering results are closer than the overall clustering results, especially for the Wdbc data set. But in general, ERCM still shows better performance than RCM for the four tested data sets.

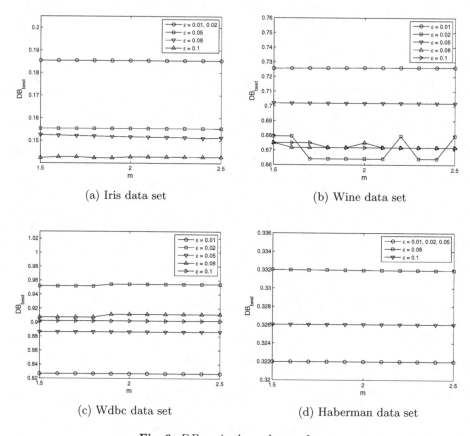

Fig. 6. DB_{best} in dependency of m

4.2 Sensitivity Studies of Parameter m

The second experiment aims at checking whether ERCM is sensitive to changes of m. To this end, we compare the performance of ERCM based on DB_{best} by using different values of m, $m = 1.5, 1.6, \cdots, 2.5$.

The results of the second experiment are depicted in Fig.6. For every dataset, the maximum variance of DB_{best} and its corresponding value of ε is listed in Table 4.

One can see that for the four data sets, the clustering results are not sensitive to small changes of the value of m. So in general, choosing an ordinary value of m (e.g. $m = 2.0$) is sufficient. In contrast, ε must be chosen more carefully.

Table 4. Maximum variance of DB_{best}

Dataset	Maximum variance
Iris	1.9298E-7 ($\varepsilon = 0.08$)
Wine	6.0699E-5 ($\varepsilon = 0.02$)
Wdbc	4.4418E-6 ($\varepsilon = 0.08$)
Haberman	7.8288E-10 ($\varepsilon = 0.08$)

5 Conclusion

Rough c-means algorithm is becoming popular because of its moderate description and restriction in representing clusters. In this paper, we propose an extension to Rough c-means clustering algorithm. In the proposed algorithm, the distinction between data points in the boundary area is captured by using the degree of a data point in the boundary area being close to its corresponding lower approximation. This information is utilized in calculating each cluster's new centroid.

We apply the extended RCM to four data sets from the UCI Machine Learning repository. Experimental results show that contrasted with the original RCM, the proposed algorithm provides superior clustering results.

There exist several issues deserving further investigation. For example, finding the optimal parameters is still a troublesome problem that needs to be solved. And we plan to further investigate the usage of local learning methods within RCM to improve the performance, as well as to test our approach with more empirical studies.

Acknowledgement. This work was supported by Fundamental Research Funds for the Central Universities (ZYGX2012J079 and ZYGX2012J086).

References

1. Mirkin, B.: Mathematical Classification and Clustering. Kluwer Academic Publishers, Dordrecht (1996)
2. Duda, R.O., Hart, P.E., Stork, D.G.: Pattern Classification. John Wiley & Sons, Inc., New York (2001)
3. Tou, J.T., Gonzalez, R.C.: Pattern Recognition Principles. Addison-Wesley, London (1974)
4. Lloyd, S.P.: Least squares quantization in pcm. IEEE Trans. Information Theory, 129–137 (1982)
5. Bezdek, J.C.: Pattern Recognition with Fuzzy Objective Function Algorithms. Plenum, New York (1981)
6. Cannon, R., Dave, J., Bezdek, J.: Efficient implementation of the fuzzy c-means clustering algorithms. IEEE Transactions on Pattern Analysis and Machine Intelligence 8(2), 248–255 (1986)
7. Ng, A.Y., Jordan, M.I., Weiss, Y.: On spectral clustering: Analysis and an algorithm. In: Advances in Neural Information Processing Systems 14, pp. 849–856. MIT Press (2001)

8. Yu, S., Shi, J.: Multiclass spectral clustering. In: Proceedings of the 9th IEEE International Conference on Computer Vision, pp. 313–319. IEEE Computer Society (2003)
9. Jain, A., Dubes, R.: Algorithms for Clustering Data. Prentice-Hall, Englewood Cliffs (1988)
10. Jain, A., Murty, M., Flynn, P.: Data clustering: a review. ACM Computing Surveys (CSUR) 31(3), 264–323 (1999)
11. Smyth, P.: Clustering sequences with hidden markov models. Advances in Neural Information Processing Systems, 648–654 (1997)
12. Rose, K.: Deterministic annealing for clustering, compression, classification, regression, and related optimization problems. Proceedings of the IEEE 86(11), 2210–2239 (1998)
13. Pawlak, Z.: Rough sets: theoretical aspects of reasoning about data. Kluwer Academic Publishers, Boston (1991)
14. Pawlak, Z.: Some Issues on Rough Sets. In: Peters, J.F., Skowron, A., Grzymała-Busse, J.W., Kostek, B., Swiniarski, R.W., Szczuka, M.S. (eds.) Transactions on Rough Sets I. LNCS, vol. 3100, pp. 1–58. Springer, Heidelberg (2004)
15. Lingras, P., West, C.: Interval set clustering of web users with rough k-means. Journal of Intelligent Information Systems 23(1), 5–16 (2004)
16. Pawlak, Z., Skowron, A.: Rough sets and boolean reasoning. Information Sciences 177(1), 41–73 (2007)
17. Bazan, J.G., Skowron, A.: Dynamic Reducts as a Tool for Extracting Laws from Decision Tables. In: Raś, Z.W., Zemankova, M. (eds.) ISMIS 1994. LNCS, vol. 869, pp. 346–355. Springer, Heidelberg (1994)
18. Polkowski, L., Skowron, A.: Rough mereology: A new paradigm for approximate reasoning. International Journal of Approximate Reasoning 15(4), 333–365 (1996)
19. Polkowski, L.: A set theory for rough sets: toward a formal calculus of vague. Fundamenta Informaticae 71(1), 49–61 (2006)
20. Peters, J.F., Szczuka, M.S.: Rough Neurocomputing: A Survey of Basic Models of Neurocomputation. In: Alpigini, J.J., Peters, J.F., Skowron, A., Zhong, N. (eds.) RSCTC 2002. LNCS (LNAI), vol. 2475, pp. 308–315. Springer, Heidelberg (2002)
21. Slowinski, R., Vanderpooten, D.: A generalized definition of rough approximations based on similarity. IEEE Transactions on Knowledge and Data Engineering 12(2), 331–336 (2000)
22. Yao, Y.Y.: Constructive and algebraic methods of theory of rough sets. Information Sciences 109, 21–47 (1998)
23. Yao, Y.Y.: Decision-Theoretic Rough Set Models. In: Yao, J., Lingras, P., Wu, W.-Z., Szczuka, M.S., Cercone, N.J., Ślęzak, D. (eds.) RSKT 2007. LNCS (LNAI), vol. 4481, pp. 1–12. Springer, Heidelberg (2007)
24. Yao, Y.Y.: Three-way decisions with probabilistic rough sets. Information Sciences 180(3), 341–353 (2010)
25. Pal, S., Mitra, P.: Case generation using rough sets with fuzzy representation. IEEE Transactions on Knowledge and Data Engineering 16(3), 292–300 (2004)
26. Zhu, W., Wang, F.: Reduction and axiomization of covering generalized rough sets. Information Sciences 152(1), 217–230 (2003)
27. Zhu, W., Wang, F.: On three types of covering rough sets. IEEE Transactions on Knowledge and Data Engineering 19(8), 1131–1144 (2007)
28. Min, F., He, H.P., Qian, Y.H., Zhu, W.: Test-cost-sensitive attribute reduction. Information Sciences 181, 4928–4942 (2011)
29. Min, F., Zhu, W.: Attribute reduction of data with error ranges and test costs. Information Sciences 211, 48–67 (2012)

30. Mitra, S.: An evolutionary rough partitive clustering. Pattern Recognition Letters 25(12), 1439–1449 (2004)
31. Mitra, S., Banka, H., Pedrycz, W.: Rough-fuzzy collaborative clustering. IEEE Transactions on Systems, Man, and Cybernetics, Part B: Cybernetics 36(4), 795–805 (2006)
32. Mitra, S., Pedrycz, W., Barman, B.: Shadowed c-means: Integrating fuzzy and rough clustering. Pattern Recognition 43(4), 1282–1291 (2010)
33. Peters, G.: Some refinements of rough k-means clustering. Pattern Recognition 39(8), 1481–1491 (2006)
34. Peters, G., Lampart, M., Weber, R.: Evolutionary Rough k-Medoid Clustering. In: Peters, J.F., Skowron, A. (eds.) Transactions on Rough Sets VIII. LNCS, vol. 5084, pp. 289–306. Springer, Heidelberg (2008)
35. Peters, G., Weber, R., Nowatzke, R.: Dynamic rough clustering and its applications. Applied Soft Computing, 3193–3207 (2012)
36. Li, F., Liu, Q.: An Extension to Rough c-Means Clustering. In: Yao, J., Ramanna, S., Wang, G., Suraj, Z. (eds.) RSKT 2011. LNCS, vol. 6954, pp. 208–216. Springer, Heidelberg (2011)
37. Lingras, P.: Applications of Rough Set Based K-Means, Kohonen SOM, GA Clustering. In: Peters, J.F., Skowron, A., Marek, V.W., Orłowska, E., Słowiński, R., Ziarko, W.P. (eds.) Transactions on Rough Sets VII. LNCS, vol. 4400, pp. 120–139. Springer, Heidelberg (2007)
38. Hassanien, A., Abraham, A., Peters, J., Schaefer, G., Henry, C.: Rough sets and near sets in medical imaging: a review. IEEE Transactions on Information Technology in Biomedicine 13, 955–968 (2009)
39. Mitra, S., Barman, B.: Rough-Fuzzy Clustering: An Application to Medical Imagery. In: Wang, G., Li, T., Grzymala-Busse, J.W., Miao, D., Skowron, A., Yao, Y. (eds.) RSKT 2008. LNCS (LNAI), vol. 5009, pp. 300–307. Springer, Heidelberg (2008)
40. Wang, G.: Rough Set Theory and Knowledge Acquisition. Xi'an Jiaotong University Press (2001) (in Chinese)
41. Bottou, L., Vapnik, V.: Local learning algorithms. Neural Computation 4(6), 888–900 (1992)
42. Wu, M., Schölkopf, B.: A local learning approach for clustering. Advances in Neural Information Processing Systems 19, 1529–1536 (2006)
43. Blake, C.L., Merz, C.J.: UCI repository of machine learning databases (1998), http://www.ics.uci.edu/~mlearn/mlrepository.html
44. Bezdek, J., Pal, N.: Some new indexes of cluster validity. IEEE Transactions on Systems, Man, and Cybernetics, Part B: Cybernetics 28(3), 301–315 (1998)

Granular Computing: Topological and Categorical Aspects of Near and Rough Set Approaches to Granulation of Knowledge

Marcin Wolski

Department of Logic and Philosophy of Science,
Maria Curie-Skłodowska University, Lublin, Poland
marcin.wolski@umcs.lublin.pl

Abstract. Knowledge (information) granulation is one of the fundamental concepts of information processing leading to a new discipline called granular computing. One of the basic problems addressed by granular computing is the higher order granulation: how to collect basic information granules into a new granule. In the paper we address this problem by purely mathematical means using two well-established methodologies of information processing: near set theory and rough set theory. We start with the simple fact that the theory of near sets and the theory of rough sets share a common metric root. Since a probe function and an equivalence relation can be regarded as a pseudometric on U, in actual fact the underlying structure of both theories is a family of pseudometrics. The same starting point one can find in metric topology: an arbitrary family of pseudometrics is called a *pregauge structure* and when this family additionally separates all points, it is called a *gauge structure*. Pregauge structures characterise all completely regular spaces, whereas gauge structures correspond to all Hausdorff completely regular spaces (often called *gauge spaces*). In consequence, a perceptual system and an information system can be regarded as both pregauge structures and as topological completely regular spaces. A perceptual system or an approximation space does usually not separate all points and thus does not form a gauge space. Therefore in the paper we introduce the concept of a *separating completion* of a pregauge structure. This notion allows us to build a non-trivial topology on the set of perceptual elementary granules of a perceptual system (or an information system); in other words, a separating completion induces the higher order granulation. The completion requirement induces also a topology on a set of objects, which is locally homeomorphic to the the topology on basic information granules. Apart from topological results, we shall also discuss both topologies using category theory. A perceptual system may be actually enriched to an abelian group or a vector space, while unchanging the original granulation. It also gives rise to a quite rich sheaf of all real-valued functions preserving the basic granulation. Summing up, our aim is to build rich mathematical structures which do not change basic granulation and may be used to solve the problem of higher order granules.

Keywords: rough set, near set, pregauge structure, gauge space, quotient topology, sheaf.

J.F. Peters et al. (Eds.): Transactions on Rough Sets XVI, LNCS 7736, pp. 34–52, 2013.
© Springer-Verlag Berlin Heidelberg 2013

1 Introduction

Generally speaking the present paper is concerned with higher order granulation of objects described by basic systems of rough set theory and near set theory: information systems and perceptual systems. The term *higher order granulation* means, intuitively, granulation of (basic) granules. It is one of the fundamental problems of a new emerging discipline of computer science called granular computing, e.g. [11]. The main scientific effort in granular computing has been devoted mainly to two areas: fuzzy sets (e.g. [19]) and rough sets (e.g. [10]); in the paper we are adding another area, namely, near sets (e.g. [12]). A basic information granule in rough set theory (and to a large extent in near set theory) is an equivalence class $[x]_E$ of some indiscernibility relation E derived from an information system (or perceptual system). The problem of higher order granulation is how to collect the basic granules into "meaningful" concepts (i.e. granules of granules). We are interested in the (purely) theoretical foundations of a such granulation.[1]

Near set theory, introduced by J. Peters [12], is an approach to processing the perceptual information about objects. In the near set approach perceptual information about objects (objects descriptions) is given with respect to probe functions, i.e. real valued functions which represent features of a physical object. Simple examples of probe functions are the size or weight of an object. A set of objects equipped with a family of probe functions is called a *perceptual system*. The rough set approach, introduced by Z. Pawlak [8], starts, in turn, with the concept of an *information system*, where objects are described by means of attributes (which are not necessarily real valued functions); for example, the colour of an object may assign to an object the value "red". Each set of attributes (from an information system) induces an indiscernibility relation E among objects: two objects are indiscernible if they have the same description in terms of these attributes. Thus, a perceptual system is a special kind of information system. On the other hand, due to the area of speciality (perceptual information), near set theory has also some conceptual and methodological autonomy: e.g., the notion of nearness not only differs from indiscernibility but is a more general concept and in consequence all basic notions of rough sets can be obtained within the near set framework. Summing up, although both theories are interrelated, in the course of time, they have also become more independent from each other.

From the perspective of metric topology, both theories share the same root. Actually, each probe function can be regarded as a pseudometric and, in consequence, the starting point of this theory is a family of pseudometrics. The same metric structure can be found in rough set theory too. Although attributes may not be real valued functions, an information system is usually converted into an approximation space: a set of objects U provided with the indiscernibility relation E induced by the set of all attributes. As is well-known, when we define $d(x, y) = 0$ if xEy (and 1 otherwise), then d will be a pseudometric induced by the equivalence relation E. Thus, rough sets are also based on a family of pseudometrics. On the other hand, by putting xEy if $d(x, y) = 0$, we define an equivalence relation induced by a pseudometric d. Thus, both theories also share the families of equivalence relations as starting points. (In what follows, we shall keep track of both types of families.) Given a family of pseudometrics, one may produce a topology

---

[1] The paper is an extension of [18] which was presented at RSKT'2011.

over a given set. This family is often called a *pregauge* structure and all completely regular spaces can be obtained from pregauges. Of course, it follows that perceptual systems and approximation spaces can be regarded as both pregauge structures and completely regular topological spaces. In actual fact, all topological results concerning near sets and rough sets can be obtained in this framework.

Interestingly, we can extend a pregauge structure (a family of pseudometrics) along with a corresponding indiscernibility relation E to richer mathematical objects e.g. abelian groups or vector spaces. As already said, E is the intersection of all equivalence relations e_i induced by pseudometrics d_i by the rule xe_iy only if $d_i(x,y) = 0$. Thus E is defined solely in terms of the value 0 and other possible values are actually regarded as a wild card. In consequence, an abelian group or a vector space extending a given pregauge structure, induces exactly the same indiscernibility relation E. Such extensions provide new means to represent some modifications of data, e.g. the change of contrast of a given picture may be represented as a scalar multiplication. Of course, such extensions no longer consist of pseudometrics, but rather of real valued functions $f : U \times U \to \mathbb{R}$. Given that, once can immediately notice that these functions give rise to a subsheaf of the standard sheaf F_D assigning to each (open) set X functions $f : X \to D$. In this way, we can give simple category theoretic representation of pregauge structures and their extensions.

For a mathematician a pregauge structure or a completely regular topological space is mainly an introduction to a richer structure. If the family of pseudometrics distinguishes all points, then it is called a *gauge structure* and the corresponding topological spaces are called gauge spaces [3]. Gauge spaces viewed as topological spaces have very strong separation properties: they are Hausdorff completely regular spaces. As is well-known, probe functions or attributes usually do not distinguish all objects and thus do not form a gauge space. The simplest solution seems to add to the family $\mathcal{D}(\mathbb{F})$ of pseudometrics induced by a set of probe functions $\mathbb{F}$ a pseudometric d (let us call it *completion*), such that $\{d\} \cup \mathcal{D}(\mathbb{F})$ will become a gauge structure. However, it is quite hard to find a working example of a such completion. On the level of equivalence relations, the things look a bit better. We can introduce a relational *completion* of a pregauge $\mathcal{D}(\mathbb{F})$. That is, the pregauge is regarded as a family $\mathcal{E}$ of equivalence relations, and a completion of $\mathcal{E}$ is a relation R which, added to $\mathcal{E}$, makes this family separating all points of U: for all $x, y \in U$, if xe_iy (for all $e_i \in \mathcal{E}$) and xRy, then $x = y$. Of course, if R is an equivalence relation, then the family of pseudometrics corresponding to $\mathcal{E} \cup R$ will be a gauge structure. However, in both cases we obtain Hausdorff topological spaces whose separating properties are very strong: if the set of objects is finite (a standard case in data analysis), then the space will be discrete.

Therefore, in order to solve this problem, we shall build two separate topologies: the first one induced by the separating completion R of a (perceptual) indiscernibility relation; the second defined on the quotient set induced by this (perceptual) indiscernibility relation. To be more precise, the second topology is defined in terms of the first one by means of a standard topological procedure. Interestingly, they are strongly linked: one can define a local homeomorphism between them (furthermore, classes of R are of special importance). Thus, a separating completion gives us a means of defining a higher order granulation of objects. Furthermore, due to the completion requirement, we can provide also a simple category theoretic description of relations between these two spaces by means of sheaves. In this way, the problem of higher order granulation would be analysed in terms of rich category theory. To complete our considerations we shall also present how these abstract topological structures can be applied to the analysis of perceptual systems provided with an additional relation, namely a preclusivity relation which has been already examined in the context of rough sets [2].

2 Mathematical Preliminaries

In this section we shortly introduce basic concepts from near set theory [12,13,14], rough set theory [8,9,10] and topology (gauge spaces). We shall present only material which is relevant to our study.

2.1 Rough Set Theory

In this section we present basic concepts from rough set theory introduced by Z. Pawlak [8,9,10]. We recall also a simple topological characterisation of approximation spaces [15,17].

Definition 1 (Information System). *A quadruple* $\mathcal{I} = (U, Att, Val, f)$ *is called an* information system, *where:*

- *U is a non-empty finite set of objects;*
- *Att is a non-empty finite set of attributes;*
- *$Val = \bigcup_{A \in Att} Val_A$, where Val_A is the value-domain of the attribute A;*
- *$f : U \times Att \rightarrow Val$ is an information function, such that for all $A \in Att$ and $x \in U$ it holds that $f(x, A) \in Val_A$.*

If f is a total function, i.e. $f(x, A)$ is defined for all $x \in U$ and $A \in Att$, then the information system $\mathcal{I}$ is called complete; *otherwise, it is called* incomplete.

In what follows, we restrict our attention to complete information systems; thus, whenever we write about an information system, we mean a complete information system. The reader interested in foundations of information systems may consult, e.g., [7].

Each subset of attributes $S \subseteq Att$ determines an equivalence relation $IND(S) \subseteq U \times U$ defined as follows:

$$IND(S) = \{(x, y) : \text{ for all } A \in S, \ f(x, A) = f(y, A)\}.$$

As usual, $IND(S)$ is called an indiscernibility relation induced by S, the partition induced by the relation $IND(S)$ is denoted by $U/IND(S)$, and $[x]_S$ denotes the equivalence class of $IND(S)$ defined by $x \in U$. Obviously, $U/IND(Att)$ refines every other partition $U/IND(S)$, where $S \subseteq Att$. So, one can start with a pair (U, E) and assume that $E = IND(Att)$ for some $\mathcal{I} = (U, Att, Val, f)$. The simple generalisation of this observation is given by:

Definition 2 (Approximation Space). *A pair (U, E), where U is a non-empty set and E is an equivalence relation on U, is called an* approximation space. *A subset $X \subseteq U$ is called* definable *if $X = \bigcup \mathcal{B}$ for some $\mathcal{B} \subseteq U/E$, where U/E is the family of equivalence classes of E.*

Definition 3 (Approximation Operators). *Let (U, E) be an approximation space. For every concept $X \subseteq U$, its E-lower and E-upper* approximations *are defined as follows, respectively:*

$$\underline{X} = \{x \in U : [x]_E \subseteq X\},$$
$$\overline{X} = \{x \in U : [x]_E \cap X \neq \emptyset\}.$$

Let $\mathcal{P}(U)$ denote the powerset of U. Then, by the usual abuse of language and notation, the operator $\underline{} : \mathcal{P}(U) \to \mathcal{P}(U)$ sending X to $\underline{X}$ will be called the *lower approximation operator*, whereas the operator $\overline{} : \mathcal{P}(U) \to \mathcal{P}(U)$ sending X to $\overline{X}$ will be called the *upper approximation operator*.

Definition 4 (Approximation Topological Space). *A topological space (U, τ_E) where $\mathcal{U}/E$, the family of all equivalence classes of E, is a subbasis of τ_E and Int is given by*

$$Int(X) = \bigcup \{[x]_E \in U/E : x \in U \text{ and } [x]_E \subseteq X\}$$

is called an approximation topological space.

On this view, a set $X \subseteq U$ is definable only if $X \in \tau_E$. It is worth emphasising that every topological approximation space satisfies the following *clopen set property*: every closed set is open and every open set is closed [15,17].

2.2 Near Set Theory

Near sets were introduced by J. Peters [12]. The algebraic properties of near sets are described in [14].

Definition 5 (Perceptual System). *A perceptual system is a pair $\langle U, \mathbb{F} \rangle$, where U is a non-empty finite set of perceptual objects and $\mathbb{F}$ is a countable set of probe functions $\phi_i : U \to \mathbb{R}$.*

The probe functions describes physical features of objects and usually are regarded as sensors. They also give rise to a number of relations between objects. Let $|\alpha - \beta|$ denote the absolute value of the difference of $\alpha, \beta \in \mathbb{R}$. Then one can formulate:

Definition 6 (Perceptual Indiscernibility Relation). *Let $\langle U, \mathbb{F} \rangle$ be a perceptual system. For every $\mathcal{B} \subseteq \mathbb{F}$, the* perceptual indiscernibility relation $\sim_\mathcal{B}$ *is defined as follows:*

$$\sim_\mathcal{B} = \{(x, y) \in U \times U : \text{for all } \phi_i \in \mathcal{B}, \ \phi_i(x) - \phi_i(y) = 0\}.$$

Of course, this relation is the counterpart of the original indiscernibility relation given by Pawlak in [8]. This induces perceptual elementary sets of the following form:

$$[x]_{\sim_\mathcal{B}} = \{x' \in X \mid x' \sim_\mathcal{B} x\}.$$

Now, one can define perceptual approximation operators as in Definition 3. However, the theory of near sets is actually focused on weaker relations.

Definition 7 (Perceptual Weak Indiscernibility Relation). *Let* $\langle U, \mathbb{F} \rangle$ *be a perceptual system. For every* $B \subseteq \mathbb{F}$, *the* perceptual weak indiscernibility relation $\simeq_B$ *is defined as follows:*

$$\simeq_B = \{(x, y) \in U \times U : \text{for some } \phi_i \in B, \ \phi_i(x) - \phi_i(y) = 0\}.$$

Definition 8 (Perceptual Tolerance Relation). *Let* $\langle U, \mathbb{F} \rangle$ *be a perceptual system and let* $\varepsilon \in \mathbb{R}$. *For every* $B \subseteq \mathbb{F}$, *the* perceptual tolerance relation $\cong_B$ *is defined as follows:*

$$\cong_{B,\epsilon} = \{(x, y) \in U \times U : \text{for all } \phi_i \in B, \ |\phi_i(x) - \phi_i(y)| \leq \varepsilon\}.$$

For notational convenience, this relation is often denoted by $\cong_B$ instead of $\cong_{B,\varepsilon}$ with the assumption that ϵ is inherent to the definition of the tolerance relation. Note that the sets of the form $x_{/\cong_B}$ cover U instead of partitioning it.

Instead of using the standard relational image of a point x, namely, $\{y \in U : xRy\}$, where R represent a tolerance relation, it is better to use preclasses and classes.

Definition 9 (Preclass, Class). $X \subseteq U$ *is called a* preclass *of* $\cong_B$ *iff, for all* $x, y \in X$, *it holds that* $x \cong_B y$. *A preclass* X *is called a* class, *if it is a maximal preclass.*

2.3 Gauge Spaces

Now, we recall those topological notions which are specific to the theory of gauge spaces [3].

Definition 10 (Pseudometric, Premetric). *A function* $d : U \times U \to [0, \infty)$ *such that:*

1. $d(x, y) \geq 0$,
2. $d(x, x) = 0$,
3. $d(x, y) = d(y, x)$,
4. $d(x, z) \leq d(x, y) + d(y, z)$,

for all $x, y \in U$, *is called a* pseudometric. *Dropping the last two axioms leads to the notion of a* premetric.

Of course, for any $\phi : U \to \mathbb{R}$, the map $d_\phi : U \times U \to \mathbb{R}$, defined by

$$d_\phi(x, y) = |\phi(x) - \phi(y)|$$

is a pseudometric. Thus, any probe function induces a pseudometric. An arbitrary family of pseudometric is called a *pregauge structure*.

Definition 11 (Topology from a Pregauge). *Let* $\mathcal{D} = \{d_i : i \in \mathcal{I}\}$ *be a pregauge structure on* U. *The topology* $\tau(\mathcal{D})$ *having for subbasis a set of balls*

$$\mathcal{B}(\mathcal{D}) = \{B(x, d_i, \epsilon) : x \in U, \ d_i \in \mathcal{D}, \ \epsilon > 0\}, \ B(x, d_i, \epsilon) = \{y : d(x, y) < \epsilon\}$$

is called the topology in U *induced by* $\mathcal{D}$.

Definition 12 (Separating Family of Premetrics). *A family* $\mathcal{D} = \{d_i : i \in \mathcal{I}\}$ *of premetrics on* U *is called* separating *if, for each pair of points* $x \neq y$, *there exists* $d_i \in \mathcal{D}$ *such that* $d_i(x, y) \neq 0$.

Definition 13 (Gauge Structure, Gauge Space). *If a pregauge structure* $\mathcal{D} = \{d_i : i \in \mathcal{I}\}$ *is separating, then* $\mathcal{D}$ *is called a* gauge *structure. A topological space* (U, τ) *that admits a gauge structure, i.e.,* $\tau = \tau(\mathcal{D})$, *is called a* gauge *space.*

Definition 14 (Regular Space). (U, τ) *is regular if, given any point* $x \in U$ *and closed set* $F \subseteq U$, *if* $x \notin F$, *then they are separated by neighbourhoods. In fact, in a regular space, any such* x *and* F *will also be separated by closed neighbourhoods.*

Definition 15 (Completely Regular Space). (U, τ) *is completely regular if, given any point* $x \in U$ *and closed set* $F \subseteq U$, *if* $x \notin F$, *then they are separated by a function.*

Of course every completely regular space is regular. The concept of a completely regular space as any other separating axiom needs a special caution. The above definitions come from [16]. However, many authors require that to call a space completely regular space it must additionally be T_0 and hence Hausdorff, e.g. [3,4]. On the other hand, a number of authors additionally call such spaces Tychonoff, e.g. [4]:

Definition 16 (Tychonoff Space). (U, τ) *is Tychonoff, or completely* T_3, *if it is both* T_0 *and completely regular.*

Let us emphasise once again: in the paper we use the latter convention and distinguish completely regular spaces from Tychonoff ones, as it is done in [16], contrary to [3,4].

Proposition 1. *A space* (U, τ) *is a gauge space if and only if it is Tychonoff.*

The proof can be found e.g. in [3]; however, as noted above, in [3] Tychonoff spaces are called completely regular spaces.

Corollary 1. *If* $\mathcal{D}$ *is pregauge structure, then* $\tau(\mathcal{D})$ *is a completely regular topological space. If* $\mathcal{D}$ *is a gauge structure, then* $\tau(\mathcal{D})$ *is additionally Hausdorff (i.e. it is Tychonoff) and when* $\mathcal{D}$ *consists of a single pseudometric* d, *then* d *must be a metric.*

For a proof see [3].

Corollary 2. *An approximation topological space* (U, τ_E) *is completely regular.*

As usual, E may be converted into a pseudometric d by the rule: $d(x, y) = 0$ iff xEy, otherwise $d(x, y) = 1$. On the one hand, for every x and $\epsilon > 1$, the ball $B(x, d, \epsilon) = U$. On the other hand, if $\epsilon \leq 1$, then for every x the ball $B(x, d, \epsilon) = [x]_E$. Thus, every approximation topological space $(U\tau_E)$ is induced by the subbasis of balls $B(x, d, \epsilon)$, and hence is completely regular.

Of course, for a given perceptual system $\langle U, \mathbb{F} \rangle$, $\mathcal{D}(\mathbb{F}) = \{d_{\phi_i} : \phi_i \in \mathbb{F}\}$, where $d_{\phi_i}(x, y) = |\phi_i(x) - \phi_i(y)|$, is a family of pseudometrics which usually does not separate all points of U. Thus, the corresponding topological space is completely regular as well.

Summing up, pregauge structures underlie both near set theory and rough set theory. In topology, this concept is strengthened by adding the condition of separating all points of a given space. In the next section, we examine how this condition can be applied to data analysis.

2.4   Bits of Category Theory

In order to make the presentation self-contained we present here the very basic definitions from category theory. However, we confine ourselves to necessary ones to make our categorical remarks intuitively clear for the reader not familiar with category theory. For a quick introduction to this area see for instance [1].

Definition 17 (Category). *A* category *C consists of:*

- *a class of objects denoted by $|C|$,*
- *a class of arrows (or morphisms) from a to b, denoted by $C(a, b)$, for all $a, b \in |C|$,*
- *a composition operation $\circ : C(b, c) \times C(a, b) \to C(a, c)$, for all $a, b, c \in |C|$,*
- *the identity arrows $id_a \in C(a, a)$, for all $a \in |C|$,*

such that for all $f \in C(a, b), g \in C(b, c), h \in C(c, d)$ the following equations are satisfied:

$$h \circ (g \circ f) = (h \circ g) \circ f$$
$$f \circ id_a = f = id_b \circ f$$

If C is a category then its *dual*, denoted by C^{op}, has the same objects as C but its morphisms are reversed. A category C is *small* if $|C|$ is a set in the sense of Gödel-Bernays set theory. C is called *locally small* if $C(a, b)$ is a set for all its objects a, b. An arrow $f \in C(a, b)$ is an *isomorphism* or *iso* if there exists $g \in C(b, a)$ such that $g \circ f = id_a$ and $f \circ g = id_b$.

A standard example of a category is Set which has sets as objects and total functions as arrows. Let us recall that for any topological space (U, τ) its open sets are partially ordered by the set inclusion $\subseteq$. Now, we can regard τ as a small category where there is an arrow from $X \in \tau$ to $Y \in \tau$ iff $X \subseteq Y$. Such a category will be denoted by $C(\tau)$.

Definition 18 (Functor). *A functor F from a category A to a category B consists of:*

- *a mapping $|A| \to |B|$ of objects; the image of $a \in |A|$ is denoted by Fa,*
- *a mapping $A(a, b) \to B(Fa, Fb)$ of arrows, for all $a, b \in |A|$; the image of $f \in A(a, b)$ is denoted Ff,*

such that for all $a, b, c \in |A|, f \in A(a, b), g \in A(b, c)$, the following conditions are satisfied:

$$F(g \circ f) = Fg \circ Ff \text{ and } Fid_a = id_{Fa}.$$

To give an example, let $C(\tau_1)$ and $C(\tau_2)$ be two categories defined above. Then a functor $F : C(\tau_1) \to C(\tau_2)$ is an order preserving function.

Definition 19 (Presheaf). *Let (U, τ) be a topological space and A be a category; a presheaf F on U with values in A is a functor from $C(\tau)^{op}$ to A.*

In other words, a presheaf F assigns to each open set X of U an object FX of A, and if $X \subseteq Y$, where X and Y are open sets of U, then there exists a morphisms in A, often denoted by ρ_X^Y and called a restriction map, which takes FY to FX, in such a way that for all open sets X, Y, Z of U if

$$X \subseteq Y \subseteq Z \text{ then } \rho_X^Z = \rho_X^Y \circ \rho_Y^Z.$$

Thus $\rho_X^Y \in \mathbf{A}(\mathrm{F}Y, \mathrm{F}X)$. For any $X \in \tau$ elements of $\mathrm{F}X$ are called sections over X and $\rho_X^Y(s)$, for $s \in \mathrm{F}X$, is usually denoted by $s_{|X}$. The standard practise is to take **Set** as **A**, and often a presheaf is defined explicitly as a functor from $\mathbf{C}(\tau)^{op}$ to **Set**.

Let $\{X_i\}_{i \in I}$ be an open covering of $X \in \tau$, that is each set X_i is an element of τ and $\bigcup_{i \in I} X_i = X$. A family of sections $\{s_i\}_{i \in I}$ such that every s_i belongs to $\mathrm{F}(X_i)$ is a coherent family on $\{X_i\}_{i \in I}$ if $s_{i|X_i \cap X_j} = s_{j|X_i \cap X_j}$.

Definition 20 (Sheaf). *A presheaf F is a sheaf if for every open covering $\{X_i\}_{i \in I}$ of X and every coherent family $\{s_i\}_{i \in I}$ on $\{X_i\}_{i \in I}$, there exists a unique element s of $\mathrm{F}X$ such that $s_{|X_i} = s_i$.*

Sheaf is a special kind of a presheaf which allows to reconstruct an object from local data about this object. Simple examples relevant to our study are as follows:

Given a topological space (U, τ) and an arbitrary (plain) set D let $\mathrm{F}_D X$ (where $X \in \tau$) be a set of all functions from X to D; the restriction functions ρ_X^Y are standard restrictions of functions. Then F_D is a sheaf. If D is a topological space, then one can define $\mathrm{F}_C X$ to be the set of all continuous functions from X to D. Once again we obtain a sheaf. If $p : V \to U$ is a continuous function of topological spaces, then one can define a sheaf F_p on U: its sections $\mathrm{F}_p X$ over X (of U) are continuous functions $s : X \to V$ such that $p \circ s = id_X$, where id_X is, as usual, the identity function on X.

3 Pregauges and Completions in the Near Set Framework

Let us recall that we are interested in *granulation of granules*. Such a higher order granulation is important when dealing with complex concepts which cannot be easily described as a subset of a given universe U. In rough set theory this higher order granulation is trivial in the following technical sense:

Definition 21 (Quotient Topology). *The quotient topology on a set V (generated by a map $f : U \to V$ and a topology τ on U) is the collection $\tau_f = \{Y \subseteq V : f^{-1}(Y) \in \tau\}$ of all subsets of V whose preimages are open in U.*

Proposition 2. *An approximation space (U, τ_E) induces a discrete quotient topology on the set U/E via the projection $p : U \to U/E$ sending $x \in U$ to $[x]_E \in U/E$.*

Firstly, $\mathbf{X} \subseteq U/E$ is open provided that $p^{-1}(\mathbf{X})$ is open. Since U/E forms a basis of τ_E, and every equivalence class $[x]_E$ is a point of U/E, the set U/E is, obviously, equipped by p with a discrete topology. In other words, we classify an undefinable concept X by means of the discrete topological space $(U/E, \mathcal{P}(U/E))$. It follows that:

Proposition 3. *Let (U, τ_E) be an approximation space. Then:*

$$\overline{X} = p^{-1}(p(X)),$$

for all $X \subseteq U$.

Since (U, τ) satisfies the clopen-set property, every closed set $Y \in \tau$ is a sum of equivalence classes (i.e. elements of the basis) induced by its elements. Here, we deal with $\overline{X}$ and $p^{-1}(p(X))$ is, in actual fact, this sum.

As usual, an indiscernibility relation E expresses the distinguishing power of our knowledge (encoded by an information system or a perceptual system), and equivalence classes form basic information granules. It would seem, that apart from the quotient topology construction used in rough set theory, there is left no mathematical means to provide U/E with other topology which in some sense would be induced or related to E. So, the process of granulation would be seen as an idempotent one:

$$granulation(granulation(U)) = granulation(U).$$

The main idea of this paper is to apply the intuitions staying behind pregauge and gauge structures to obtain a non-discrete topology on U/E which would be (in some way) induced by E. Of course, U/E "consumes" all pieces of knowledge encoded by E and therefore to define a non-discrete topology on U/E we must use some sophisticated approach, which is dealing with E in an indirect way. As one can guess, a pregauge structure and its completions bring solution to this problem. Interestingly, the main theoretical results presented in the paper nicely fit the framework of near set theory. As presented in Section 2.4, near set theory is based on relational structures, therefore, we shall mainly consider equivalence relations corresponding to pseudometrics. Although in this way we "forget" some information conveyed by pseudometrics, we gain however "freedom" to extend a given perceptual system to richer mathematical structures such as an abelian group or a vector space. So, as usual, this approach has some *pros* and *cons*.

Given a perceptual system $\langle U, \mathbb{F} \rangle$ and $\mathcal{D}(\mathbb{F}) = \{d_{\phi_i} : \phi_i \in \mathbb{F}\}$, two points $x \neq y$ are not separated only if, for all $\phi_i \in \mathbb{F}$, $d_{\phi_i}(x, y) = 0$. As an easy consequence one obtains:

Proposition 4. *For a perceptual system $\langle U, \mathbb{F} \rangle$, the family $\mathcal{D}(\mathbb{F}) = \{d_{\phi_i} : \phi_i \in \mathbb{F}\}$ of pseudometrics on U does not distinguish two points $x \neq y$ iff $x \sim_\mathbb{F} y$.*

Let us recall that family of pseudometrics distinguishes two points x and y if it includes a pseudometric d such that $d(x, y) > 0$. So, it does not distinguish these points only if, for all its pseudometrics d, it holds that $d(x, y) = 0$, which is the very definition of $\sim_\mathbb{F}$. In other words, if the family $\mathcal{D}(\mathbb{F})$ distinguishes all points, then $\sim_\mathbb{F}$ is the identity. As said in the Introduction, our idea is to add to $\mathcal{D}(\mathbb{F})$ a gauge (a pseudometric) d such that $\mathcal{D}(\mathbb{F}) \cup \{d\}$ will become a separating family.

Definition 22 (Corresponding Relation). *Let d be a premetric, then:*

$$E = \{(x, y) : d(x, y) = 0\}$$

is called the relation corresponding *to d.*

Proposition 5. *Let $\mathcal{D} = \{d_i : i \in \mathcal{I}\}$ be a separating family of pseudometrics on U, and let e_i denote the relation corresponding to d_i. Then*

$$E_\mathcal{D} = \bigcap \{e_i : i \in \mathcal{I}\}$$

is the identity.

The result is quite straightforward. Of course, for every $d_i \in \mathcal{D}$, we have $d_i(x, x) = 0$ and $(x, x) \in e_i$, for all $x \in U$. Since $\mathcal{D}$ is separating, for every x and y, such that $x \neq y$, there exists d_j, for which $d_j(x, y) > 0$. Hence, $(x, y) \notin e_j$.

Definition 23 (Corresponding Pseudometric). *Let E be an equivalence relation defined on U, then*

$$d(x, y) = \begin{cases} 0, & \text{if } xEy, \\ 1, & \text{otherwise,} \end{cases}$$

for any $x, y \in U$, is called the pseudometric corresponding *to E.*

Proposition 6. *Let $\mathcal{E} = \{e_i : i \in \mathcal{I}\}$ be a family of equivalence relations on U such that*

$$E = \bigcap \{e_i : i \in \mathcal{I}\}$$

is the identity. Then $\mathcal{D} = \{d_i : i \in \mathcal{I}\}$, where d_i is the pseudometric corresponding to e_i, is a separating family of pseudometrics on U.

It is also quite easy to prove. Since E is the identity, for every x and y, such that $x \neq y$, there must be e_i, which does not include (x, y). So, for the corresponding d_i, it must hold $d_i(x, y) > 0$, which means that $\mathcal{D}$ is a separating family.

Proposition 7. *Let $\langle U, \mathbb{F} \rangle$ be a perceptual system and let $\mathcal{D}(\mathbb{F})$ not be separating. If $\mathcal{D}(\mathbb{F}) \cup \{d\}$ is separating then it holds that:*

$$\text{if } x \sim_{\mathbb{F}} y \text{ and } xey \text{ then } x = y, \tag{1}$$

for all $x, y \in U$, where e is the relation corresponding to d.

As earlier, if $\mathcal{D}(\mathbb{F})$ is not separating, then $\sim_{\mathbb{F}}$ is not the identity relation. Since $\mathcal{D}(\mathbb{F}) \cup \{d\}$ is a separating family, by previous propositions, it holds that $\sim_{\mathbb{F}} \cap e$ is the identity, which amounts to (1).

Thus, we can speak about families of pseudometrics in terms of equivalence relations (what, as noted, has some *pros* and *cons*). Let us for a moment discuss *pros*. Each pseudometric $d \in \mathcal{D}(\mathbb{F})$ is a function from $U \times U$ to the set of non-negative real numbers $\mathbb{R}^+$. Since U is finite, each d can be regarded as a square $n \times n$ matrix $\mathfrak{m}_d$, where n is the number of elements in U, and the entry d_{ij} in each matrix $\mathfrak{m}_d$ is equal to $d(x_i, x_j)$, for $x_i, x_j \in U$ and $d \in \mathcal{D}(\mathbb{F})$. For a finite set U, there is a one-to-one correspondence between functions $f : U \times U \to \mathbb{R}$ and matrices $\mathfrak{m}_f$. Therefore, in what follows, by the standard abuse of language and notation, we do not make any distinction between a function $f : U \times U \to \mathbb{R}$ and its matrix $\mathfrak{m}_f$. When a matrix $\mathfrak{m}_d$ of a pseudometric d is viewed from the perspective of $\sim_{\mathbb{F}}$, then the only important entries are 0s, whereas other values actually do not contribute to E and can be represented by a wild card $*$. For simplicity, suppose that $n = 3$, $U = \{x_1, x_2, x_3\}$ and $\sim_{\mathbb{F}} = \{(x_1, x_1), (x_2, x_2), (x_3, x_3), (x_2, x_3), (x_3, x_2)\}$; then each matrix $\mathfrak{m}_d$, $d \in \mathcal{D}(\mathbb{F})$, viewed through $\sim_{\mathbb{F}}$ has the form depicted by Eq. 2.

$$\mathfrak{m}_d = \begin{pmatrix} d_{11} = 0 & d_{12} = * & d_{13} = * \\ d_{21} = * & d_{22} = 0 & d_{23} = 0 \\ d_{31} = * & d_{32} = 0 & d_{33} = 0 \end{pmatrix} \tag{2}$$

Since $*$ is a wild card, which actually can represent any value from $\mathbb{R}$, the relation $\sim_\mathbb{F}$ can see no difference between pregauge structure $\mathcal{D}(\mathbb{F})$ and the set $\mathfrak{M}$ of all functions $f : U \times U \rightarrow \mathbb{R}$ such that $f(x, y) = 0$ iff $x \sim_\mathbb{F} y$.

As one can guess, $\mathfrak{M}$ is closed under some operations. It is obvious that the matrix addition $\oplus$ of two matrices from $\mathfrak{M}$ returns a matrix from $\mathfrak{M}$. Furthermore, the inverse $-m_d$ of a matrix m_d from $\mathfrak{M}$, which is obtained by replacing each non-zero entry $r \in \mathbb{R}$ of m_d by $-r \in \mathbb{R}$, is also a member of $\mathfrak{M}$. Thus, the set $\mathfrak{M}$ is closed under the operations of addition and taking inverses.

Definition 24 (Abelian Group). *An* abelian group $(U, +)$ *is a set U equipped with a binary operation $+ : U \times U \rightarrow U$ such that for all $x, y \in U$:*

- *Closure: if x and y are two elements in U, then the product $x + y$ is also in U.*
- *Associativity: the operation $+$ is associative, i.e., for all x, y, x in U, $(x + y) + z = x + (y + z)$.*
- *Identity: There is an identity element I that $I + x = x + I = x$ for every element x in U.*
- *Inverse: There must be an inverse of each element: for every x in U, the set U contains an element $y = x^{-1}$ such that $x + x^{-1} = x^{-1} + x = I$.*

Hence $(\mathfrak{M}, \oplus)$ is an abelian group, where $\oplus$ is the standard matrix addition and I is an $n \times n$ matrix m^0 whose all entries are 0's. Another simple operation, which $\mathfrak{M}$ is closed under, is multiplication by scalars.

Definition 25 (Vector Space). *Let F be a field (of which addition and multiplication is denoted by $+$ and $*$ respectively). A vector space is an abelian group $(U, \oplus)$, whose operation $\oplus$ is called vector addition, equipped with the scalar multiplication $\odot : U \times U \rightarrow F$, such that*

- *Distributivity of scalar multiplication with respect to vector addition:*

$$r \odot (v \oplus w) = r \odot v \oplus r \odot w.$$

- *Distributivity of scalar multiplication with respect to field addition:*

$$(r + k) \odot v = r \odot v + k \odot v.$$

- *Compatibility of scalar multiplication with field multiplication:*

$$r * (b \odot v) = (a * b) \odot v.$$

- *Identity element of scalar multiplication $1 \odot v = v$, where 1 is the multiplicative identity in F.*

As the reader may know, matrices provide classic examples of vector spaces: vector addition $\oplus$ is just matrix addition and scalar multiplication $\odot$ is defined in the obvious way (by multiplying each entry by the same scalar). The zero vector is just the zero matrix m^0. Thus, $(\mathfrak{M}, \oplus, \odot)$ is a vector space over the field $\mathbb{R}$. As said earlier, $\sim_\mathbb{F}$ does regard $\mathcal{D}(\mathbb{F})$ and $\mathfrak{M}$ as the same structure, therefore we can enrich $\mathfrak{M}$ to a vector space safely.

Define $x_i \sim_m x_j$ iff for every matrix m in $\mathfrak{M}$ the entry $d_{ij} = 0$.

Proposition 8. *For a perceptual system $\langle U, \mathbb{F} \rangle$ and $\mathfrak{M}$, or its abelian group $(\mathfrak{M}, \oplus)$, or its vector space $(\mathfrak{M}, \oplus, \odot)$, it holds that*

$$\sim_{\mathbb{F}} = \sim_m .$$

Straightforward: by the definition of $\mathfrak{M}$, for every $f \in \mathfrak{M}$ it holds that $f(x, y) = 0$ iff $x \sim_{\mathbb{F}} y$; and by the definition of $\sim_m$, we get $x \sim_m y$ iff for every $f \in \mathfrak{M}$ it holds that $f(x, y) = 0$. Thus $x \sim_{\mathbb{F}} y$ iff $x \sim_m y$.

Summing up, the indiscernibility relation $\sim_{\mathbb{F}}$ is preserved under the extending of $\mathcal{D}(\mathbb{F})$ to some richer structures. Of course, not every m in $\mathfrak{M}$ represents a pseudometric. Such an extension can be useful when dealing with problems for which mathematical means offered by near set theory or rough set theory are too restrictive. E.g. the change of contrast of a given picture could be represented by a multiplication of (some) probe functions by scalars. It is worth to observe that such a change would infect the perceptual tolerance relation and in consequence change the classification induced by this relation. However, the perceptual indiscernibility relation is preserved under a such operation, which (the operation) can additionally be represented in $(\mathfrak{M}, \oplus, \odot)$.

From category theory perspective the family $\mathfrak{M}$ is just a set of functions from $U \times U$ to $\mathbb{R}$, which additionally satisfies the condition that $f(x, y) = 0$ iff $x \sim_{\mathbb{F}} y$. Similar families where considered in Section 2.4 as examples of sheaves: F_D assigning to every open set X all functions from X to D, or $F_{\mathbb{R}}$ assigning to every open set X the set of all continuous functions $f : X \to \mathbb{R}$. So, in order to define a functor $M : \mathbf{C}(\tau)^{op} \to \mathbf{Set}$ in terms of $\mathfrak{M}$ we need a topology τ on $U \times U$. Now, we assume the discrete topology τ; later, having defined some other topologies, we shall return to this sheaf.

Proposition 9. *Let $(U \times U, \tau)$ be a discrete topological space. Define $M(U \times U)$ to be $\mathfrak{M}$, and for every $X \subseteq U \times U$, $M(X)$ to be the set of standard restrictions of functions from $\mathfrak{M}$ to X. Then M is a functor from $\mathbf{C}(\tau)^{op}$ to $\mathbf{Set}$, which is a sheaf.*

Actually, there is not much to prove: M is obviously a subsheaf of $F_D : \mathbf{C}(\tau)^{op} \to \mathbf{Set}$, where $F(X)$ is a set of all functions from X to $\mathbb{R}$. Since τ is discrete, M is also a subsheaf of F_C, assigning to each open set X a set of continuous functions $f : X \to \mathbb{R}$. Hence $\mathfrak{M}$, for a category theorist, is a simple sheaf. As said, it would be good to find more interesting topology on $U \times U$ than the discrete one. In order to do so, we must return to a pregauge $\mathcal{D}(\mathbb{F})$ and its completions.

So far we have been dealing with pseudometrics and corresponding equivalence relations. However, it is very difficult for a given pregauge $\mathcal{D}(\mathbb{F})$ to find a completion being an equivalence relation E, which would still be applicable to data. (It's not by accident that new extensions of rough set theory deal with weaker relations than equivalence relations.) Therefore, we shall generalise this relation to a tolerance relation.

Definition 26 (Separating Completion). *Let $\langle U, \mathbb{F} \rangle$ be a perceptual system such that $\mathcal{D}(\mathbb{F})$ is not a separating family on U. Then a tolerance relation R on U satisfying (1) from Proposition 7 will be called a separating completion of $\mathcal{D}(\mathbb{F})$.*

However, there are other (theoretical) shortcomings concerning both equivalence relations and tolerance relations.

Corollary 3. *Let $\langle U, \mathbb{F} \rangle$ be a perceptual system and let $\mathcal{D}(\mathbb{F})$ not be separating. If an equivalence relation E is a separating completion of $\mathcal{D}(\mathbb{F})$, then $\mathcal{D}(\mathbb{F}) \cup \{d_E\}$, where d_E is the pseudometric corresponding to E, is a gauge structure. Of course, the induced gauge space is discrete.*

Since E is a separating completion of $\mathcal{D}(\mathbb{F})$, whenever $\mathcal{D}(\mathbb{F})$ does not distinguish two different points x and y ($x \neq y$), that is, $x \sim_\mathbb{F} y$, by (1) we get $(x, y) \notin E$, which means $d_E(x, y) > 0$. Otherwise, (x, y) would be an element of $\sim_\mathbb{F} \cap E$ and in consequence $\sim_\mathbb{F} \cap E$ would differ from the identity. Thus, by the very definition, $\mathcal{D}(\mathbb{F}) \cup \{d_E\}$ is a gauge space.

Corollary 4. *Let $\langle U, \mathbb{F} \rangle$ be a perceptual system defined as above. Assume also that R is a separating completion of $\mathcal{D}(\mathbb{F})$, a topology τ is induced by a subbasis $\{e_i(x) : x \in U\}$, $e_i(x) = \{y \in U : x e_i y\}$, where e_i is either the relation corresponding to $d_i \in \mathcal{D}(\mathbb{F})$ or R. Then τ is discrete.*

This is actually trivial. Since R is a separating completion of $\mathcal{D}(\mathbb{F})$, $\bigcap e_i$ is the identity and hence $\bigcap e_i(x) = \{x\}$. Given that from a subbasis we make a basis by taking all finite intersections (let us recall that U is finite), $\{x\}$ belongs to the basis, for every $x \in U$, and in consequence τ is discrete.

Thus, finite topologies induced by separating families of relations (or pseudometrics) are discrete, and hence not applicable to data analysis. Therefore, in what follows, we consider separately a topology induced by a pregauge structure $\mathcal{D}(\mathbb{F})$ and its separating completion. So, we actually pose a new problem: given a perceptual system (or an information system) to find a separating completion which fits the data. It is worth to observe that the separating capabilities of a pregauge structure $\mathcal{D}(\mathbb{F})$, induced by a given perceptual system, are encoded by $\sim_\mathbb{F}$ (which is an analogue of E). Thus, a separating completion is indirectly and non-deterministically induced by $\sim_\mathbb{F}$ (or E). Below we focus on pure theory and present properties fulfilled by any completion. Interestingly, due to the completion requirement, the two topologies are strictly connected. Furthermore, this relationship may be expressed by means of concepts from near set theory.

As usual, when one deals with an equivalence relation E on U, one is interested in the canonical projection $p : U \to U/E$ sending $x \in U$ to $[x]_E$.

Proposition 10. *Let be given a perceptual system $\langle U, \mathbb{F} \rangle$, such that $\mathcal{D}(\mathbb{F})$ is not separating family on U, and let R be a separating completion of $\mathcal{D}(\mathbb{F})$. For every class X of R, it holds that the restriction of $p : U \to U/E_{\mathcal{D}(\mathbb{F})}$ to X, denoted by $p|_X$, is a $1-1$ correspondence.*

The proposition follows from the completion requirement which enforces that every member of a class must be assigned a unique equivalence class of $E_\mathcal{D}$. Otherwise, $\sim_\mathbb{F} \cap R$ would differ from the identity. It also means that for a class X and its image $p(X)$ under p, the function p is bijective.

The standard practice is to use the canonical map p to obtain a quotient topology. Let us recall that in rough set theory this construction brings a discrete topological space. Our aim here is to obtain a non-trivial topology.

Due to Proposition 10 it would be good to have classes of R as open sets.

Proposition 11. *Let X be a class for R. Define $x \leq y$ iff $R(x) \subseteq R(y)$. Then $\leq$ is a preorder and X is an up-set of $\leq$:*

$$X = \{y \in U : x \leq y \text{ for some } x \in X\}.$$

Since X is a class and $x \in X$, we must have $X \subseteq R(x)$. Assume that $x \leq y$, then, by previous observation and the definition of $\leq$, it holds that $X \subseteq R(y)$; that is, y is R-related to all members of X. Given that R is a tolerance relation and that X is a maximal preclass, we obtain $y \in X$. Of course, not every up-set of $\leq$ is a class of R.

Definition 27 (Alexandrov Topological Space). *A topological space (U, τ) is Alexandrov only if its topology is closed under arbitrary intersections.*

As is well-known, given a preorder one can form an Alexandrov topological space in the standard way.

Definition 28 (Alexandrov Topology). *The Alexandrov topology τ on a preordered set $(U, \leq)$ is the family of up-sets of $\leq$:*

$$\tau = \{X \subseteq U : \text{for all } x, y \in U \text{ if } x \in X \text{ and } x \leq y \text{ then } y \in X\}.$$

Corollary 5. *Let be given a perceptual system $\langle U, \mathbb{F} \rangle$, such that $\mathcal{D}(\mathbb{F})$ is not a separating family on U, and let R be a separating completion of $\mathcal{D}(\mathbb{F})$. As above, define $x \leq y$ iff $R(x) \subseteq R(y)$. Then every class of R is an open set in the Alexandrov topology $\tau_{\leq}$ induced by $\leq$.*

Straightforward: every class is an up-set, and every up-set is open.

Thus, we have just obtained a topology τ on U for which classes of R are open sets. Now, we can define the quotient topology on $U/E_{\mathcal{D}(\mathbb{F})}$ (see Definition 21).

Proposition 12. *Let be given a perceptual system $\langle U, \mathbb{F} \rangle$ as defined in Proposition 5, and let X be a class of R. Then $p(X) \in \tau_p$, where τ_p is the quotient topology on $U/E_{\mathcal{D}(\mathbb{F})}$ induced by $\tau_{\leq}$ and the canonical projection $p : U \to U/E_{\mathcal{D}(\mathbb{F})}$.*

As noted earlier, any class X is an open set and $p_{|X}$ is a $1 - 1$ correspondence, thus $p(X)$ must be open in the quotient topology.

Definition 29 (Local Homeomorphism). *A map $f : U \to W$ of topological spaces U and W is a local homeomorphism, if each $x \in U$ has an open neighbourhood Y such that f homeomorphically maps Y onto an open subspace $f(Y)$ of W.*

Proposition 13. *Let be given a perceptual system $\langle U, \mathbb{F} \rangle$ defined as above. Furthermore, let $(U, \tau_{\leq})$ be the Alexandrov topological space generated by $\leq$ defined as above, and $(U/E_{\mathcal{D}(\mathbb{F})}, \tau_p)$ be the quotient topology induced by $\tau_{\leq}$ and $p : U \to U/E_{\mathcal{D}(\mathbb{F})}$. Then $p : U \to U/E_{\mathcal{D}(\mathbb{F})}$ is a local homeomorphism.*

Of course, an open neighbourhood is given by a class of R.

Let us briefly comment the above results in the context of the main topic of this paper: granulation of granulation. As in the framework of rough sets, we deal with a an approximation space $(U, \sim_{\mathbb{F}})$ induced however by a perceptual system $(U, \mathbb{F})$. As in the case of rough sets, we build the quotient set $U/\sim_\mathbb{F}$. Since $\sim_\mathbb{F}$ was "eaten" by $U/\sim_\mathbb{F}$, it cannot be directly used to define a topology on $U/\sim_\mathbb{F}$. That is why in rough set theory $U/\sim_\mathbb{F}$ is given a discrete topology. Our proposal is to take a separating completion R of $\sim_\mathbb{F}$. Firstly, R is not arbitrary relation, but it must fit $\sim_\mathbb{F}$. Thus to an extent our knowledge encoded by $\sim_\mathbb{F}$ restricts possible candidates for R. Secondly, since R is not defined a priori, but must be "discovered" from data, the being offered solution seems flexible enough to be applicable.

Before we go further into theoretical considerations, let us consider a simple application of the above concepts. As already noted, the main construction can be reduced to the case of two relations: $\sim_\mathbb{F}$ and R. In the rough set literature, apart from the indiscernibility (or similarity) relation, there are also considered dissimilarity relations and measures, e.g. A. Gomolińska [5] extends similarity-based approximation spaces by a relation of dissimilarity of objects, or G. Cattaneo [2] introduces an irreflexive and symmetric binary relation called a discernibility or preclusivity relation. In other words, instead of working with a single indiscernibility relation, it becomes more common to use at least two separate relations (as we do). Suppose that we are given a perceptual system $\langle U, \mathbb{F}\rangle$ and independently we are given a preclusivity relation P. Now we would like to show how to apply the concept of separating completion (together with above topologies) in order to reason about data using both relations. Firstly, a preclusivity relation is the complement of a reflexive and symmetric relation called a compatibility or tolerance relation. In what follows, R will denote the complement of P.

The perceptual indiscernibility relation $\sim_\mathbb{F}$ induces a pregauge $\{d\}$, where d is the pseudometric corresponding to $\sim_\mathbb{F}$. To define a separating completion R^c of $\{d\}$, we use the complement of the preclusivity relation P, that is R:

$$C = \{(x, y) : x, y \in U,\ [x]_{\sim_\mathbb{F}} \neq [y]_{\sim_\mathbb{F}}\ \text{and}\ xRy\} \tag{3}$$

Now, define R^c as the reflexive closure of C, i.e., $R^c = C \cup \{(x, x) : x \in U\}$.

Proposition 14. *Let be given a perceptual system $\langle U, \mathbb{F}\rangle$ together with a preclusivity relation P; as above, R is the complement of P. Then the reflexive closure R^c of the relation C defined by (3) is a separating completion of $\mathcal{D}(\mathbb{F})$.*

Of course, R^c is a tolerance relation. Assume that for x and y, such that $x \neq y$, we have $x \sim_\mathbb{F} y$. It means that $[x]_{\sim_\mathbb{F}} = [y]_{\sim_\mathbb{F}}$. Then by (3) $(x, y) \notin C$. Thus, wherever $\mathcal{D}(\mathbb{F})$ does not distinguishes two points x and y, d_{R^c} does it.

Let us recall that starting from a tolerance relation R^c, we can obtain a preorder $\leq$ as follows: $x \leq y$ iff $R^c(x) \subseteq R^c(y)$. As said earlier, given a preorder $\leq$, we can produce the Alexandrov topological space $(U, \tau_\leq)$, which allows us to define the quotient topology on $U/\sim_\mathbb{F}$.

Definition 30 (Approximation Operators). *Let $\langle U, \mathbb{F}\rangle$ be perceptual system equipped with a preclusivity relation P, and let $(U, \tau_{\sim_\mathbb{F}})$ be the approximation space induced by the perceptual indiscernibility relation $\sim_\mathbb{F}$. Define:*

$$\overline{\overline{X}} = p^{-1}(Cl(p(X))),$$

for all $X \subseteq U$, where $p : U \to U/\sim_\mathbb{F}$, Cl is the closure operator of $(U/\sim_\mathbb{F}, \tau)$, and τ is induced by the Alexandrov topological space $(U, \tau_\leq)$ and p.

Proposition 15. *Let $\langle U, \mathbb{F} \rangle$ be a perceptual system equipped with a total preclusivity relation $P = \{(x, y) : x, y \in U\}$. Then $\overline{\overline{X}} = \overline{X}$ for all $X \subseteq U$.*

If P is total, then R will be the empty relation and R^c will become the identity and the induced Alexandrov space will be discrete. In consequence, $Cl(p(X)) = p(X)$, for all $X \subseteq U$, which gives Proposition 3. In other words, in the case of a complete system, we shall obtain the standard approximation operators. Thus, these operators can be regarded as generalisations of operators introduced by Z. Pawlak.

Needless to say, R^c can be replaced by any relation which satisfies the completion requirement. We decided to use the relations already used in rough set theory to show that here and there two relations had been already applied to the framework of rough sets, and so, our proposal is actually an application of this idea by means of completions of pregauge structures.

Now, let us come back to the sheaf $\mathsf{M} : \mathbf{C}(\tau)^{op} \to \mathbf{Set}$, which was defined for the discrete topology τ on $U \times U$. As said then, it would be good to change this topology for a non-trivial one. We have just defined above an Alexandrov topology $\tau_{\leq}$ for U, which is locally homeomorphic to a quotient topology τ on U/E. Let us recall that for a Cartesian product $U_1 \times U_2$ of two spaces (U_1, τ_1) and (U_2, τ_2), the product topology ν on $U_1 \times U_2$ is the topology in which a subset $X \subseteq U_1 \times U_2$ is open only if $p_i(X)$ is open for each i, where $p_i(x_1, x_2) = x_i$. So, the final sheaf would have the following form: $\mathsf{M} : \mathbf{C}(\nu)^{op} \to \mathbf{Set}$, where ν is the product topology of two copies of $(U, \tau_{\leq})$, and

$$\mathsf{M}(U \times U) = \{f : U \times U \to \mathbb{R} : f(x, y) = 0 \text{ iff } x \sim_{\mathbb{F}} y\},$$

where $\mathbb{R}$ (for simplicity) is regarded as a set (i.e. without a topology). The other components of definition work as in $\mathbb{F}_D$; that is, M is a subsheaf of F_D, where D is $\mathbb{R}$. Interestingly, the underlying topological space $(U \times U, \nu)$, has a subspace which is locally homeomorphic to $(U/\sim_{\mathbb{F}}, \tau)$. It comes straightforwardly from the basic topological characteristic of product topologies that each space (U_i, τ_i) is homeomorphic to a subspace of $(U \times U, \nu)$. So, we have passed to a richer structure which still in some sense has a linkage to $\sim_{\mathbb{F}}$.

On the other hand, we can also define a sheaf on $(U/\sim_{\mathbb{F}}, \tau)$. As is well known, e.g. [6], given a local homeomorphism $p : U \to V$ of topological spaces (U, τ_1) and (V, τ_2), we can associate with it a sheaf F of cross-sections:

$$\Gamma(X, U) = \{s : X \to U | p \circ s = id_X\}$$

where s is a continuous function and $X \subseteq V$. Now, the assignment F defined by $\mathsf{F}X = \Gamma(X, U)$ is a functor from $\mathbf{C}(\tau_2)^{op}$ to $\mathbf{Set}$, where for $Y \subseteq X$, $s \in \mathsf{F}X$ is taken to the standard restriction of the function s, i.e. $s_{|Y} \in \mathsf{F}Y$. In actual fact, up to isomorphism, every sheaf on V is of the form $\Gamma(-, U)$, for some local homeomorphism $p : U \to V$. As stated above, for a perceptual system $\langle U, \mathbb{F} \rangle$ (or for an information system) the canonical projection $p : U \to U/E_{\mathcal{D}(\mathbb{F})}$ sending $x \in U$ to the corresponding perceptual elementary set is a local homeomorphism between the topological space $(U, \tau_{\leq})$, where $\leq$ is defined in terms of separating completion R of $\sim_{\mathbb{F}}$, and the family of perceptual elementary sets equipped with the quotient topology $(U/E_{\mathcal{D}(\mathbb{F})}, \tau_p)$.

Summing up, we can construct two sheaves: the sheaf M on the product topological space $(U \times U, \nu)$ and the sheaf F on the quotient topological space $(U/\sim_{\mathbb{F}}, \tau)$. Both shaves take into account the granulation induced by $\sim_{\mathbb{F}}$ and the Alexandrov topology obtained from a tolerance relation R, which is a separating completion of $\mathcal{D}(\mathbb{F})$. As said at the beginning of the paper, we are interested in theoretical foundations of higher order granulations; therefore our aim is to give mathematical structures which provide a means to perform such granulations.

4 Conclusions

The starting point of both near set theory and rough set theory is given by a family of pseudometrics (a pregauge structure) or, considered from a different angle, by a family of equivalence relations. In metric topology, pregauge structures are used to describe the class of completely regular spaces. When a pregauge separates all points, then it is called a gauge structure and the corresponding Hausdorff completely regular space is called a gauge space. In the paper, we have discussed the idea of a completion of a pregauge structure. To make the consideration more flexible (and thus applicable), we have introduced the concept of a relational completion. In the case of a completion being an equivalence relation, we can obtain a gauge space. However, it is quite hard to find a working example of a gauge space in data analysis. In the paper, we have shown how a completion (defined as a tolerance relation) can be used to provide a topology on the family of perceptual elementary sets (that is a higher order granulation of objects). Due to the completion requirement such a granulation can be analysed in terms of sheaves. The paper has also presented a simple application of these very abstract concepts to a perceptual system (or an information system) equipped with an additional preclusivity relation which was already examined in the context of rough sets.

Acknowledgments. Many thanks to Roman Rędziejowski who brought to my attention the topic of gauge spaces. I am also indebted to anonymous referees for the valuable comments and suggestions. The author's research was partially supported by the grant N N516 077837 from Ministry of Science and Higher Education of the Republic of Poland.

References

1. Blute, R., Scott, P.: Category Theory for Linear Logicians. In: Ehrhard, T., Jirard, J., Ruet, P., Scott, P. (eds.) Linear Logic in Computer Science, pp. 3–64. Cambridge University Press (2005)[2]
2. Cattaneo, G., Ciucci, D.: A Quantitative Analysis of Preclusivity vs. Similarity Based Rough Approximations. In: Alpigini, J.J., Peters, J.F., Skowron, A., Zhong, N. (eds.) RSCTC 2002. LNCS (LNAI), vol. 2475, pp. 69–76. Springer, Heidelberg (2002)
3. Dugundji, J.: Topology. Allyn and Bacon, Boston (1966)
4. Engelking, R.: General Topology. Heldermann Verlag (1989)

---

[2] The paper is available at www.cs.bham.ac.uk/ vdp/pscott.pdf

5. Gomolińska, A.: Approximation Spaces Based on Relations of Similarity and Dissimilarity of Objects. Fundamenta Informaticae 79, 319–333 (2007)
6. Mac Lane, S., Moerdijk, I.: Sheaves in Geometry and Logic: A First Introduction to Topos Theory. Springer (1992)
7. Marek, W., Pawlak, Z.: Information storage and retrieval systems, mathematical foundations. Theoretical Computer Science 1(4), 331–354 (1976)
8. Pawlak, Z.: Information systems – Theoretical Foundations. International Journal of Computer and Information Sciences 11, 341–356 (1981)
9. Pawlak, Z.: Rough sets. Int. J. Computer and Information Sci. 11, 341–356 (1982)
10. Pawlak, Z.: Rough Sets: Theoretical Aspects of Reasoning About Data. Kluwer, Dordrecht (1991)
11. Pedrycz, W., Skowron, A., Kreinovich, V. (eds.): The Handbook of Granular Computing. Wiley (2008)
12. Peters, J.F.: Near Sets. Special Theory About Nearness of Objects. Fundamenta Informaticae 75, 407–433 (2007)
13. Peters, J.F., Skowron, A., Stepaniuk, J.: Nearness of Objects: Extension of Approximation Space Model. Fundamenta Informaticae 79, 497–512 (2007)
14. Peters, J.F., Wasilewski, P.: Foundations of Near Sets. Information Sciences 179, 3091–3109 (2009)
15. Rasiowa, H.: Algebraic Models of Logics. Warsaw University Press, Warsaw (2001)
16. Willard, S.: General Topology. Addison-Wesley Publishing Company (1970)
17. Wiweger, A.: On topological rough sets. Bulletin of the Polish Academy of Sciences, Mathematics 37, 89–93 (1989)
18. Wolski, M.: Gauges, Pregauges and Completions: Some Theoretical Aspects of Near and Rough Set Approaches to Data. In: Yao, J., Ramanna, S., Wang, G., Suraj, Z. (eds.) RSKT 2011. LNCS, vol. 6954, pp. 559–568. Springer, Heidelberg (2011)
19. Zimmermann, H.J.: Fuzzy Set Theory - and its Applications. Springer (2001)

The Concept of Reducts in Pawlak Three-Step Rough Set Analysis

Yiyu Yao and Rong Fu

Department of Computer Science, University of Regina
Regina, Saskatchewan, Canada S4S 0A2
{yyao,fu207}@cs.uregina.ca

Abstract. Rough set approaches to data analysis involve removing redundant attributes, redundant attribute-value pairs, and redundant rules in order to obtain a minimal set of simple and general rules. Pawlak arranges these tasks into a three-step sequential process based on a central notion of reducts. However, reducts used in different steps are defined and formulated differently. Such an inconsistency in formulation may unnecessarily affect the elegancy of the approach. Therefore, this paper introduces a generic definition of reducts of a set, uniformly defines various reducts used in rough set analysis, and examines several mathematically equivalent, but differently formulated, definitions of reducts. Each definition captures a different aspect of a reduct and their integration provides new insights.

Keywords: Pawlak three-step analysis, reducts, rough set analysis.

1 Introduction

In his seminal book, *Rough Sets: Theoretical Aspects of Reasoning About Data*, Pawlak [13] provided a simple and elegant method for analyzing data represented in a tabular form. The method can be applied to decision table simplification and rule learning. In our previous paper [21], with a slightly different formulation, we reviewed Pawlak approach and examined several of its variations. More specifically, we introduced a generic notion of a reduct of a set and an explicit expression of a concept by a pair of intension and extension of the concept. We formulated Pawlak approach as a three-step method for analyzing an information table. Our objective was to show that the three steps use three types of reducts, namely, attribute reducts of the table with respect to decision attributes, attribute reducts of an object with respect to decision attributes, and rule reducts. However, due to space limitation, we were only able to provide an outline of our argument. The objective of this paper is to expand our outline into a more complete and thorough investigation.

This paper is different from and complementary to many other studies. It is not intent on proposing a new method nor comparing different methods. The main contribution is to provide a new interpretation of Pawlak approach to data analysis. Our formulation starts with a generic notion of reducts of a set and an

J.F. Peters et al. (Eds.): Transactions on Rough Sets XVI, LNCS 7736, pp. 53–72, 2013.
© Springer-Verlag Berlin Heidelberg 2013

explicit expression of a concept as a pair of a logic formula (i.e., intension of the concept) and a set of objects (i.e., the extension of the concept) [18]. We hope that a reformulation and reinterpretation of Pawlak three-step approach will offer several new insights. The generic notion of reducts unifies the three steps and demonstrates the simplicity and reflexibility of Pawlak approach. Instead of using several forms and definitions of reducts, we use only one general form and one definition. The explicit expression of concepts, in terms of logic formulas as intensions and subsets of objects as extensions, offers new understanding of reducts and rules. In summary, our reformulation, based on a single notion and a uniform exposition, aims at showing the simplicity, elegancy, and flexibility of Pawlak three-step approach at a conceptual level, rather than demonstrating its efficiency at an implementation level. This allows us to focus on the definition of reducts, instead of designing methods for constructing reducts.

For simplicity and clarity, we restrict our discussion to the basic notion of reducts in Pawlak's book. As future work, it will be interesting and worthwhile to investigate if the same argument, with some modifications, can be applied to various generalized notions of reducts, including, dynamic reducts [1], association reducts [16], approximate reducts [11], decision bireducts [17], and many others [5,8,9,10,19,24]. The restriction allows us to concentrate on the basic issues without being distracted by minute details of various generalizations. The results of this paper can be used to relate Pawlak approach to other standard rule learning algorithms, such as partition-based decision-tree methods [15,25] and covering-based sequential covering methods [2,3,4,6,7,26]. While other methods focus mainly on rule learning algorithms, Pawlak approach emphasizes on a study of intrinsic properties of rules independent of a particular rule learning algorithm [14].

The rest of this paper is organized as follows. Section 2 presents an overview of rough set analysis and explicitly expresses such an analysis into a sequential three-step process. Section 3 introduces a general definition of a reduct of a set, examines a simpler definition of a reduct when a monotonic evaluation is used, and investigates an ∩-reduct and an ∪-reduct of a family of subsets of a set. Section 4 is a critical analysis of Pawlak three-step approach. Based on the generic definition of a reduct of a set introduced in Section 3, we study about twenty different definitions of reducts used in rough set analysis. Each definition interprets a reduct from an unique angle and, pooling together, all interpretations provide new insights.

2   An Overview of Pawlak Rough Set Analysis

Rough set analysis (RSA) deals with a finite set of objects called the universe, in which each object is described by values of a finite set of attributes. In his book, Pawlak first used a subset of objects to represent a concept and a partition of the universe to represent a classification at an abstract level. More specifically, he called subsets categories, a partition or equivalently an equivalence relation (classification) knowledge and a family of equivalence relations a knowledge base.

Those notions were later explained by using an information table. Although such a formulation, from abstract notions to concrete examples, provides a more general framework, the meanings of various notions are not entirely clear when they are introduced. For this reason, we start our formulation by directly referring to an information table.

Definition 1. *An information table is the following tuple:*

$$S = (U, At, \{V_a \mid a \in At\}, \{I_a \mid a \in At\}),$$

where U is a finite nonempty set of objects called the universe, At is a finite nonempty set of attributes, V_a is a nonempty set of values for $a \in At$, and $I_a : U \longrightarrow V_a$ is a complete information function that maps an object of U to exactly one value in V_a.

For an object $x \in U$, $I_a(x)$ denotes the value of x on attribute $a \in At$. For notational simplicity, for a subset of attributes $A \subseteq At$, $I_A(x)$ denotes the vector value of x on A.

Definition 2. *A classification table or a decision table is an information table $S = (U, At = C \cup D, \{V_a \mid a \in At\}, \{I_a \mid a \in At\})$, where C is a set of condition attributes and D is a set of classification or decision attributes. If for all objects $x, y \in U$, $I_C(x) = I_C(y)$ implies that $I_D(x) = I_D(y)$, the table is called a consistent classification table, and is called an inconsistent table otherwise.*

An information table provides all available information about a set of objects. We analyze attributes and objects based on the information functions in the table. Pawlak investigated three main tasks of rough set analysis and presented them in a sequential three steps [13], as shown in Figure 1. We use a naming system that is slightly different from the one used in Pawlak's book. More specifically, we use "attribute reduction" and "attribute reduct" instead of "knowledge reduction" and "reduct of knowledge," respectively, and use "attribute-value-pair reduction" and "attribute-value-pair reduct" instead of "reduction of categories" and "value reducts," respectively.

The first step analyzes attribute dependencies with an objective to simplifying a table. The main tasks involving identifying superfluous (i.e., dispensable) attributes and finding a minimal subset of attributes that preserves the same information as the entire set of attributes for the purpose of classification. Such a minimal set of attributes is called an attribute reduct of the table or a relative attribute reduct of a classification table. There may exist more than one reduct for each table. With respect to a reduced table with a minimal set of attributes in a decision table, we can construct a set of decision rules. The left-hand-side of each decision rule is a conjunction of a set of attribute-value pairs.

The second step analyzes dependencies of attribute values with an objective to simplifying a decision rule. Similar to the notion of superfluous attribute in a table, there may exist superfluous attribute-value pairs in the left-hand-side of a decision rule. The main tasks of the second step are to identify superfluous

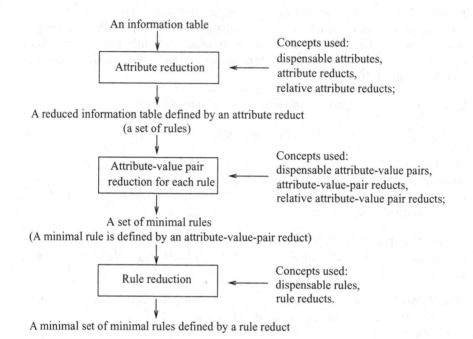

Fig. 1. Pawlak three-step rough set analysis

attribute-value pairs and to derive a minimal set of attribute-value pairs for each decision rule. A minimal set of attribute-value pairs is called a relative attribute-value-pair reduct. Again, there may exist more than one reduct. The result of the second step is a set of minimal decision rules.

The third step analyzes dependencies of decision rules with an objective to simplifying a set of decision rules. There may exist superfluous (i.e., dispensable) rules in the set of decision rules obtained in the second step. By removing superfluous rules, one can obtain a minimal set of rules called a rule reduct.

In Pawlak's book, the three steps are clearly separated. As pointed out by a reviewer of this paper, the steps of attribute reduction, attribute-value reduction and rule reduction do not need to follow each other. In applications they may occur optionally or independently. For example, attribute-value reduction may be treated as a special case of attribute reduction, which leads toward merging the first two above-mentioned steps together. In some cases, rule reduction may be avoided; one may simply use techniques based on voting to deal with redundant or conflicting rules.

Although each of the three steps involves different entities or subjects, they share high-level similarities. All analyze relationships between entities with an objective to make simplification by removing superfluous entities. More importantly, the result of simplification is a reduct, namely, an attribute reduct of a table, an attribute-value-pair reduct of a rule, and a rule reduct of a set of

rules. This observation suggests that one may unify the three steps. However, in rough set literature, different forms and definitions are given for different types of reducts. In rest of this paper, we present a generic definition of reducts and show that Pawlak three-step analysis may be uniformly formulated based on a generic definition of reducts.

3   A General Definition of Reducts

We introduce a general definition of reducts of a set and examine basic properties of reducts.

3.1   Reducts of a Set

Reducts are a fundamental notion of rough set analysis. As showed in the last section, different types of reducts have been proposed and studied. Consider an attribute reduct of a table, intuitively speaking, an attribute reduct is a subset of attributes that preserve the same information or property as the entire set of attributes (i.e., the sufficiency condition) and at the same time contains no superfluous attributes (i.e., non-redundancy condition). This interpretation of an attribute reduct can be generalized into a generic definition of a reduct of any set. First, we specify a property such that the entire set has the property. Then, we state the sufficiency and non-redundancy conditions on a subset of the set for it to be a reduct. The sufficiency condition suggests that a reduct has the same property as the entire set. The non-redundancy condition requires that a reduct must be a minimal subset having the property.

Definition 3. *Suppose S is a finite set and 2^S is the power set of S. Let $\mathbb{P}$ denote a unary predicate on subsets of S, that is, for $X \subseteq S$, $\mathbb{P}(X)$ stands for the statement that "subset X has the property $\mathbb{P}$." An evaluation e of $\mathbb{P}$ is understood as a truth assignment for every subset of S: $\mathbb{P}_e(X)$ is true if X has property $\mathbb{P}$, otherwise, it is false.*

An evaluation typically depends on a particular data set. For example, an evaluation of subsets of attributes is determined by a particular information table. A reduct of S is therefore defined with respect to a given evaluation. We use a subscript e to explicitly denote the evaluation.

Definition 4. *Given an evaluation e of $\mathbb{P}$, A subset $R \subseteq S$ is called a reduct of S if it satisfies the following conditions:*

$$(w) \quad \mathbb{P}_e(S),$$
$$(s) \quad \mathbb{P}_e(R),$$
$$(n) \quad \forall B \subset R, (\neg \mathbb{P}_e(B)).$$

Condition (w) requires that the whole set S must have the property $\mathbb{P}$. In many studies, this condition is typically implicitly assumed or embedded in $\mathbb{P}$. It ensures that a reduct of S exists. Condition (s) is a sufficiency condition, stating that a reduct R of S is sufficient for preserving property $\mathbb{P}$ of S. Condition (n) is a non-redundancy condition, indicating that none of the proper subsets of R has the property.

According to Definition 4, it is necessary to check all proper subsets of R in order to verify if R is a reduct. This imposes an unpractical computational constraint. In many situations, one can study a special class of property $\mathbb{P}$ that satisfies the monotonicity with respect to set inclusion.

Definition 5. *A predicate $\mathbb{P}$ is said to be monotonic with respect to set inclusion if it satisfies the following property:*

$$\forall A, B \subseteq S, (A \subseteq B \Longrightarrow (\mathbb{P}(A) \Longrightarrow \mathbb{P}(B))). \tag{1}$$

The monotonicity states that if a subset has a property, then a superset of it also has the property. It is important to point out that, unlike the definition of a reduct, the monotonicity of $\mathbb{P}$ is defined based on all possible evaluations. That is, the monotonicity must hold for all possible evaluations. In terms of information tables, each table determines an evaluation and all possible tables determine all possible evaluations. The monotonicity of a predicate must hold for all possible information tables. The monotonicity can be equivalently re-expressed as

$$\forall A, B \subseteq S, (A \subseteq B \Longrightarrow (\neg \mathbb{P}(B) \Longrightarrow \neg \mathbb{P}(A))). \tag{2}$$

That is, if a set does not have the property, then none of its subsets has the property. Thus, once we know that a set does not have the property, we do not need to check its subsets. This leads to a simplified definition of reducts.

Definition 6. *Suppose $\mathbb{P}$ satisfies monotonicity. Given an evaluation e of $\mathbb{P}$, a subset $R \subseteq S$ is called a reduct of S if it satisfies the following properties:*

$$(w) \quad \mathbb{P}_e(S)$$
$$(s) \quad \mathbb{P}_e(R)$$
$$(n) \quad \forall a \in R, (\neg \mathbb{P}_e(R - \{a\}))$$

Condition (n) shows that each element $a \in R$ is necessary. That is, elements of R are individually necessary. With the monotonicity, a verification of a reduct becomes easier, one only needs to check individual elements from S based on condition (n) instead of all subsets of R. A reduct is always defined with respect to a particular evaluation. In the rest of this paper, for notational simplicity we sometimes omit the subscript e by simply writing $\mathbb{P}_e(X)$ as $\mathbb{P}(X)$ for subset $X \subseteq S$. It may be commented that many definitions of reducts in rough set theory obey the monotonicity.

In the study of reducts, there are two additional important notions. The first one is superfluous or redundant elements and the second one is core elements. The concept of superfluous element is only applicable when considering monotonic evaluations. One can also define generic notions of redundant elements and core elements.

Definition 7. *Suppose $\mathbb{P}$ satisfies monotonicity. Given an evaluation e of $\mathbb{P}$, an element a is called a redundant element if it satisfies the following properties:*

$$\text{(r1)} \quad \mathbb{P}_e(S)$$
$$\text{(r2)} \quad \mathbb{P}_e(S - \{a\})$$

Condition (r2) states that when removing an element a from the set S, the rest elements of the set still satisfy the property $\mathbb{P}$. That is to say, the element a is unnecessary and dispensable for preserving the property $\mathbb{P}$ and one can have a same result without considering the element a in S. Therefore, we say a is redundant in S.

Definition 8. *Suppose $\mathbb{P}$ satisfies monotonicity. Given an evaluation e of $\mathbb{P}$, an element a is called a core element if it satisfies the following properties:*

$$\text{(c1)} \quad \mathbb{P}_e(S)$$
$$\text{(c2)} \quad \neg\mathbb{P}_e(S - \{a\})$$

That is, for a set S satisfying property $\mathbb{P}$, if we remove element a from S, the rest elements can no longer preserve property $\mathbb{P}$. Therefore, the element a is necessary and indispensable for keeping property $\mathbb{P}$.

Definition 9. *Given a set S, let $RED(S)$ denote the family of all reducts of S, the set of core elements of S can be defined as follows:*

$$CORE(S) = \bigcap RED(S). \tag{3}$$

The $CORE$ is the intersection of all reducts, in other words, elements in $CORE$ are included in every reduct. Therefore, the $CORE$ is the most important subset that none of its elements can be eliminated for preserving a specific property.

3.2 Reducts of a Family of Subsets of a Set

The proposed definition of reducts is flexible. As an example, we consider a set S whose elements are subsets of a set. Given a set W, suppose $S \subseteq 2^W$ is a family of subsets of W. According to definition of reducts in Definition 6, we introduce $\cap$-reducts and $\cup$-reducts of S.

Definition 10. [13] *Suppose W is a finite set and $S \subseteq 2^W$. A set $R \subseteq S$ is called an $\cap$-reduct of S if it satisfies the following conditions:*

$$\text{(w)} \quad \cap S = \cap S,$$
$$\text{(s)} \quad \cap R = \cap S,$$
$$\text{(n)} \quad \forall a \in R, (\neg(\cap(R - \{a\}) = \cap S)).$$

A set $Q \subseteq S$ is called an $\cup$-reduct of S if it satisfies the following conditions:

$$\text{(w')} \quad \cup S = \cup S,$$
$$\text{(s')} \quad \cup Q = \cup S,$$
$$\text{(n')} \quad \forall a \in Q, (\neg(\cup(Q - \{a\}) = \cup S)).$$

Conditions (w) and (w') simply state that S has the property. We explicit list them to show the connection to Definition 6.

In some situations, we want to use a family of subset S to represent other subsets of W. This leads to a definition of relative reducts.

Definition 11. [13] *Suppose $S \subseteq 2^W$ is a family of subsets of a finite set W and $T \subseteq W$ is a subset of W. An $\cap$-reduct of S relative to T, or simply an $\cap$-relative-reduct is defined by the following conditions,*

$$(w) \quad \cap S \subseteq T,$$
$$(s) \quad \cap R \subseteq T,$$
$$(n) \quad \forall a \in R, (\neg(\cap(R - \{a\}) \subseteq T)).$$

Condition (w) states that the family S has the property of $\cap S \subseteq T$. We will show later that those reducts form a basis of rough set analysis.

4   A Critical Analysis of Pawlak Three-Step Approach

In this section, we provide a critical analysis of Pawlak three-step approach based on the notion of reducts introduced in the last section.

4.1   Rough Set Approximations

Rough set theory analyzes an information table based on equivalence relations (i.e., reflexive, symmetric and transitive relations) induced by subsets of attributes [12,13].

Definition 12. [13] *Given an information table, a subset of attributes $A \subseteq At$ defines an equivalence relation on U as follows:*

$$xE_A y \Longleftrightarrow \forall a \in A, (I_a(x) = I_a(y))$$
$$\Longleftrightarrow I_A(x) = I_A(y). \tag{4}$$

That is, x and y are equivalent if and only if they have the same values on all attributes in A. The equivalence relation E_A induces a partition of the universe and is denoted by $U/E_A = \{[x]_{E_A} \mid x \in U\}$, where $[x]_{E_A} = \{y \mid xE_A y\}$ is the equivalence class containing x.

There is a one-to-one correspondence between all equivalence relations on U and all partitions of U. Therefore, we use equivalence relations and partitions interchangeably. An equivalence relation E is a set of pairs, that is, $E \subseteq U \times U$, where $U \times U$ is the cartesian product of U and U. One can apply set-theoretic operations and relations on equivalence relations. If E_1 and E_2 are two equivalence relations, then $E_1 \cap E_2$ is also an equivalence relation.

The standard set inclusion of equivalence relations defines a partial order on partitions as follows: for two equivalence relations E_1 and E_2,

$$U/E_1 \preceq U/E_2 \Longleftrightarrow E_1 \subseteq E_2. \tag{5}$$

If $U/E_1 \preceq U/E_2$, each block of U/E_2 must be a union of some blocks of U/E_1, and the partition U/E_1 is called a refinement of U/E_2 and U/E_2 a coarsening of U/E_1.

With respect to different subsets of attributes, we can establish the following relationships, for $A, B \subseteq At, x \in U$,

$$(1) \qquad E_A = \bigcap_{a \in A} E_{\{a\}},$$

$$E_{A \cup B} = E_A \cap E_B,$$

$$(2) \qquad [x]_{E_A} = \bigcap_{a \in A} [x]_{E_{\{a\}}},$$

$$[x]_{E_{A \cup B}} = [x]_{E_A} \cap [x]_{E_B},$$

$$(3) \qquad A \subseteq B \Rightarrow E_B \subseteq E_A,$$

$$A \subseteq B \Rightarrow U/E_B \preceq U/E_A. \qquad (6)$$

Properties (1) and (2) show that the equivalence relation, or the corresponding partition, defined by a subset of attributes can be constructed from the individual equivalence relations or partitions defined by singleton subsets of attributes. Property (3) states that the refinement-coarsening relation $\preceq$ is monotonic with respect to set inclusion of sets of attributes.

Consider the equivalence relation E_A defined by a subset of attributes $A \subseteq At$. For a subset $X \subseteq U$, its lower and upper approximations are defined by [12,13]:

$$\underline{apr}(X) = \bigcup \{ [x]_{E_A} \in U/E_A \mid [x]_{E_A} \subseteq X \};$$

$$\overline{apr}(X) = \bigcup \{ [x]_{E_A} \in U/E_A \mid [x]_{E_A} \cap X \neq \emptyset \}. \qquad (7)$$

That is, the lower approximation $\underline{apr}(X)$ is the union of those equivalence classes that are subsets of X, and the upper approximation $\overline{apr}(X)$ is the union of those equivalence classes that have nonempty intersection with X. By the lower and upper approximation, one can divide the universe U into three pair-wise disjoint regions [12], namely, the positive region $\mathrm{POS}(X)$, the boundary region $\mathrm{BND}(X)$, and the negative region $\mathrm{NEG}(X)$:

$$\mathrm{POS}(X) = \underline{apr}(X),$$

$$\mathrm{BND}(X) = \overline{apr}(X) - \underline{apr}(X),$$

$$\mathrm{NEG}(X) = U - \overline{apr}(X) = (\overline{apr}(X))^c, \qquad (8)$$

where $(\cdot)^c$ denotes the set complement. Some of these regions may be empty. The pair of lower and upper approximations and the three regions uniquely define each other. One can formulate the theory of rough sets by using any one of them.

Based on these notions, we are ready to review Pawlak three-step approach to data analysis.

4.2   Step 1: Analysis of Attribute Dependencies

Pawlak refers to partitions, or equivalently equivalence relations, defined by subsets of attributes as classification knowledge or simply classification. Analysis of attribute dependencies is performed through equivalence relations defined by subsets of attributes.

Reducts of an Information Table. Consider first the notion of reducts of an information table. Pawlak introduces the notion of a reduct of a family of partitions or equivalence relations. Since each attribute defines an equivalence relation, we can use a Pawlak reduct of a family of equivalence relations to define an attribute reduct of an information table.

Definition 13. *Given an information table, consider the family of equivalence relations defined by singleton subsets of attributes* $S = \{E_{\{a\}} \mid a \in At\}$. *A reduct of S is defined as a subset $R \subseteq S$ satisfying the following conditions:*

$$(s1) \quad \cap R = \cap S,$$
$$(n1) \quad \forall E \in R, (\neg(\cap(R - \{E\}) = \cap S)). \tag{9}$$

In this definition, the condition $\cap S = \cap S$ is not explicitly given. Recall that an equivalence relation is a set of pairs. It follows that $S \subseteq 2^{U \times U}$. Therefore, according to Definition 10, a reduct as defined by Definition 13 is in fact an $\cap$-reduct of S.

There is only a small problem when characterizing an information table by the family of equivalence relations $\{E_{\{a\}} \mid a \in At\}$. Two different attributes $a, b \in At$ may define the same equivalence relation, that is, $E_{\{a\}} = E_{\{b\}}$. To resolve the problem, Pawlak treats all those attributes that define the same equivalence relation as one attribute. According to Definition 6, the following definition resolves this problem by directly referring to the set of attributes At.

Definition 14. *In an information table, an attribute reduct is a subset of attributes $R \subseteq At$ satisfying each of the following equivalent pairs of conditions:*

equivalence relation based conditions :
$$(s2) \quad E_R = E_{At},$$
$$(n2) \quad \forall a \in R, (\neg(E_{R-\{a\}} = E_{At}));$$
partition based conditions :
$$(s3) \quad U/E_R = U/E_{At},$$
$$(n3) \quad \forall a \in R, (\neg(U/E_{R-\{a\}} = U/E_{At}));$$
equivalence class based conditions :
$$(s4) \quad \forall x \in U, ([x]_{R_R} = [x]_{E_{At}}),$$
$$(n4) \quad \forall a \in R \exists x \in U, (\neg([x]_{E_{R-\{a\}}} = [x]_{E_{At}})). \tag{10}$$

The definition contains both commonly used conditions based on equivalence relations or partitions and new conditions based on equivalence classes. Each pair of conditions provides a different characterization and understanding of a reduct. That is, a reduct is a minimal set of attributes that defines the same equivalence relation as E_{At}. The last pair of conditions is particularly interesting

and is closely related to the conditions for defining attribute-value-pair reducts. Again, the first condition of a general reduct is not explicitly stated. For example, we omit the condition $E_{At} = E_{At}$. By Definition 6, an attribute reduct is an exapmle of a reduct of the set of attributes At.

Relative Reducts of a Consistent Classification Table. For analyzing a classification table, Pawlak introduces the notion of a relative reduct. Based on an equivalence relations defined by subsets of attributes, a classification table with $At = C \cup D$ is consistent if

$$E_C \subseteq E_D, or\ equivalently\ U/E_C \preceq U/E_D. \tag{11}$$

For a consistent classification table, similar to Definition 13, a relative reudct can be defined according to Definition 11.

Definition 15. *Consider the set of all equivalence relations defined by singleton subsets of condition attributes $S = \{E_{\{a\}} \mid a \in C\}$. A reduct of S relative to E_D is a subset $R \subseteq S$ satisfying the following properties:*

$$\begin{aligned}
&\text{(w)} \quad \cap S \subseteq E_D, \\
&\text{(s)} \quad \cap R \subseteq E_D, \\
&\text{(n)} \quad \forall E \in R, (\neg(\cap(R - \{E\}) \subseteq E_D)).
\end{aligned} \tag{12}$$

Similar to Definition 14, a relative attribute reduct can also be equivalently defined by using equivalence relations, partitions, and equivalence classes, respectively.

Definition 16. *Given a consistent classification table $S = (U, At = C \cup D, \{V_a \mid a \in At\}, \{I_a \mid a \in At\})$, a subset $R \subseteq C$ is called a reduct of C relative to D, or simply a relative reduct, if R satisfies one of the following equivalent pairs of conditions:*

equivalence relation based conditions :
$$\begin{aligned}
&\text{(s5)} \quad E_R \subseteq E_D, \\
&\text{(n5)} \quad \forall a \in R, (\neg(E_{R-\{a\}} \subseteq E_D));
\end{aligned}$$
partition based conditions :
$$\begin{aligned}
&\text{(s6)} \quad U/E_R \preceq U/E_D;, \\
&\text{(n6)} \quad \forall a \in R, (\neg(U/E_{R-\{a\}} \preceq U/E_D));
\end{aligned}$$
equivalence class based conditions :
$$\begin{aligned}
&\text{(s7)} \quad \forall x \in U, ([x]_{E_R} \subseteq [x]_{E_D}), \\
&\text{(n7)} \quad \forall a \in R \exists x \in U, (\neg([x]_{E_{R-\{a\}}} \subseteq [x]_{E_D})).
\end{aligned} \tag{13}$$

Conditions in the definition suggest that a relative reduct R is a minimal set of attributes whose partition U/E_R is the same or finer than E_D. For example, condition (s7) states that the equivalence class of E_R containing x is a subset of the equivalence class of E_D containing x. That means that one can infer the equivalence class $[x]_{E_D}$ from the equivalence class $[x]_{E_R}$ so that it preserves the descriptive ability for classification. Condition (n7) states that any attribute in R is necessary for inferring $[x]_{E_D}$.

Relative Reducts of an Inconsistent Classification Table. For an inconsistent classification table, Pawlak defines a relative reduct by using the positive region of the classification U/E_D induced by E_C:

$$\text{POS}_{E_C}(U/E_D) = \bigcup \{\text{POS}_{E_C}(K) \mid K \in U/E_D\}$$
$$= \bigcup \{\underline{apr}_{E_C}(K) \mid K \in U/E_D\}. \tag{14}$$

More specifically, a relative reduct of an inconsistent classification table is a minimal set of attributes that preserves the positive region of U/E_D; there is not consideration of objects not in the positive region.

Definition 17. [13] *Given a consistent classification table $S = (U, At = C \cup D, \{V_a \mid a \in At\}, \{I_a \mid a \in At\})$, a subset $R \subseteq C$ is called a reduct of C relative to D if R satisfies the two conditions:*

$\quad$ (s8) $\quad \text{POS}_{E_R}(U/E_D) = \text{POS}_{E_C}(U/E_D)$,

$\quad$ (n8) $\quad \forall a \in R, (\neg(\text{POS}_{E_{R-\{a\}}}(U/E_D) = \text{POS}_{E_C}(U/E_D)))$.

For a consistent classification table, we have $\text{POS}_{E_C}(U/E_D) = U$. Pawlak's definition is therefore applicable to both consistent and inconsistent classification tables.

$\quad$ Although Pawlak's definition is an example of a relative reduct of the set of condition attribute C, it is very different in form from the definition of a relative reduct of a consistent classification table as given by Definitions 15 and 16. By insisting on having the same positive region, a relative reduct does not care about objects in the boundary region. This observation provides a hint: one can transform an inconsistent table into a consistent table by focusing on individual positive regions of equivalence classes of E_D so that the definition of a relative reduct of a consistent table can be used. According to the positive regions of equivalence classes in U/E_D, we can form the following partition:

$$\{\text{POS}_{E_C}(K) \neq \emptyset \mid K \in U/E_D\} \cup (\{\cup\{\text{BND}_{E_C}(K) \mid K \in U/E_D\}\} - \{\emptyset\}).\text{(15)}$$

Suppose $E_{D'}$ is the equivalence relation corresponding to this partition, and the partition can be denoted by $U/E_{D'}$. According to the partition $U/E_{D'}$, a relative reduct of an inconsistent table can be defined based on Definition 16.

Definition 18. *Given a consistent classification table $S = (U, At = C \cup D, \{V_a \mid a \in At\}, \{I_a \mid a \in At\})$, a subset $R \subseteq C$ is called a reduct of C relative to D if R satisfies any of the following equivalent pairs of conditions:*

equivalence relation based conditions :

(s9) $E_R \subseteq E_{D'}$,

(n9) $\forall a \in R, (\neg(E_{R-\{a\}} \subseteq E_{D'}))$;

partition based conditions :

(s10) $U/E_R \preceq U/E_{D'}$,

(n10) $\forall a \in R, (\neg(U/E_{R-\{a\}} \preceq U/E_{D'}))$;

equivalence class based conditions :

(s11) $\forall x \in U, ([x]_{E_R} \subseteq [x]_{E_{D'}})$,

(n11) $\forall a \in R \exists x \in U, (\neg([x]_{R-\{a\}} \subseteq [x]_{E_{D'}}))$.   (16)

A consistent table is a special case of inconsistent tables. For a consistent table, $U/E_{D'} = U/E_D$ and, hence, the definition is also valid for a consistent table. An advantage of Definition 18 is that it is in a uniform format as relative of a consistent table. However, we need to construct a partition $U/E_{D'}$ originally not in the table. If we examine the Pawlak definition again, we find that keeping the positive region is equivalent to saying that $\forall x \in U, ([x]_{E_C} \subseteq [x]_{E_D} \implies [x]_{E_R} \subseteq [x]_{E_D})$. Based on this observation, we can have another definition of a relative reduct of an inconsistent table.

Definition 19. *Given an inconsistent classification table $S = (U, At = C \cup D, \{V_a \mid a \in At\}, \{I_a \mid a \in At\})$, a subset $R \subseteq C$ is called a reduct of C relative to D if R satisfies the two conditions:*

(s12) $\forall x \in U, ([x]_{E_C} \subseteq [x]_{E_D} \implies [x]_{E_R} \subseteq [x]_{E_D})$,

(n12) $\forall a \in R \exists x \in U, (\neg([x]_{E_C} \subseteq [x]_{E_D} \implies [x]_{E_{R-\{a\}}} \subseteq [x]_{E_D}))$.   (17)

For a consistent table, $[x]_{E_C} \subseteq [x]_{E_D}$ is true for all $x \in U$. In this case, conditions (s12) and (n12) are equivalent to conditions (s7) and (n7). Therefore, Definition 19 is also valid for a consistent table. An advantage of this definition is that we do not need to introduce any extra structures not given in the table.

Relative Reducts and Attribute Dependencies. The relationship between the set of condition attributes C and the set of decision attributes D can be easily extended to a study of dependency of any two sets of attributes in an information table.

Consider two arbitrary subsets of attributes $A, B \subseteq At$ in an information table. The two subsets may have an nonempty intersection. If $E_A \subseteq E_B$ holds, we say that B depends on A. In this paper, we only consider two sets of attributes with a full dependency. This is similar to a consistent classification table with $E_C \subseteq E_D$. By applying the results of a relative reduct of a consistent classification table, it is straightforward to define a reduct of A relative to B for a simplified attribute dependency.

Definition 20. *For a pairs of subsets of attributes $A, B \subseteq At$ in an information table with $E_A \subseteq E_B$, a subset $R \subseteq A$ is called a reduct of A relative to B if it satisfies the following conditions:*

(s13) $E_R \subseteq E_B$,

(n13) $\forall a \in R, (\neg(E_{R-\{a\}} \subseteq E_B))$.   (18)

With this definition, we can interpret a reduct $R \subseteq At$ of an information table as a relative reduct with respect to the entire set of attributes At. Relative reducts of a consistent classification table are also special cases.

Attribute dependencies can be formally studied through attribute-level rules in a table [23]. In this way, we can unify notions of attribute reducts and attribute-value-pair reducts in a common framework of rules. For such a purpose, we need to introduce a decision logic language $\mathcal{L}_A$ similar to the one used in Pawlak's book.

Definition 21. *In an information table, a decision logic language $\mathcal{L}_A$ is recursively defined as follows: an atomic formula is given by $=_a$, where $a \in At$. If p and q are formulas, then $p \wedge q$ is a formula.*

By using language $\mathcal{L}_A$, we can express attribute dependency $E_A \subseteq E_B$ as,

$$\bigwedge_{a \in A} =_a \quad \rightarrow \quad \bigwedge_{b \in B} =_b,$$ (19)

or simply,

$$=_A \quad \rightarrow \quad =_B,$$ (20)

where both the left-hand-side and right-hand-side of $\rightarrow$ are formulas of $\mathcal{L}_A$. Consequently, finding a relative reduct can be viewed as searching for a minimal set of atomic formulas on the left-hand-side of an attribute-dependency rule.

The meaning of formulas of $\mathcal{L}_A$ are given by pairs of objects. More specifically, a pair of objects (x, y) is said to satisfy an atomic formula $=_a$ if and only if $I_a(x) = I_a(y)$. In general, the meanings of formulas can be recursively defined.

Definition 22. *The meanings of formulas of $\mathcal{L}_A$ are recursively computed as follows:*

$$m(=_a) = \{(x, y) \in U \times U \mid I_a(x) = I_a(y)\},$$
$$m(p \wedge q) = m(p) \cap m(q).$$ (21)

A formula may be interpreted as the intension of a concept and the meanings set is the extension of the concept. In this way, we express a concept jointly by a pair of a formula and a set. A concept in the context of attribute-level rules is an equivalence relation. By definition, it follows that,

$$m(=_a) = E_{\{a\}},$$
$$m(\bigwedge_{a \in A} =_a) = E_A.$$ (22)

With respect to the left-hand-side of an attribute dependency rule given by equation (20), we can define a set of atomic formulas and a set of the meaning sets of atomic formulas:

$$S = \{=_a | \ a \in A\},$$
$$m(S) = \{m(=_a) \mid a \in A\}. \tag{23}$$

In this way, an attribute reduct of A relative to B can be interpreted as a) a reduct of the set of atomic formulas S relative to the formula $=_B$, and b) a reduct of the family of equivalence relations $m(S)$ relative to the equivalence relation E_B. According to Definition 6 and Definition 10, we have two more definitions of a ruduct of an attribute dependency rule.

Definition 23. *For an attribute dependency rule* $=_A \ \rightarrow \ =_B$, *a subset* $R \subseteq S$ *is reduct of the set of atomic formulas* S *relative to* $=_B$ *if* R *satisfies the following conditions:*

(s14) $\cap \{m(p) \mid p \in R\} \subseteq m(=_B)$,

(n14) $\forall q \in R, (\neg(\cap\{m(p) \mid p \in (R - \{q\})\} \subseteq m(=_B)))$. $\qquad$ (24)

Definition 24. *For an attribute dependency rule* $=_A \ \rightarrow \ =_B$, *a subset* $R' \subseteq m(S)$ *is reduct of the set of equivalence relations* $m(S)$ *relative to the equivalence relation* $m(=_B)$ *if* R' *satisfies the following conditions:*

(s15) $\cap R' \subseteq m(=_B)$,

(n15) $\forall E \in R, (\neg(\cap(R' - \{E\}) \subseteq m(=_B)))$. $\qquad$ (25)

Recall that different attributes may define the same equivalence relation, like Definition 13, Definition 24 is not a very accurate characterization of a reduct of an attribute dependency rule.

4.3   Step 2: Analysis of Attribute-Value Dependencies

For a classification table with $At = C \cup D$, the result of Step 1 analysis is an attribute reduct $R \subseteq C$. For an equivalence class $[x]_{E_R}$ satisfying the condition $[x]_{E_R} \subseteq [x]_{E_D}$, Pawlak constructs a classification rule showing a dependency between values of x on attributes R and D, respectively. To represent formally such classification rules, we consider a sub-language of the decision logic language used Pawlak [13].

Definition 25. *In an information table, a decision logic language* $\mathcal{L}_V$ *is recursively defined as follows: an atomic formula is given by* $a = v$, *where* $a \in At$ *and* $v \in V_a$. *If* ϕ *and* ψ *are formulas, then* $\phi \wedge \psi$ *is a formula.*

An atomic formula $a = v$ is commonly known as an attribute-value pair, written (a, v), or a descriptor. By restricting to the logic connective $\wedge$, we only consider a formula that is the conjunction of a family of atomic formulas. The meaning of a formula is defined by the set of objects satisfying the formula.

Definition 26. *The meanings of formulas of* $\mathcal{L}_V$ *are recursively computed as follows:*

$$m(a = v) = \{x \in U \mid I_a(x) = v\},$$
$$m(\phi \wedge \psi) = m(\phi) \cap m(\psi). \tag{26}$$

With the introduced logic language $\mathcal{L}_V$, a classification rule can be defined as:

$$\bigwedge_{a \in R} a = I_a(x) \to \bigwedge_{d \in D} d = I_d(x), \tag{27}$$

or simply,

$$R = I_R(x) \to D = I_D(x), \tag{28}$$

The left-hand-side of the rule can be understood as a set of attribute value pairs, namely, atomic formulas. A classification rule is therefore called an attribute-value-level rule. Like an attribute-level rule, there may exist superfluous attribute-value pairs on the left-hand-side of the rule. Pawlak calls $m(a = v)$ a category and introduces the notion of a reduct of categories to simplify a classification rule.

By using the same argument for defining a reduct of an attribute dependency rule, we can define a reduct of an attribute-value dependency rule. For an object $x \in U$, we have:

$$m(a = I_a(x)) = [x]_{E_{\{a\}}},$$
$$m(\bigwedge_{a \in A} a = I_a(x)) = [x]_{E_A}. \tag{29}$$

Based on these results, for rule $R = I_R(x) \to D = I_D(x)$, we introduce two definitions of an attribute-value-pair reduct relative to $[x]_{E_D}$.

Definition 27. *For a classification rule $R = I_R(x) \to D = I_D(x)$, a subset of attributes $R(x) \subseteq R$ is called an attribute reduct of x relative to D if $R(x)$ satisfies the two conditions:*

(s16) $[x]_{E_{R(x)}} \subseteq [x]_{E_D}$;

(n16) $\forall a \in R(x), (\neg([x]_{E_{R(x)-\{a\}}} \subseteq [x]_{E_D}))$.

Note that (s16) and (n16) are related to (s7) and (n7) of Definition 16. By comparison, an attribute reduct of an information table must be defined with respect to all objects in the table and an attribute reduct of a classification rule is defined with respect to only objects equivalent to x.

Given a classification rule $R = I_R(x) \to D = I_D(x)$, we can construct a set of attribute-value pairs (i.e., atomic formulas) and the set of their meaning sets, respectively, as follows:

$$S(x) = \{a = I_a(x) \mid a \in R\},$$
$$m(S(x)) = \{m(a = I_a(x)) \mid a \in R\}. \tag{30}$$

We can use reducts of the two sets to define reducts of of a classification rule in a similar manner as in Definitions 23 and 24.

Definition 28. *For a classification rule $R = I_R(x) \to D = I_D(x)$, a subset $R(x) \subseteq S(x)$ is called an attribute-value-pair reduct relative to $D = I_D(x)$ if $R(x)$ satisfies the conditions:*

(s17) $\cap\{m(\phi) \mid \phi \in R(x)\} \subseteq m(D = I_D(x))$;

(n17) $\forall \psi \in R(x), (\neg(\cap\{m(\phi) \mid \phi \in (R(x) - \{\psi\})\} \subseteq m(D = I_D(x))))$. (31)

Definition 29. *For a classification rule $R = I_R(x) \rightarrow D = I_D(x)$, a subset $R'(x) \subseteq m(S(x))$ is called a reduct of $m(S(x))$ relative to $m(D = I_D(x))$ if $R'(x)$ satisfies the conditions:*

$$\text{(s18)} \quad \cap R'(x) \subseteq m(D = I_D(x));$$
$$\text{(n18)} \quad \forall K \in R'(x), (\neg(\cap(R'(x) - \{K\}) \subseteq m(D = I_D(x)))). \tag{32}$$

Definition 29 is in fact an $\cap$-reduct in Definition 11. Different attribute-value pairs may have the same meaning set, Definition 29 as used by Pawlak is not a very accurate characterization of a relative attribute-value-pair reduct.

In general, similar to the study of attribute-level rules in Section 4.2, we can study attribute-value dependencies for any pair of sets of attributes $A, B \subseteq At$. For example, one can consider attribute-value dependencies by using the set of condition attributes C and the set of decision attributes D in a classification table, instead of using a reduct $R \subseteq C$ from Step 1.

4.4   Step 3: Analysis of Rule Dependencies

After Step 2 analysis, for an attribute reduct $R(x) \subseteq R$ for an object x, we have $[x]_{E_{R(x)}} \subseteq [x]_{E_D}$, which produces a classification rule:

$$R(x) = I_{R(x)}(x) \rightarrow D = I_D(x). \tag{33}$$

The third step of Pawlak data analysis consists of constructing a rule set and simplifying the rule set by removing redundant rules. Pawlak compiles a set of simplified rules by choosing one rule defined by an attributive-value-pair reduct $R(x)$ for each equivalence class $[x]_R$, where R is a relative attribute reduct obtained in Step 1. Let RS denote the rule set obtained in Step 2. There may exist redundant rules in RS. It is therefore necessary to introduce the notion of a rule reduct of RS.

For a classification rule $c \rightarrow d$, we define its meaning as the set of correctly classified objects:

$$m(c \rightarrow d) = m(c \wedge d) = m(c) \cap m(d). \tag{34}$$

Pawlak only considers certain rules derived from the lower approximations. In this case, we have $m(c) \subseteq m(d)$ and $m(c \rightarrow d) = m(c)$. In general, this may not be true. Based on the meaning sets of rules, a rule reduct is related to an $\cup$-reduct of the following family of subsets of U:

$$m(RS) = \{m(c \rightarrow d) \mid (c \rightarrow d) \in RS\}. \tag{35}$$

For an inconsistent table, we have $\cup m(RS) = \text{POS}_{E_C}(U/E_D)$; for a consistent table we have $\cup m(RS) = U$. According to Definition 6, we introduce the notion of a reduct of a rule set.

Definition 30. *A subset of rules $R \subseteq RS$ is called a rule reduct of a set of rules RS if R satisfies the condition:*

      (s19) $\cup m(R) = \cup m(RS)$,

      (n19) $\forall(c \rightarrow d) \in R, (\neg(\cup m(R - \{c \rightarrow d\}) = \cup m(RS)))$,

where $m(R) = \{m(c \rightarrow d) \mid (c \rightarrow d) \in R\}$.

Condition (s19) states that rules in R are sufficient for correctly classifying all objects as the entire rule set RS and condition (n19) states that each rule in R is necessary.

According to Definition 10, we can directly compute an $\cup$-reduct of the family $m(RS)$ to interpret a reduct of the rule set RS. However, since two rules may have the same meaning set, such an interpretation is not precise.

5   Discussions and Conclusion

In our interpretation and formulations of Pawlak three-step approach, we use a more general and generic notion of reducts. By exploring the monotonicity of evaluations, a reduct of a set is defined as a subset of a set satisfying a pair of conditions, namely, a jointly sufficient condition (s) and an individually necessary condition (n). In total, we consider about twenty definitions of various reducts.

The unified framework based on reducts has a number of advantages. One can apply a generic reduct construction algorithm for constructing any of the three types of reducts. In particular, one may use any of the three classes of algorithms, deletion, addition-deletion, and addition algorithms [22]. All three steps of Pawlak analysis can be viewed as different applications of the same data reduction method. The same framework can be further applied to new situations where a reduct of a set is of interest.

In our formulation, we explicitly express intension and extension of a concept. A classification rule is expressed as a pair of two rules, one for extension and the other for intension: for $x \in U, R \subseteq C$,

$$\begin{cases} [x]_{E_R} \subseteq [x]_{E_D}, \\ \bigwedge_{a \in R} a = I_a(x) \rightarrow \bigwedge_{d \in D} d = I_d(x). \end{cases} \tag{36}$$

It enables us to see additional insights into Pawlak three-step approach. Steps 1 and 2 use both intensions and extensions. Step 3 only uses extensions.

The Pawlak three-step approach can be modified in several ways. It can be observed that the first step is not necessary. Thus, a two-step approach can be derived based only on Steps 2 and 3. In Step 2, attribute-value-pair reducts are constructed based on both intensions and extensions. One may consider only extensions of concepts without reference to intensions. This can be formulated as a search for a reduct of the family of subsets of U given by:

$$\{[x]_{E_A} \mid A \subseteq C, [x]_{E_A} \subseteq [x]_{E_D}\}. \tag{37}$$

The results are a new rule learning method [20]. In Step 3, a rule reduct is defined independent of how a rule set is formed.

This paper contributes to Pawlak three-step rough set analysis by introducing a generic notion of reducts, providing multiple interpretations of reducts, and unifying different definitions of reducts in a common framework. We demonstrate that rough set analysis can be formulated based on the central notion of reducts. With some modifications, it is possible to investigate various generalized notions of reducts by using the results from this paper.

Acknowledgements. This work is partially supported by a Discovery Grant from NSERC Canada. The authors thank reviewers for their constructive comments.

References

1. Bazan, J.G., Skowron, A., Synak, P.: Dynamic Reducts as a Tool for Extracting Laws from Decisions Tables. In: Raś, Z.W., Zemankova, M. (eds.) ISMIS 1994. LNCS (LNAI), vol. 869, pp. 346–355. Springer, Heidelberg (1994)
2. Blaszczynski, J., Slowinski, R., Szelag, M.: Sequential covering rule induction algorithm for variable consistency rough set approaches. Information Sciences 181, 987–1002 (2011)
3. Cendrowska, J.: PRISM: An algorithm for inducing modular rules. International Journal of Man-Machine Studies 27, 349–370 (1987)
4. Fürnkranz, J.: Separate-and-conquer rule learning. Artificial Intelligence Review 13, 3–54 (1999)
5. Janusz, A., Ślęzak, D.: Utilization of attribute clustering methods for scalable computation of reducts from high-dimensional data. In: Federated Conference on Computer Science and Information Systems, pp. 307–313 (2012)
6. Grzymala-Busse, J.: LERS - A system for learning from examples based on rough sets. In: Slowinski, R. (ed.) Intelligent Decision Support: Handbook of Applications and Advances of the Rough Sets Theory, pp. 3–18. Kluwer Academic Publishers, Dordrecht (1992)
7. Grzymala-Busse, J., Rzasa, W.: Approximation space and LEM2-like algorithms for computing local coverings. Fundamenta Informaticae 85, 205–217 (2008)
8. Mi, J.S., Leung, Y., Wu, W.Z.: Dependence-space-based attribute reduction in consistent decision tables. Soft Computing 15, 261–268 (2011)
9. Miao, D.Q., Zhao, Y., Yao, Y.Y., Li, H.X., Xu, F.F.: Relative reducts in consistent and inconsistent decision tables of the Pawlak rough set model. Information Sciences 179, 4140–4150 (2009)
10. Moshkov, M.J., Piliszczuk, M., Zielosko, B.: Partial Covers, Reducts and Decision Rules in Rough Sets: Theory and Applications. Springer, Berlin (2008)
11. Nguyen, H.S., Ślęzak, D.: Approximate Reducts and Association Rules. In: Zhong, N., Skowron, A., Ohsuga, S. (eds.) RSFDGrC 1999. LNCS (LNAI), vol. 1711, pp. 137–145. Springer, Heidelberg (1999)
12. Pawlak, Z.: Rough sets. International Journal of Computer and Information Sciences 11, 341–356 (1982)
13. Pawlak, Z.: Rough Sets: Theoretical Aspects of Reasoning About Data. Kluwer Academic Publishers, Dordrecht (1991)

14. Pawlak, Z., Skowron, A.: Rough sets and Boolean reasoning. Information Sciences 177, 41–73 (2007)
15. Quinlan, J.R.: Induction of decision trees. Machine Learning 1, 81–106 (1986)
16. Ślęzak, D.: Association Reducts: A Framework for Mining Multi-attribute Dependencies. In: Hacid, M.-S., Murray, N.V., Raś, Z.W., Tsumoto, S. (eds.) ISMIS 2005. LNCS (LNAI), vol. 3488, pp. 354–363. Springer, Heidelberg (2005)
17. Ślęzak, D., Janusz, A.: Ensembles of Bireducts: Towards Robust Classification and Simple Representation. In: Kim, T.-H., Adeli, H., Slezak, D., Sandnes, F.E., Song, X., Chung, K.-I., Arnett, K.P. (eds.) FGIT 2011. LNCS (LNAI), vol. 7105, pp. 64–77. Springer, Heidelberg (2011)
18. van Mechelen, I., Hampton, J., Michalski, R.S., Theuns, P. (eds.): Categories and Concepts: Theoretical Views and Inductive Data Analysis. Academic Press, New York (1993)
19. Wu, W.Z.: Knowledge reduction in random incomplete decision tables via evidence theory. Fundamenta Informaticae 115, 203–218 (2012)
20. Yao, Y., Deng, X.: A Granular Computing Paradigm for Concept Learning. In: Ramanna, S., Howlett, R.J. (eds.) Emerging Paradigms in ML and Applications. SIST, vol. 13, pp. 307–326. Springer, Heidelberg (2013)
21. Yao, Y., Fu, R.: Partitions, Coverings, Reducts and Rule Learning in Rough Set Theory. In: Yao, J.T., Ramanna, S., Wang, G., Suraj, Z. (eds.) RSKT 2011. LNCS (LNAI), vol. 6954, pp. 101–109. Springer, Heidelberg (2011)
22. Yao, Y., Zhao, Y., Wang, J.: On Reduct Construction Algorithms. In: Wang, G.-Y., Peters, J.F., Skowron, A., Yao, Y. (eds.) RSKT 2006. LNCS (LNAI), vol. 4062, pp. 297–304. Springer, Heidelberg (2006)
23. Yao, Y., Zhou, B., Chen, Y.H.: Interpreting Low and High Order Rules: A Granular Computing Approach. In: Kryszkiewicz, M., Peters, J.F., Rybiński, H., Skowron, A. (eds.) RSEISP 2007. LNCS (LNAI), vol. 4585, pp. 371–380. Springer, Heidelberg (2007)
24. Zhang, W.X., Mi, J.S., Wu, W.Z.: Approaches to knowledge reductions in inconsistent systems. International Journal of Intelligent Systems 18, 989–1000 (2003)
25. Zhao, Y., Yao, Y.Y., Yao, J.T.: Level construction of decision trees for classification. International Journal Software Engineering and Knowledge Engineering 16, 103–126 (2006)
26. Zhu, W., Wang, F.Y.: Reduction and axiomization of covering generalized rough sets. Information Sciences 152, 217–230 (2003)

Nearness of Subtly Different Digital Images[*]

Leszek Puzio[1,2] and James F. Peters[1,3]

[1] Computational Intelligence Laboratory
Department of Electrical & Computer Engineering, University of Manitoba
Winnipeg, Manitoba R3T 5V6 Canada
jfpeters@ee.umanitoba.ca
[2] Department of Information Systems and Applications
University of Information Technology and Management
35-225 Rzeszow, ul. H.Sucharskiego 2, Poland
lpuzio@wsiz.rzeszow.pl
[3] School of Mathematics & Computer/Information Sciences,
University of Hyderabad, Central Univ. P.O., Hyderabad 500046, India

Abstract. The problem considered in this article is how to measure the
nearness or apartness of digital images in cases where it is important to
detect subtle changes in the contour, position, and spatial orientation of
bounded regions. The solution of this problem results from an application of anisotropic (direction dependent) wavelets and a tolerance near
set approach to detecting similarities in pairs of images. A wavelet-based
tolerance Nearness Measure (tNM) makes it possible to measure fine-
grained differences in shapes in pairs of images. The application of the
proposed method focuses on image sequences extracted from hand-finger
movement videos. Each image sequence consists of hand-finger movements recorded during rehabilitation exercises. The nearness of pairs of
images from such sequences is measured to check the extent that normal hand-finger movement differs from arthritic hand-finger movement.
Experimental results of the proposed approach are reported, here. The
contribution of this article is an application of an anisotropic wavelet-
based tNM in classifying arthritic hand-finger movement images in terms
of their degree of nearness to or apartness from normal hand-finger movement images.

Keywords: anisotropic wavelets, arthritis, digital image sequence, nearness measure, tolerance near sets.

1 Introduction

This paper considers the problem of how to measure the nearness or apartness
of digital images in cases where it is important to detect subtle changes in the

[*] This research was supported by Natural Sciences & Engineering Research Council
of Canada (NSERC) grant 185986, Manitoba Center of Excellence Fund (MCEF)
grant, Canadian Network of Excellence (CNE), and Canadian Arthritis Network
(CAN) grant SRI-BIO-05.

J.F. Peters et al. (Eds.): Transactions on Rough Sets XVI, LNCS 7736, pp. 73–82, 2013.
© Springer-Verlag Berlin Heidelberg 2013

contour, position, and spatial orientation of bounded regions. The solution to this problem is given in terms of an anisotropic wavelet-based, tolerance nearness measure in classifying arthritic hand-finger movement images relative to their degree of *nearness* to or *apartness* from normal hand-finger movement images.

In this work, we utilise near set theory informally introduced in 2002 [1] and formally introduced in 2007 [2,3]. Near sets were inspired by a study of the perceptual resemblance of objects during a collaboration between Z. Pawlak and J.F. Peters [1]. Recent research proves that near set theory can be used effectively to define distance functions that measure the nearness of digital images [4–20]. The anisotropic wavelet-based tolerance nearness measure [9] is based on recent work by C. Henry and J.F. Peters [4, 12, 13]. The contribution of this article is an application of an anisotropic wavelet-based tNM in classifying arthritic hand-finger movement images in terms of their degree of nearness to or apartness from normal hand-finger movement images.

This paper has the following organization. Sect. 2 gives the basic mathematics underlying the proposed classification method. Sect. 3 briefly presents the nearness measurement method and sample experimental results.

2   Preliminaries: Anisotropic Wavelets and Tolerance Nearness Measure

An anisotropic wavelet (*i.e.*, dependent on the direction (angle) that is used to define a wavelet) is constructed in a polar coordinate system as a product of the Hann window function and the Gaussian wavelet [21]. The Hann window function is given in (1).

$$\rho(\alpha) = 0.5(1 - \cos(\alpha)), \ \alpha \in [0, 2\pi), \tag{1}$$

$$\psi(r) = -2r \left(\frac{2}{\pi}\right)^{1/4} e^{-r^2}. \tag{2}$$

An anisotropic wavelet $\psi(\alpha, r)$ is a product of a Hann window $\rho(\alpha)$ and translated by n_r Gaussian wavelet $\psi(r)$ represented in (3). By putting (1) and (2) into (3), we obtain a so-called 'mother wavelet', *i.e.*, a wavelet function (4) that is used to construct a wavelet set. Each wavelet in our set we calculate in (5).

$$\psi(\alpha, r) = \rho(\alpha)\psi(r), \tag{3}$$

$$\psi(\alpha, r) = 0.5(1 - \cos(\alpha)) \, (-2r) \left(\frac{2}{\pi}\right)^{1/4} e^{-r^2}, \tag{4}$$

$$\psi_{\mathcal{I}}(\alpha, r) = \left(1/\sqrt{2\pi n_r/2^{s_\alpha+1}}\sqrt{2^{-s_r}}\right) \cdot \psi\left(2^{s_\alpha}\alpha - \pi(n_\alpha - 1), 2^{-s_r}(r - n_r)\right), \tag{5}$$

$$C_\psi\{f\}(\dots) = \int \int f(\alpha, r)\psi^*_{s_\alpha, s_r, n_\alpha, n_r}(\alpha, r) \, d\alpha \, dr. \tag{6}$$

where $C_\psi\{f\}(\ldots)$ denotes $C_\psi\{f\}(s_\alpha, s_r, n_\alpha, n_r)$, ψ denotes a wavelet with (α, r) a polar coordinates and where $\mathcal{I} = \{s_\alpha, s_r, n_\alpha, n_r\}$ denotes an index set used in (5) to define a wavelet with an angular scale s_α, radial scale s_r, an angular translation n_α and a radial translation n_r. In particular, it is n_α that makes (5) anisotropic, while n_r is a radial distance from the pole (origin of a polar coordinate system).

Perception-based description of an object x in near set theory is in the form of feature vectors $\phi(x)$ containing *probe function* values [7,14]

$$\phi(x) = (\phi_1(x), \phi_2(x), \ldots, \phi_i(x), \ldots, \phi_l(x))^T \tag{7}$$

where $\phi_i : O \longrightarrow [0, \infty)$ is a probe function that represents a single object feature. This leads to the notion of a perceptual information system.

Definition 1. Perceptual information system [7]

A perceptual information system $\langle O, \mathbb{F} \rangle$ or more concisely, perceptual system is a real-valued, total deterministic information system where O is a non-empty set of perceptual objects, while $\mathbb{F}$ a countable set of probe functions.

Definition 2. Perceptual tolerance relation [22,23]
Let $\langle O, \mathbb{F} \rangle$ be perceptual system and put $\varepsilon \in [0, \infty)$. For every $\mathcal{B} \subseteq \mathbb{F}$, the perceptual tolerance relation $\cong_{\mathcal{B}, \varepsilon}$ is defined as (8).

$$\cong_{\mathcal{B}, \varepsilon} = \{(x, y) \in O \times O, \| \phi(x) - \phi(y) \| \leq \varepsilon\} \tag{8}$$

where $\| \cdot \|_2$ is the L_2 norm, $\phi(x) = [\phi_1(x) \ldots \phi_i(x) \ldots \phi_l(x)]^T$ is a feature vector obtained using all probe functions $\phi_i \in \mathcal{B}$. For simplicity, we write $x \cong_{\mathcal{B}} y$ instead of $x \cong_{\mathcal{B}, \varepsilon} y$.

Relations with the same formal properties as similarity relations of sensations considered by Poincaré [24] are nowadays, after Zeeman [25], called *tolerance relations.*

A *tolerance* τ on a set O is a relation $\tau \subseteq O \times O$ that is reflexive and symmetric. Transitive tolerance relations are equivalence relations. A set O together with a tolerance τ is called a *tolerance space* (denoted $\langle O, \tau \rangle$). The useful notion of a tolerance preclass was first introduced by M.J. Schroeder and M.H. Wright [26]. A set

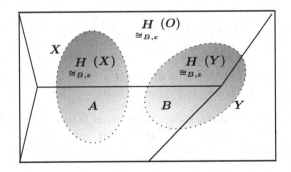

Fig. 1. Sample Tolerance Near Sets

$A \subseteq O$ is a τ-*preclass* (or briefly *preclass* when τ is understood) if and only if for any $x, y \in A$, $(x, y) \in \tau$. The family of all preclasses of a tolerance space is naturally ordered by set inclusion and preclasses that are maximal with respect to a set inclusion are called τ-*classes* or just *classes*, when τ is understood. The family of all classes of the space $\langle O, \tau \rangle$ is denoted by $H_\tau(O)$. The family $H_\tau(O)$ is a covering of O.

Definition 3. Tolerance Near Sets [22,23]

Let $\langle O, \mathbb{F} \rangle$ be a perceptual system and let $X, Y \subseteq O$. A set X is perceptually near a Y within the perceptual system $\langle O, \mathbb{F} \rangle$ (i.e., $(X \bowtie_{\mathbb{F}} Y)$) iff there are $x \in X$ and $y \in Y$ and there is $\mathcal{B} \subseteq \mathbb{F}$ such that $x \cong_{\mathcal{B},\varepsilon} y$. We than say that X, Y are perceptually near *each other in the tolerance sense of nearness in Def.2.*

Fig. 1 points to candidate tolerance near sets. Let O denote a set of pixels inside the rectangle. Further, assume $\underset{\cong_{\mathcal{B},\varepsilon}}{H}(O)$ denotes the family of all tolerance classes of the space $\langle O, \cong_{\mathcal{B},\varepsilon} \rangle$ determined by the tolerance relation $\cong_{\mathcal{B},\varepsilon}$ in a covering of the non-empty set O. Let $X, Y \subset O$ be represented by the shaded ellipses in Fig. 1. In this Figure, tolerance class $A \in \underset{\cong_{\mathcal{B},\varepsilon}}{H}(X)$ and tolerance class $B \in \underset{\cong_{\mathcal{B},\varepsilon}}{H}(Y)$. For simplicity, let the set of probe functions $B = \{\phi_{gr}\}$, where $\phi_{gr}(o) = $ intensity for pixel $o \in O$. It is apparent from the greylevel intensities in classes A and B, that these classes contain pixels with similar descriptions, *i.e.*, pixels with similar intensities. To determine nearness of tolerance spaces, we consider the tolerance nearness measure tNM.

Definition 4. Tolerance Nearness Measure (tNM) [4]

The distance $D_{tNM} : \mathcal{P}(O) \times \mathcal{P}(O) :\rightarrow [0, \infty)$ is defined by

$$D_{tNM}(X, Y) = \begin{cases} 1 - tNM_{\cong_{\mathcal{B},\varepsilon}}(A, B), & \text{if } X \text{ and } Y \text{ are not empty,} \\ \infty, & \text{if } X \text{ or } Y \text{ is empty,} \end{cases}$$

where

$$tNM_{\cong_{\mathcal{B},\varepsilon}}(X, Y) = \left(\sum_{C \in H_{\cong_{\mathcal{B},\varepsilon}}(Z)} |C| \right)^{-1} \cdot \sum_{C \in H_{\cong_{\mathcal{B},\varepsilon}}(Z)} |C| \frac{\min(|C \cap X|, |[C \cap Y|)}{\max(|C \cap X|, |C \cap Y|)}.$$

For simplicity, tNM is abbreviated NM. The details concerning NM are given in [4, 8, 9, 13] and not repeated here.

Nearness measure values range from 0 to 1 ($D_{tNM}(X, Y) = 0$ means that sets X, Y are near (*i.e.*, X, Y have similar descriptions), while $D_{tNM}(X, Y) = 1$ means that sets X, Y are far apart (*i.e.*, X, Y have dissimilar descriptions)).

Example 1. Sample image features extraction using wavelet method

We study images resemblance using features obtained using a wavelet-based edge extraction method [27]. This method is based on a anisotropic wavelet [21]. Each edge is described by localization, orientation, wavelet coefficient proportional to

2.1: Original image   **2.2:** Detected edges

Fig. 2. Wavelet algorithm application result

edge gradient value, object contour number and contour lengths. Fig. 2 presents example of the use of the wavelet algorithm on an image containing a hand (Fig. 2.1). Fig. 2.2 contains points where edges were detected using wavelet algorithm.

*Table 1 contains wavelet algorithm results presented in Fig. 2.2. The algorithm reveals that hand contour contains 132 edges. We present only part of the results just to show what value they could take. Each edge has its order number #, position **X**, **Y**, spatial orientation given in radians, and wavelet coefficient value.*

Table 1. Results obtained for sample hand image in scale $s_r=0$

#	X	Y	Orient.	Coef.
1	13	5	-0,363	1,113
2	14	8	-0,198	1,007
3	14	11	-0,061	1,012
4	14	14	0,028	0,881
5	14	17	0,031	0,866
⋮	⋮	⋮	⋮	⋮
132	95	74	-0,588	0,459

3   Anisotropic Wavelet-Based Tolerance Nearness

This section introduces an an application of the anisotropic wavelet algorithm from [27] considered in the context of tolerance near sets.

3.1   Image Comparison Methodology

A method that combines the original anisotropic wavelet algorithm [27] and tolerance nearness measure tNM is summarised in Alg. 1. In this work, hand-finger movement video recording are made during rehabilitation exercise. Sequences of

3.1: A normal hand   **3.2:** A rheumatic hand

Fig. 3. Single images from two video hand movement sequences

images are extracted from those videos. For every image in an image sequence, we extract features such us edge localization, edge spatial orientation, wavelet coefficient proportional to edge gradient value, objects contour number and contour lengths using a wavelet algorithm. Those features was utilized to nearness measures evaluation of two images from sequence, *i.e.*, first image with second, second with third, and so on. Alg. 1 step (4.1) is time consuming because at this step τ classes are determined. One could find solution to this problem in [28].

Algorithm 1. tNM calculation algorithm of digital images

Input : $Img1,Img2$ (pair of images), s_r (wavelet algorithm scale),
ε (tolerance).
Output: tNM (Tolerance Nearness Measure value).

1 Initialize algorithm parameters:
(1.1) $s_r \leftarrow$ wavelet scale value;
(1.2) $\varepsilon \leftarrow$ Nearness Measure tolerance value;
2 Extract $Img1, Img2$ features using anisotropic wavelets from Sect. 2:
(2.1) $Feat1 \leftarrow WavAlg(s_r, Img1)$;
(2.2) $Feat2 \leftarrow WavAlg(s_r, Img2)$;
3 Obtain edge positions from images features:
(3.1) $X \leftarrow Feat1(x, y)$;
(3.2) $Y \leftarrow Feat2(x, y)$;
4 Compute tNM from Def. 4:
(4.1) $tNM \leftarrow tNM_{\cong_{B,\varepsilon}}(X, Y)$;

3.2   Experimental Results

To measure the nearness of a pair of digital images, we utilize the tNM measure from Def. 4. Image features are obtained using the wavelet algorithm from Sect. 2 and [27]. tNM was based on edge localization (*i.e.*, X and Y edge position), wavelet coefficient value, edge spatial orientation, and object contour length features.

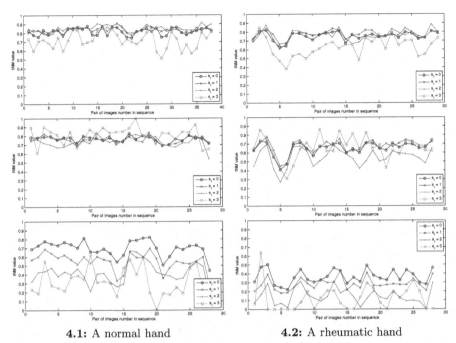

4.1: A normal hand   **4.2:** A rheumatic hand

Fig. 4. From top: tNM values with $\varepsilon=0.1$, $\varepsilon=0.05$, and $\varepsilon=0.01$ based on edge localization

As was expected tNM values decreases with ε, but $D_{t_{NM}}$ should increase with decreasing ε. This is illustrated in Fig 4, with the X axis marked a succession of pairs of images from video sequence. On the Y axis, tNM values are given for pairs of images. Each plot consists of four data series, because image features was extracted for a different scale parameter of the wavelet algorithm ($s_r = 0, 1, 2, 3$). In this example tNM was calculated based on edges localization feature from rheumatic hand images sequence. Obtained tNM values are only different in Fig. 4 for normal and arthritic hands when ε equals to 0.01.

Fig. 5 presents calculated tNM values with different ε values based on wavelet coefficient values. From this figure, one could find that tNM values for normal and rheumatic hands are very similar.

Fig. 6 shows tNM values when utilized edge spatial orientation as a image feature. Surprisingly, we obtained the same tNM values for all tolerance values ($\varepsilon=0.1$, 0.05, 0.01). That is why Fig. 6 contains one figure for normal and one figure for arthritic hand.

We conclude by our research (some of which we presented in Fig. 4, Fig. 5, and Fig. 6) that applying such image features as edge spatial orientation, edge wavelet coefficient value, contour number, or contour length to tNM calculation results in small difference in obtained tNM values for given images sequences. Fig. 5 and Fig. 6 illustrates this. Both, normal and arthritic hand images sequences with calculated tNM values are at almost the same level.

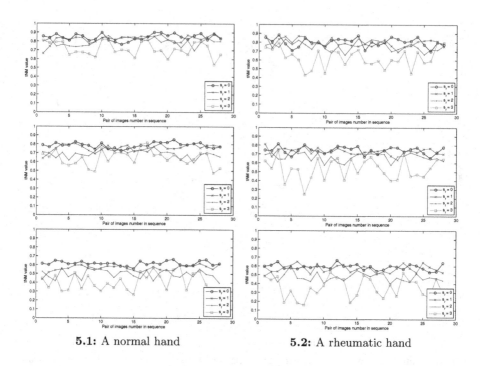

5.1: A normal hand   **5.2:** A rheumatic hand

Fig. 5. From top: tNM values with $\varepsilon=0.1$, $\varepsilon=0.05$, and $\varepsilon=0.01$ based on coefficient values

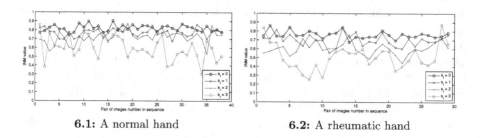

6.1: A normal hand   **6.2:** A rheumatic hand

Fig. 6. The tNM values with $\varepsilon=0.01$ obtained using spatial orientation feature

In sum, we conclude that the best distinction between normal and arthritic hand-finger sequences with tNM is based on edge localization as a image feature in s_r scale equals to 0, and with tolerance value ε equals to 0.01.

Figure 7 presents nearness measures values for hand-finger image pairs. It is clear that tNM values for normal hand sequences are two times bigger (on average) than arthritic hand sequences. This suggests that this tNM function is able to distinguish between normal and arthritic hand-finger movements.

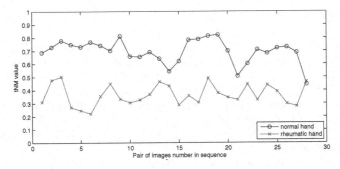

Fig. 7. The tNM of normal and rheumatic hand sequences with $s_r=0$, and $\varepsilon=0.01$

4 Conclusion

This paper presents a number of research results, namely,

(result.i) We are able to apply near sets theory and the tNM measure with wavelets in image analysis.

(result.ii) The Puzio-Walczak wavelet algorithm has utility in image edge extraction for a number of parameters: position, spatial orientation, number and length of objects contours.

(result.iii) It is possible to distinguish between normal and arthritic hand-finger movements using the tNM distance function based on edge position with $\varepsilon = 0.01$.

Future work will include further work on a family of wavelet-based nearness distance functions and classification of images containing subtly different shapes.

References

1. Pawlak, Z., Peters, J.: How near. Systemy Wspomagania Decyzji I, 57 (2002)
2. Peters, J.: Near sets. Special theory about nearness of objects. Fund. Inform. 75(1-4), 407–433 (2007)
3. Peters, J.: Near sets. General theory about nearness of objects. Applied Math. Sci. 1(53), 2609–2629 (2007)
4. Henry, C.: Near Sets: Theory and Applications, supervisor: J.F. Peters. PhD thesis, Department of Electrical & Computer Engineering (2010)
5. Henry, C., Peters, J.: Arthritic hand-finger movement similarity measurements: Tolerance near set approach. Comp. & Math. Methods in Medicine, Article ID 569898, 1–14 (2011), doi:10.1155/2011/569898
6. Pal, S., Peters, J.: Rough Fuzzy Image Analysis. Foundations and Methodologies. Chapman & Hall/CRC Press Mathematical & Computational Imaging Sciences, London (2010)
7. Peters, J., Wasilewski, P.: Foundations of near sets. Info. Sci. 179, 3091–3109 (2009)
8. Hassanien, A., Abraham, A., Peters, J., Schaefer, G., Henry, C.: Rough sets and near sets in medical imaging: A review. IEEE Trans. Info. Tech. in Biomedicine 13(6), 955–968 (2009), doi:10.1109/TITB.2009.2017017

9. Peters, J., Puzio, L.: Anisotropic wavelet-based image nearness measure. Int. J. Computational Intell. Sys. 2(3), 168–183 (2009)

10. Henry, C.: Near set evaluation and recognition (near) system. In: Pal, S., Peters, J.F. (eds.) Rough Fuzzy Image Analysis. Foundations and Methodologies, ch. 7, pp. 7.1–7.22. CRC Press, Boca Raton (2010)

11. Ramanna, S., Meghdadi, A.: Measuring resemblances between swarm behaviours: A perceptual tolerance near set approach. Fundamenta Informaticae 95(4), 533–552 (2009), doi:10.3233/FI-2009-163.

12. Henry, C., Peters, J.: Perceptual image analysis. Int. J. Bio-Inspired Comput. 2, 271–181 (2010),
http://inderscience.metapress.com/link.asp?id=r5w2662q8m18rg23

13. Henry, C., Peters, J.: Perception-based image classification. Int. J. Intell. Comput. Cybern. 3(3), 410–430 (2010), http://www.emeraldinsight.com/1756-378X.htm

14. Peters, J.: Metric spaces for near sets. Ap. Math. Sci. 5(2), 73–78 (2011)

15. Peters, J., Naimpally, S.: Approach spaces for near filters. Gen. Math. Notes 2(1), 159–164 (2011)

16. Peters, J., Tiwari, S.: Approach merotopies and near filters. Gen. Math. Notes 2(2), 1–15 (2011)

17. Henry, C., Peters, J.: Near sets. Wikipedia (2011),
http://en.wikipedia.org/wiki/Near_sets

18. Peters, J.F.: Visual Perception in Image Analysis. In: Kwaśnicka, H., Jain, L.C. (eds.) Innovations in Intelligent Image Analysis. SCI, vol. 339, pp. 105–125. Springer, Heidelberg (2011)

19. Peters, J.F., Wasilewski, P.: Tolerance spaces: Origins, theoretical aspects and applications. Information Sciences 195, 211–225 (2012)

20. Henry, C.: Perceptual Indiscernibility, Rough Sets, Descriptively Near Sets, and Image Analysis. In: Peters, J.F., Skowron, A. (eds.) Transactions on Rough Sets XV. LNCS, vol. 7255, pp. 41–121. Springer, Heidelberg (2012)

21. Puzio, L., Walczak, A.: 2-d wavelet with position controlled resolution. In: SPIE Medical Imaging, vol. 5959, p. 59590S (2005), doi:10.2478/s11772-007-0040-6

22. Peters, J.F.: Tolerance near sets and image correspondence. Int. J. of Bio-Inspired Comput. 1(4), 239–245 (2009)

23. Peters, J.F.: Corrigenda and addenda: Tolerance near sets and image correspondence. Int. J. of Bio-Inspired Comput. 2, 310–318 (2010)

24. Poincaré, J.: Dernières pensées, trans. by J.W. Bolduc as Mathematics and Science: Last Essays. Flammarion & Kessinger Pub., Paris & NY (1913 & 2009)

25. Zeeman, E.: The topology of the brain and visual perception. In: Fort Jr., M.K. (ed.) Topology of 3-Manifolds and Related Topics. University of Georgia Institute Conference Proceedings, pp. 240–256. Prentice-Hall, Inc. (1962)

26. Schroeder, M., Wright, M.: Tolerance and weak tolerance relations. Journal of Combinatorial Mathematics and Combinatorial Computing 11, 123–160 (1992)

27. Puzio, L., Walczak, A.: Adaptive edge detection method for images. Opto-Electronics Review 16(1), 60–67 (2008)

28. Henry, C.J., Ramanna, S.: Parallel Computation in Finding Near Neighbourhoods. In: Yao, J., Ramanna, S., Wang, G., Suraj, Z. (eds.) RSKT 2011. LNCS, vol. 6954, pp. 523–532. Springer, Heidelberg (2011)

Semantic Clustering of Scientific Articles Using Explicit Semantic Analysis*

Marcin Szczuka and Andrzej Janusz

Faculty of Mathematics, Informatics, and Mechanics,
The University of Warsaw, Banacha 2, 02-097 Warsaw, Poland
szczuka@mimuw.edu.pl, andrzejjanusz@gmail.com

Abstract. This paper summarizes our recent research on semantic clustering of scientific articles. We present a case study which was focused on analysis of papers related to the Rough Sets theory. The proposed method groups the documents on the basis of their content, with an assistance of the DBpedia knowledge base. The text corpus is first processed using Natural Language Processing tools in order to produce vector representations of the content. In the second step the articles are matched against a collection of concepts retrieved from DBpedia. As a result, a new representation that better reflects the semantics of the texts, is constructed. With this new representation the documents are hierarchically clustered in order to form a partitioning of papers into semantically related groups. The steps in textual data preparation, the utilization of DBpedia and the employed clustering methods are explained and illustrated with experimental results. A quality of the resulting clustering is then discussed. It is assessed using feedback form human experts combined with typical cluster quality measures. These results are then discussed in the context of a larger framework that aims to facilitate search and information extraction from large textual repositories.

Keywords: Text mining, semantic clustering, DBpedia, document grouping, rough sets.

1 Introduction

In this paper we present results that are an extension of the RSKT 2011 conference paper [20] and the book chapter [21]. We demonstrate how theses results have been augmented and extended for the purposes of a larger (SONCA) system, that is being developed. The original method was modified and tested on more extensive data sets, as described in [18] and [8].

---

* This work was supported by the grant N N516 077837 from the Ministry of Science and Higher Education of the Republic of Poland, the Polish National Science Centre grant 2011/01/B/ST6/03867 and by the Polish National Centre for Research and Development (NCBiR) under Grant No. SP/I/1/77065/10 in frame of the strategic scientific research and experimental development program: "Interdisciplinary System for Interactive Scientific and Scientific-Technical Information".

J.F. Peters et al. (Eds.): Transactions on Rough Sets XVI, LNCS 7736, pp. 83–102, 2013.

The original RSKT article [20] is focused on a case study of semantic clustering of scientific articles related to the area of Rough Sets. Our efforts are meant to answer the demand for developing semantic methods for document processing, expressed in a major project (SYNAT), that we are involved in.

The SYNAT project (abbreviation of Polish "**SY**stem **NA**uki i **T**echniki", see [2]) is a large, national R&D program of Polish government aimed at establishment of a unified network platform for storing and serving digital information in widely understood areas of science and technology. The project is composed of nearly 50 modules developed by research teams at 16 leading research institutions in Poland.[1] Within the framework of the larger project we want to design and implement a solution that will make it possible for a user to search within repositories of scientific information (articles, patents, biographical notes, etc.) using their semantic content. Our prospective system for doing that is called SONCA (abbreviation for **S**earch based on **ON**tologies and **C**ompound **A**nalytics, see [12,14,15] and Fig. 5).

Ultimately, SONCA should be capable of answering the user query by listing and presenting the resources (documents, Web pages, et cetera) that correspond to it *semantically*. In other words, the system should have some *understanding* of the intention of the query and of the contents of documents stored in the repository as well as the ability to retrieve relevant information with high efficacy. The system should be able to use various knowledge sources related to the investigated areas of science. It should also allow for independent sources of information about the analyzed objects, such as, e.g., information about scientists who may be identified as the stored articles' authors.

The idea that we pursue in this study is to perform semantic grouping (clustering) of documents based on their associations with concepts drawn from the DBpedia knowledge base. If done right, such clustering should make a good start point for, e.g., a system with extended search features, capable of returning results that are topically close to the search terms, not just those that actually contain the terms from the query (semantic vs. syntactic). It would also make it possible to associate (tag) documents with meaningful concepts. This approach is in line with the general trend of finding semantic similarities between documents with assistance of additional knowledge sources (ontologies, thesauri, taxonomies, Wikipedia) in order to obtain more meaningful and useful results. In order to be able to provide semantical relationships between concepts and documents we employ a method called Explicit Semantic Analysis (ESA) [5]. This method associates elementary data entities with concepts coming from knowledge base.

In our initial RSKT 2011 article we have presented a case study using a text corpus consisting of scientific papers related to Rough Sets. In this way we have gained some additional insight into our own field of research, verify (positively or negatively) some hypotheses and common beliefs, and possibly find some new. At the same time, since we know the document corpus well, we have used our own expertise to judge the quality of clustering and tagging solution. The experience

---

[1] http://www.synat.pl

gained from the case study made it possible to modify, extend and improve the ESA approach to semantic tagging of scientific documents (articles).

The article is organized as follows. Section 2 describes the methodology and motivation behind our approach. Then we describe our data set (Section 3) and DBpedia knowledge base (Section 4), providing some details about their characteristics and the way they were collected and prepared for experiments. Section 5 contains description of the actual experiment on rough-set-related articles and explanation of its results. The results and conclusions from the experiments (Section 5) led us to constructions of one of the base components of the SONCA system, that works on other sets of data. An overview of experiments on various collections of scientific articles, made with modified ESA method, is provided in Section 6. We finish with conclusions and directions for further work in Section 7.

2   The Purpose and Methodology of the Study

The purpose of this experimental study is two-fold.

1. We want to test and verify methodology for document grouping (clustering) based on their semantic content and using a knowledge base. In particular, we want to identify the best configuration for various steps in the process, one that is both computationally feasible and produces meaningful clusters of documents. The goal is to establish a procedure that we will be able to apply semi-automatically to various future text corpora. Since the area of Rough Sets is close to us, we are able to better evaluate the results of experiments on the corpus of texts collected in this field of research. As a consequence, we can identify strengths and weaknesses of the method under scope.

2. We want to learn as much as the methodology permits about our corpus of documents (research papers) related to the area of Rough Sets. Since the individual documents used for this case study (Section 3) are familiar to us, we want to discover the semantic structure of the corpus as a whole and draw some conclusions regarding the features of publications in this scientific area. In particular, we are interested in identifying the most prevalent concepts that characterize this corpus.

Figure 1 shows the general layout of the method that we employ in our case study. The methodology of our was inspired by the Explicit Semantic Analysis (ESA) approach presented in IJCAI paper [5]. Since this article is quite involved, the method which it discusses requires more detailed explanation. The data sources, i.e., collection of documents and DBpedia knowledge base, together with the NLP[2] methods for their pre-processing ("Initial text processing" box in Fig. 1) are described in Sections 3 and 4, respectively.

In our approach, after initial processing, both collections of texts (the corpus and the DBpedia abstracts) are converted to the *bag-of-words* (word-vector)

---

[2] Natural Language Processing (NLP) tools as in [4].

representation. The bag-of-words representation of a text (document) is a vector based on vocabulary, i.e., the collection of unique words (stems) in the corpus.

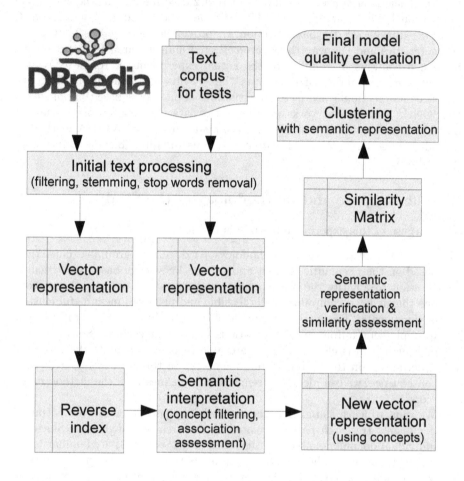

Fig. 1. The general scheme of the experiment

Assume, that after initial processing of a text corpus $D = \{T_1, \ldots, T_M\}$ we have collected a vocabulary consisting of N unique terms (e.g. words, stems, n-grams) $w_1, \ldots, w_N$. Then, any text (document) T_i $(i = 1, \ldots, M)$ in the corpus can be represented by a vector $\langle v_1, \ldots, v_N \rangle \in \mathbb{R}_+^N$, where each coordinate v_j expresses a value of some relatedness measure for j-th term in vocabulary (w_j), relative to this document. The most common measure used to calculate the weights v_j is the *tf-idf* (term frequency-inverse document frequency) index (see [4,13]) defined as:

$$v_j = tf_{i,j} \times idf_j = \frac{n_{i,j}}{\sum_{k=1}^{N} n_{i,k}} \times \log\left(\frac{M}{|\{i : n_{i,j} \neq 0\}|}\right), \tag{1}$$

where $n_{i,j}$ is the number of occurrences of the considered word w_j in the document T_i.

Next, the bag-of-words representation of concept definitions is transformed into an inverted index that maps words into lists of K concepts described in the knowledge base. The inverted index is used to speed up the semantic interpretation of documents (*Semantic Interpretation* in Fig.1). Given a text from a corpus, it iterates over words from the text, retrieves the corresponding entries and merges them into a weighted vector of concepts that represents this text.

Let $W_i = \langle v_1, \ldots, v_j, \ldots, v_N \rangle$ be a bag-of-words representation of an input text T_i, where v_j is the tf-idf index of w_j described in (1). Let $inv_{j,k}$ be an inverted index entry for w_j. It quantifies the strength of association of the term w_j with a knowledge base concept c_k, $k \in \{1, \ldots, K\}$. For convenience, all the weights $inv_{j,k}$ can be arranged in a matrix with N rows and K columns, denoted by INV, such that $INV[j, k] = inv_{j,k}$ for any pair (j, k). This matrix has a sparse structure, since usually only a relatively small number of words have any significance in describing a given concept. The new vector representation of T_i will be denoted by $C_i = \langle c_1, \ldots, c_K \rangle$ where:

$$c_k = \sum_{j:w_j \in T_i} v_j \times inv_{j,k} = W_i * INV[\cdot, k]. \tag{2}$$

In the above equation $*$ stands for the standard scalar product and $INV[\cdot, k]$ indicates k-th column of the sparse matrix INV. We will refer to this new vector representation using a notion of *bag-of-concepts* of a text T_i.

The new vector (bag-of-concepts) representation makes it possible to examine relations between concepts and documents, identify and filter key concepts for the given document corpus, and calculate semantic similarity between texts by comparing their bag-of-concepts representations. It can also be utilized by a search engine as a kind of a semantic index for efficient retrieval of relevant documents.

For technical reasons, in this experimental study, we chose to store all semantic similarity values for pairs of texts in a structure called *Similarity Matrix*. Entries in this matrix are used to numerically represent the conformity between documents (their bag-of-concept representations), which in our case is calculated using the cosine similarity.

$$Sim_{cos}(T_i, T_j) = \frac{c_i * c_j}{\|c_i\| \times \|c_j\|}, \tag{3}$$

where $*$ is the scalar product of two vectors and $\| \cdot \|$ is the L^2 norm of a vector. This particular measure is very often used for comparison of texts due to its robust behaviour in highly dimensional and sparse data domains [4,13].

The fact that we can calculate semantic similarity between documents gives us the means to perform clustering. Considering that we want to obtain a meaningful grouping of documents we decided to use an agglomerative hierarchical clustering. In order to decide for how many clusters we should divide our data we use a cluster quality measure, in particular the silhouette coefficient.

For detailed description of agglomerative clustering, silhouette coefficient, and cluster distance refer to [22].

The quality of resulting clusters is evaluated manually with help of experts in the area of Rough Sets as well as compared with results of a different clustering method. In order to have a reference point we perform a "classical" agglomerative clustering on the bag-of-words representation of documents, without any use of knowledge base. The resulting partition of documents is then compared with our approach using various measures for cluster consistency as well as manual evaluation of cluster meaningfulness.

3 Data Acquisition and Preparation

For our case study we have used 349 documents in PDF format. These documents are selected from the collection of papers published by the members and associates of the Group of Logic at the University of Warsaw. The subset used for our experiment is significantly smaller than the entire collection, which consists of over 600 publications. While choosing documents for this subset we have used the following criteria:

- We restricted publications to those published in last 15 years (between 1996 and 2011) and written in English.
- We have only chosen "regular" articles, i.e., standard journal, book, and conference papers. They roughly correspond to BibTeX categories: `article`, `inproceedings`, and `incollection`.
- Papers that are very short (extended abstracts) or unusually long (mini-monographs) have been left out.
- Some articles have been removed from the study due to technical difficulties they posed. This was mostly due to problems with incorrect PDF format and usually concerned older (pre-2003) publications.

There were several reasons for using the above criteria in the process of constructing initial data sample for our study. The most important are as follows:

- We wanted the corpus of documents to be relatively regular. Since our ultimate goal is the grouping (clustering), we tried to eliminate outliers early on. The idea is to have well-comparable documents and then do the clustering on the basis of their semantic content rather than attributes of their syntactic composition, such as size, level of complication or number of words.
- We have chosen the collection of documents that were created over the years in our group in order to have good understanding of the corpus from the very beginning. Since we know the field and in many cases have direct contact with authors, we can evaluate the outcome with greater ease and confidence. This is a big advantage, especially for an initial, explorative study such as the one that we conduct. It gives us the ability to clearly identify strong and weak points in our methodology.

- We have decided to use this particular number of documents (349) because we wanted to construct a corpus which would be as representative for the area of Rough Sets as it is possible in the given circumstances. The 349 documents in our collection correspond to roughly 10% of all documents of this kind listed in the Rough Set Database system (RSDS [6]). At the moment of writing, the RSDS contains 3641 bibliographical notes that belong to categories that we are interested in.
- Last, but not the least, we selected this particular document corpus because we have both access to their PDF versions and the limited copyrights that allow us to re-use (but not re-distribute) them.

The original PDF documents were first converted to a pure text format with the use of Python script based on PDFMiner library [17]. All documents were divided into blocks of plain text. Based on certain text statistics the script extracted only the text contained in paragraphs, sections and their titles. It has to be underlined that author names article and page headers, footers, tables, equations and other parts of text which were irrelevant and could bias further analysis were discarded. The purpose of this step was to remove various artifacts and clarify text files before attempting to calculate word frequencies and clustering.

This step, although it may appear simple, proved to be troublesome at times. Typical problems at this stage are associated with conversion of hyphenated (broken between lines) words and ligatures (e.g., fi in "classification") back to their original (textual) form. These problems were partly resolved with use of an English dictionary which made it possible to guess the right encoding of some characters by determining whether words created after substitution of missing characters were proper English terms. Articles contained also a great amount of mathematical symbols which were encoded in PDF files in various, sometimes very unexpected way. These unusual characters were filtered out as well. Additionally, the bibliography section (references) was removed from each of selected text files. It was done in order to assure that we perform analysis on actual semantic content of the document and to reduce the influence of certain words contained in references, like: publisher, journal name, etc.

The corpus of 349 plain text files was then processed in order to calculate word-vector (bag of words) representations in the next step. First, stop words were removed and then we have performed stemming on the set of words contained in these documents. For stemming of both documents and DBpedia abstracts (as described in Section 4) we use a version of popular Porter's algorithm (cf. [11]). Initially, the corpus contained 35507 unique words (excluding stop words). After stemming we have obtained 26800 unique words (stems) to work with. On average a single document in the collection contains 3524 stems, with minimum of 362 and maximum of 13640.

4   The DBpedia Knowledge Base

According to its creators, the DBpedia (cf. [23,3]) is a community effort to extract structured information from Wikipedia (cf. [25]) and to make this information available on the Web. DBpedia allows to ask queries against Wikipedia data and structure, and to link other data sets to Wikipedia data. In layman terms, DBpedia is a snapshot of the original Wikipedia with mostly preserved structure, but reduced content.

For the purpose of our study we needed to use DBpedia as an enriched dictionary. The version of DBpedia that we use (version 3.5.1 for English Wikipedia) contains 3,257,133 notions (so called *things*). Each DBpedia *thing* represents a single Wikipedia concept (a single Wikipedia page including disambiguation pages and lists). Due to the distributed and asynchronous nature of the process in which the Wikipedia is created by members of its community, there are some consistency and regularity issues with it. Much of these issues are inherited by DBpedia, which results in some problems related to conflicting or expired names for concepts and categories.

In DBpedia, pages from the original Wikipedia are represented only by their abstracts. For most of the DBpedia concepts there is also additional information derived from Wikipedia, such as classification to Wikipedia categories. There are 3,144,262 abstracts available in DBpedia 3.5.1, but they are very diverse in their length and quality. The length of abstracts vary from empty (0 words) to quite long ones (the longest has 16850 words), with an average of 101 words per abstract. Most of those texts are well formatted and structured but there are exceptions, e.g., some contain only LaTeX-styled source code of tables or figures which were, probably unintentionally, placed in the abstract section of the corresponding Wikipedia article. There are also cases when a whole text of Wikipedia page is placed in the abstract which results in considerably longer DBpedia representations.

Taken altogether, DBpedia 3.5.1 entries constitute a text corpus consisting of 316,631,010 words (after filtration). The number of unique words, before stemming and filtering, is 2,818,483. There are 560,049 *categorical notions* (Wikipedia categories) of which 449,140 are *direct*, i.e., contain some concepts and the rest are *indirect*, i.e., they contain only other categories.

5   Experimental Evaluation of the Approach

Our experiment was conducted in three main steps which we implemented in R System ([16]). First, DBpedia and the selected text collection were preprocessed. Each DBpedia entry and a document was was cleaned, in particular: its encoding was changed to UTF-8, words that contained special characters (!@#$%_&*+-=) or numbers were removed, the most common shortcuts were expanded, and the most common words from a special *stop word list*[3] were removed. The Porter's algorithm [11], implemented in the *Rstem* library, was used for finding stems of words. The stems that occurred less than three times in DBpedia were also eliminated from the texts. Finally, the concepts that were represented by less than 10 unique stems were removed from the knowledge base.

---

[3] A standard stop word list from *openNLP* library was extended by the 100 most common words from DBpedia abstracts.

As a result, the size of the knowledge base was reduced to around 2.5 million concepts described by approximately 850 thousands of unique stems.

Table 1. List of ten most relevant DBpedia concepts for three exemplary documents, with degree of association included

(LTF-C): Architecture, Training Algorithm and Applications of New Neural Classifier		
[1]	9.19	"Neural_Lab"
[2]	9.17	"Echo_state_network"
[3]	8.75	"Auto-encoder"
[4]	8.30	"Interneuron"
[5]	8.09	"Oja's_rule"
[6]	8.08	"Multilayer_perceptron"
[7]	8.06	"Biological_neural_network"
[8]	8.06	"Artificial_neural_network"
[9]	8.00	"Artificial_neuron"
[10]	7.84	"Neuroevolution"
Judgment of satisfiability under incomplete information		
[1]	8.21	"Definable_set"
[2]	8.08	"Schaefer's_dichotomy_theorem"
[3]	7.96	"Formal_semantics_of_programming_languages"
[4]	7.85	"Empty_domain"
[5]	7.78	"Tautology_(logic)"
[6]	7.68	"Equisatisfiability"
[7]	7.54	"Method_of_analytic_tableaux"
[8]	7.38	"Conditional_quantifier"
[9]	7.36	"Model_checking"
[10]	7.32	"Satisfiability_and_validity"
Combination of Metric-Based and Rule-Based Classification		
[1]	8.92	"K-nearest_neighbor_algorithm"
[2]	6.19	"Backmarking"
[3]	6.08	"Wolfe_conditions"
[4]	5.90	"Evolutionary_data_mining"
[5]	5.66	"Event_condition_action"
[6]	5.64	"Transduction_(machine_learning)"
[7]	5.63	"Soft_independent_modelling_of_class_analogies"
[8]	5.63	"Ground_truth"
[9]	5.56	"Proximity_problems"
[10]	5.50	"Dominating_decision_rule"

In the second step, the bag-of-concepts representations of texts from the rough set corpus were created using the method described in Section 2. A modified tf-idf index was used to assess the relevance of words (stems) to documents and to concepts. For each text, the frequencies of words, i.e., the tf component in tf-idf formula (1), were smoothed by taking their square root. This modification was dictated by a fact, that many of the documents which we use are of technical nature and as such contain many repetitions of specific terms (or single words).

The strength of bounds between the concepts and the rough set articles was computed using the equation (2). Following the intuition, that it is meaningless to associate any document with a large number of specific concepts, we have restricted the number of concepts associated with each document. We have decided to use no more than 35 most related concepts for characterization of any given text. This number (35) was selected because it corresponds to around 1% of the average number of stems appearing in the single document in the corpus, which in turn gives more compact and comprehensive representation. Table 1 presents associations of top 10 concepts to three exemplary articles from the corpus.

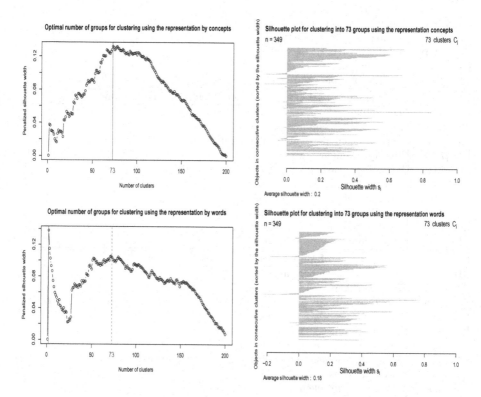

Fig. 2. The plot of silhouette coefficient values across clusters used to establish the optimal number of groups (on the left) and the silhouette coefficient values of individual documents for the selected clustering (on the right). The results obtained from the representation by concepts (on the top) is compared to representation by words (at the bottom).

The last step of our experiment involved computation of distances between documents from the rough set corpus which we then use for clustering. Due to an extremely sparse representation of our texts (only 35 non-zero values out of ≈ 2.5 million) the *cosine distance* was employed, which is a commonly used measure in high-dimensional information retrieval tasks ([4],[22]). For the clustering we utilized an agglomerative hierarchical approach with the "average" as a linking function (see [22]). The optimal number of groups (clusters) was decided using the *silhouette width coefficient* which was additionally penalized for selecting larger number of clusters. Figure 2 illustrates the values of average silhouette coefficient w.r.t. the growing number of clusters, along with the silhouette coefficient values of individual documents for the selected clustering. From this picture one can see that the highest cluster separability is achieved when we use 73 of them.

Figure 3 presents the clustering tree corresponding to the partition into 73 groups. Apart from using the silhouette coefficient, quality of the 73 clusters was also assessed with the aid of human experts. Mutual relatedness of documents from several groups has been evaluated. In order to gain another point of reference we have also performed clustering using the original bag-of-words[4] representation of texts in the corpus (Figure 4).

The results of this comparison are encouraging. The consecutive partitions obtained using the bag-of-concepts representation yielded much more stable silhouette coefficients than those for the original word-vector (bag-of-words) one. The optimal number of groups for the latter is two, which corresponds to meaningless grouping of documents. This number is also not very stable as it may vary wildly between 2 and 157 if we alter the penalty for producing excessive clusters. Moreover, if with the bag-of-words representation we make the clustering algorithm produce 73 groups (optimal number for the bag-of-concepts), then brief analysis of this partition reveals a significant imbalance in the size of clusters (Figure 4). The largest cluster obtained in this way contained 60 papers and there were 29 singletons (clusters that contained only a single document). To make things worse, many of the larger groups constructed in this manner contain semantically unrelated documents and are very difficult to label. In contrast, size of the largest group resulting from utilization of the bag-of-concepts representation was 27 and there were only 19 singletons.

The observation, that employment of domain knowledge improves the quality of clustering was confirmed by domain experts. For instance, Table 2 shows members of three exemplary clusters taken from distinct branches in the clustering tree (Figure 3). Labels that briefly summarize contents of those groups were given by experts. Among 13 papers that belong to the cluster 21, 12 were recognized by experts as related to the notion of neural computing and artificial neural networks. The same subset of papers, partitioned based on the bag-of-words representation, was broken between three different clusters of which only one was semantically homogeneous and meaningful.

It it also worth mentioning that, even though information about authors and bibliography was removed from the corpus during the preprocessing phase, 12 out of 14 articles grouped in the cluster 39 were written by a single author (Anna Gomolińska). In those papers, the author consider a problem of partial

---

[4] For consistency, we used the smoothed tf-idf vector representation.

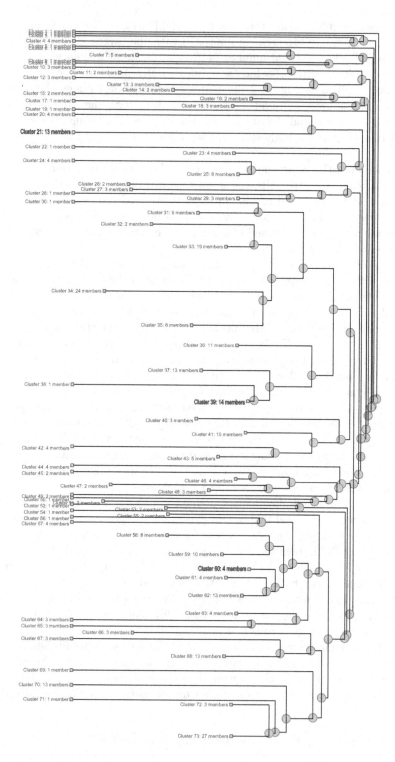

Fig. 3. Truncated tree of clusters (dendrogram) with 73 leafs, based on the bag-of-concepts representation. Clusters 21,39, and 60 that are detailed in Table 2 are marked with different font.

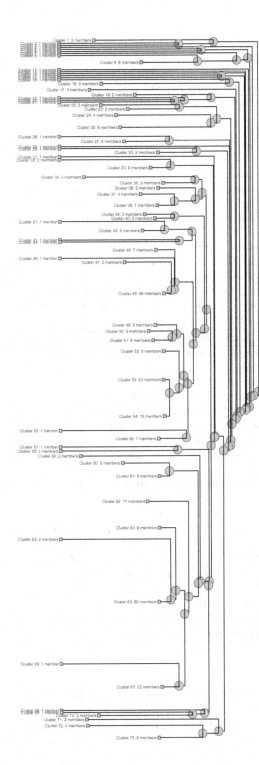

Fig. 4. A clustering tree based on the bag-of-words representation, truncated to 73 leafs

Table 2. Members of exemplary partitions, resulting from clustering with the bag-of-concepts representation. The IDs of branches from the clustering tree are given along with the labels assigned by domain experts and titles of corresponding documents.

Cluster 21: Neurocomputing and Artificial Neural Networks
[1] (LTF-C): Architecture, Training Algorithm and Applications of New Neural Classifier
[2] Rough Neurons: Petri Net Models and Applications
[3] Rough-Neural Computing: An Introduction
[4] Toward Rough Neural Computing Based on Rough Membership Functions: Theory and Application
[5] Rough Neurocomputing: A Survey of Basic Models of Neurocomputation
[6] Design of rough neurons: Rough set foundation and Petri net model
[7] Constructing Extensions of Bayesian Classifiers with use of Normalizing Neural Networks
[8] Refining decision classes with neural networks
[9] Harnessing Classifier Networks - Toward Hierarchical Concept Construction
[10] Feedforward concept networks
[11] Neural network design: Rough set approach to real-valued data
[12] Hyperplane-based neural networks for real-valued decision tables
[13] Rough Sets and Artificial Neural Networks

Cluster 39: Logical Satisfiability and Validity of Formulas
[1] Judgment of satisfiability under incomplete information
[2] A graded applicability of rules
[3] Toward rough applicability of rules
[4] Satisfiability and meaning in approximation spaces
[5] Satisfiability Judgment Under Incomplete Information
[6] Reasoning Based on Information Changes in Information Maps
[7] Rough validity, confidence, and coverage of rules in approximation spaces
[8] Satisfiability and meaning of formulas and sets of formulas in approximation spaces
[9] On rough judgment making by socio-cognitive agents
[10] Rauszer's R-logic for multiagent systems
[11] Rough rule-following by social agents
[12] Satisfiability of formulas from the standpoint of object classification
[13] Construction of rough information granules
[14] Patterns in Information Maps

Cluster 60: Instance-based Learning
[1] Combination of Metric-Based and Rule-Based Classification
[2] Rough Set Approach to CBR
[3] Local Attribute Value Grouping for Lazy Rule Induction
[4] Granulation in Analogy-based Classification

satisfiability and validity of formulas (such as decision rules) under incomplete or uncertain information.

It seems that with the bag-of-concept representation, the clustering algorithm was able to conceptually discern them from other research topics of this particular author. The articles of the same author that belong to other research direction, the theory of approximation spaces, are located in another cluster. These articles (six of them) are placed in the cluster 36 (Figure 3). In comparison, when the representation by bag-of-words is used, almost all publications of Anna Gomolińska from our corpus (21 out of 22) are placed in a single group. That last fact, in our opinion, is probably due to usage of a characteristic and highly specialized vocabulary that inadvertently biases the bag-of-words representation.

Table 3. Tags (concept labels) for three examples of clusters

Cluster 21: Neurocomputing and Artificial Neural Networks
[1] "ADALINE"
[2] "Artificial_neural_network"
[3] "Artificial_neuron"
[4] "Auto-encoder"
[5] "Delta_rule"
[6] "Multilayer_perceptron"
[7] "Universal_approximation_theorem"
[8] "Echo_state_network"
[9] "Neural_Lab"
Cluster 39: Logical Satisfiability and Validity of Formulas
[1] "Empty_domain"
[2] "Formal_theorem"
[3] "Limit-preserving_function_(order_theory)"
[4] "Satisfiability_and_validity"
[5] "Schaefer's_dichotomy_theorem"
[6] "Tautology_(logic)"
[7] "Well-definition"
Cluster 60: Instance-based Learning
[1] "Attribute_(computing)"
[2] "Attribute_(network_management)"
[3] "Integrity_constraints"
[4] "K-nearest_neighbor_algorithm"
[5] "Online_machine_learning"
[6] "Relation_(database)"
[7] "Structured_SVM"

We have also investigated whether the bag-of-concepts representation may be used for the purpose of automated tagging (labeling) of clusters. For this purpose we associated each group (cluster) of articles with DBpedia concepts that appear in representations of at least 80% of its members. Table 3 presents these associations for the three exemplary clusters.

From this example one can see that the selected concepts (cluster tags) are well in line with cluster labels assigned by the experts. Unfortunately, they seem to be too specific to express the semantic relatedness of the documents in the cluster by themselves. To overcome this issue, in the future we plan to employ knowledge about DBpedia categories and the structure of concepts to construct more general tags.

6 A More General Framework

The research on automatic tagging and semantic clustering algorithms is a part of a much larger SYNAT project [2]. The SONCA system [14,15], which we are developing, is designed to store and process scientific articles acquired from a wide range of domains and sources. As shown in Fig. 5, the system comprises of several major components and is organized into several layers.

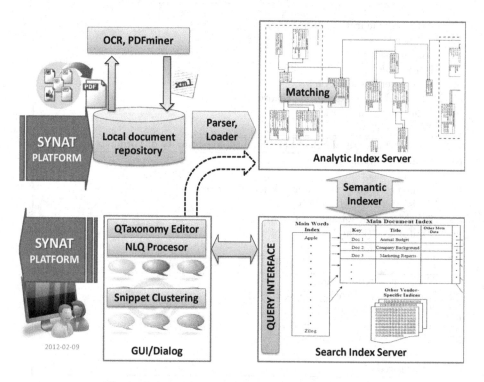

Fig. 5. A general architecture of the SONCA system

Our work on semantic clustering and tagging is instrumental in designing a bridge between the analytic data warehouse used to store structural information about scientific documents and a collection of indexes stored on a dedicated server, which are used to quickly retrieve answers (links to documents) that are semantically relevant to the user query. Information provided by our algorithms also play an important role in the process of preparing the result for presentation to the user.

Using our experience with the corpus analyzed in the presented case study we were able to design an automatic tagging module for semantic indexing of scientific articles and assessment of their relatedness. We used this module in several large-scale experiments on papers from the biomedicine and life sciences domains obtained from the PubMed Central database[5] [18,10]. As a knowledge base we used the Medical Subject Headings (MeSH [6]) – a comprehensive controlled vocabulary for the purpose of indexing journal articles and books in the life sciences [24].

A large share of documents from the PubMed Central repository was labeled with pairs of MeSH headings/subheadings by experts from U.S. National Library of Medicine. For the purpose of our experiments we obtained those labels. It gave us an opportunity to quantitatively evaluate a quality of tags produced by our implementation of ESA within the SONCA engine. Table 4 shows three exemplary documents from PMC with tags assigned by SONCA.

Table 4. Three exemplary PMC documents with tags assigned by SONCA's implementation of ESA. The matching tags are bolded.

Document title	MeSH tags by MEDLINE	MeSH tags by SONCA
Cockroaches (Ectobius vittientris) in an intensive care unit, Switzerland.	**Cockroaches***, Insect Control*, **Intensive Care Units***, **Cross Infection**, Insect Vectors	**Cockroaches**, **Intensive Care Units**, Klebsiella Infections, Pest Control, **Cross Infection**
Serotonin transporter genotype, morning cortisol and subsequent depression in adolescents.	**Depressive Disorder***, Genetic Predisposition to Disease*, Serotonin Plasma Membrane Transport Proteins*, Genotype, **Multilevel Analysis**	**Depressive Disorder**, Genome-Wide Association Study, **Multilevel Analysis**, Cohort Studies, Adolescent Psychiatry
Capacity of Thailand to contain an emerging influenza pandemic.	Disaster Planning*, Health Policy*, Disease Outbreaks, **Health Resources**, Influenza Human	Health Care Rationing, **Health Resources**, Epidemics, Evidence-Based Medicine, Influenza B virus

On average, *recall* of MeSH headings assigned by experts in top 30 labels returned by our system was 0.26. Although this quantity itself is not very high, it is noticeable that many of the non-conforming labels were still quite reasonable – such as the *Klebsiella Infections* for the first title and *Influenza B virus* for the third one. Additionally, out further studies show that there is a vital possibility for improving this result by using supervised learning algorithms for adaptive tuning the weights from the semantic inverted index [10].

Since our module is fully integrated with the SONCA engine, we were able to overcome the scalability problem by running the tagging algorithm on a column-oriented database solution, which is optimized for executing large analytical SQL

---

[5] PMC, see [1].

[6] http://www.nlm.nih.gov/pubs/factsheets/mesh.html

queries [12,19]. During the initial tests we found out that, although conceptually the method performs really well, the underlying data processing, especially the part performed inside RDBMS, requires introduction of some new techniques.

We have also conducted a different series of experiments related to the utilization of the semantic tagging for clustering of research papers. Their aim was to verify what impact the automatically generated labels have on performance of different similarity measures and similarity learning methods, in general [9]. Form those experiments we have learned that tags generated by methods such as ESA can be successfully used in combination with similarity learning techniques as semantic features of documents.

Finally, the experience in using ESA for enhancing the document representation led us toward consideration of a different problem, namely multi-label classification of textual data [8]. In this task, a categorization of documents is done in a supervised manner. We wanted to verify whether the limitation of the semantic representation to K top associated concepts still allows to conduct accurate multi-label classification. Indeed, this representation appears to be useful for certain document corpora, such as PubMed, for which the corresponding labels (MeSH sub-headings) can be predicted[7].

7   Conclusions and Plans for the Future

The conclusions drawn from this case study, just like the motivations presented in Section 2, are of two kinds.

Firstly, we can draw conclusions regarding the structure and characteristics of the corpus of 349 rough set related documents that were used as the basis for the study. The experimental results confirm that our text corpus is fairly uniform and focused. It is quite clear, that our articles share a lot of common concepts at the same time being separable from other research areas. Within the area of rough sets, the papers can be arranged into groups (clusters) in a really meaningful manner.

The second conclusion is that the proposed approach to clustering, based on ESA approach, has a significant potential and shall be seriously considered as an element in the future studies. During the experiments it was possible to establish some ground knowledge about features of the method used. That gives us some confidence about the viability of this approach and its potential to become an element of the prototype software solution that we are eyeing in the frame of our main (SYNAT) project. The results of other experiments, as discussed in Section 6, support the claim that ESA is a well-suited tool for our needs.

As usual with this kind of experimental study, there is a plethora of things we can do next. In shorter perspective, next steps should include testing more clustering methods and playing with parameters to obtain more optimal and versatile solution for the same corpus of documents. Also, it would be very

---

[7] This was exemplified by the results of the data mining competition associated with the JRS'2012 conference (JRS'2012 Data Mining Contest: Topical Classification of Biomedical Research Papers) http://tunedit.org/challenge/JRS12Contest, [8].

interesting to investigate whether including more knowledge from DBpedia, such as some structural information about categories, helps to improve the overall results. Another natural next step is to extend the investigated corpus and check if our findings remain valid for a larger data set. This step, however, requires an access to a larger sources of PDF documents related to the theory of rough sets.

Another possible direction for a continuation of this study may regard different methods for assessment of similarity between pairs of scientific documents. Currently, only the cosine similarity is being used. This fact restrains our ability to detect semantically similar texts since it enforces potentially undesirable properties of a metric on the similarity measure. We believe, that in order to capture more semantic resemblance of articles, the similarity measure should be more dependent on a domain from which the documents come from. One way to achieve that is through utilization of some similarity learning methods, such as the Rule-based Similarity model described in [7]. Some preliminary experiments with a adaptation of this similarity model to the unsupervised case have already been performed and their results seem very promising (cf. [9]).

In a long run, the follow-up of this study should produce a software module that will serve as a part of SONCA system, supporting development of information platform in the SYNAT project. The prototype of such module is currently being integrated with SONCA platform and tested against various document corpora and various knowledge bases.

References

1. Beck, J., Sequeira, E.: PubMed Central (PMC): An archive for literature from life sciences journals. In: McEntyre, J., Ostell, J. (eds.) The NCBI Handbook, ch. 9. National Center for Biotechnology Information, Bethesda (2003), http://www.ncbi.nlm.nih.gov/books/NBK21087/
2. Bembenik, R., Skonieczny, Ł., Rybiński, H., Niezgódka, M. (eds.): Intelligent Tools for Building a Scientific Information Platform. SCI, vol. 390. Springer, Heidelberg (2012)
3. Bizer, C., Lehmann, J., Kobilarov, G., Auer, S., Becker, C., Cyganiak, R., Hellmann, S.: DBpedia – a crystallization point for the web of data. Journal of Web Semantics: Science, Services and Agents on the World Wide Web 7, 154–165 (2009)
4. Feldman, R., Sanger, J. (eds.): The Text Mining Handbook. Cambridge University Press (2007)
5. Gabrilovich, E., Markovitch, S.: Computing semantic relatedness using Wikipedia-based explicit semantic analysis. In: Proceedings of the 20th International Joint Conference on Artificial Intelligence, pp. 6–12 (2007)
6. Grochowalski, P., Suraj, Z.: RSDS - the Rough Set Database System - a bibliographic database on wide aspects of rough sets (2009), http://rsds.univ.rzeszow.pl/
7. Janusz, A.: Dynamic Rule-Based Similarity Model for DNA Microarray Data. In: Peters, J.F., Skowron, A. (eds.) Transactions on Rough Sets XV. LNCS, vol. 7255, pp. 1–25. Springer, Heidelberg (2012)

8. Janusz, A., Nguyen, H.S., Ślęzak, D., Stawicki, S., Krasuski, A.: JRS 2012 Data Mining Competition: Topical Classification of Biomedical Research Papers. In: Yao, J.T., Yang, Y., Słowiński, R., Greco, S., Li, H., Mitra, S., Polkowski, L. (eds.) RSCTC 2012. LNCS (LNAI), vol. 7413, pp. 422–431. Springer, Heidelberg (2012)
9. Janusz, A., Ślęzak, D., Nguyen, H.S.: Unsupervised similarity learning from textual data. Fundamenta Informaticae 119(3)
10. Janusz, A., Świeboda, W., Krasuski, A., Nguyen, H.S.: Interactive Document Indexing Method Based on Explicit Semantic Analysis. In: Yao, J.T., Yang, Y., Słowiński, R., Greco, S., Li, H., Mitra, S., Polkowski, L. (eds.) RSCTC 2012. LNCS (LNAI), vol. 7413, pp. 156–165. Springer, Heidelberg (2012)
11. Jones, K.S., Willet, P.: Readings in Information Retrieval. Morgan Kaufmann, San Francisco (1997)
12. Kowalski, M., Ślęzak, D., Stencel, K., Pardel, P., Grzegorowski, M., Kijowski, M.: RDBMS model for scientific articles analytics. In: Bembenik, et al. [2], ch. 4, pp. 49–60
13. Manning, C., Raghavan, P., Schütze, H.: Introduction to information retrieval (2007) (online edition), http://nlp.stanford.edu/IR-book/
14. Nguyen, A.L., Nguyen, H.S.: On designing the SONCA system. In: Bembenik et al. [2], ch. 2, pp. 9–35
15. Nguyen, H.S., Ślęzak, D., Skowron, A., Bazan, J.: Semantic search and analytics over large repository of scientific articles. In: Bembenik, et al. [2], ch. 1, pp. 1–8
16. R Development Core Team: R: A Language and Environment for Statistical Computing. R Foundation for Statistical Computing, Vienna, Austria (2009), http://www.R-project.org
17. Shinyama, Y.: PDFMiner: Python PDF parser and analyzer (2010), http://www.unixuser.org/~euske/python/pdfminer/
18. Ślęzak, D., Janusz, A., Świeboda, W., Nguyen, H.S., Bazan, J.G., Skowron, A.: Semantic Analytics of PubMed Content. In: Holzinger, A., Simonic, K.-M. (eds.) USAB 2011. LNCS, vol. 7058, pp. 63–74. Springer, Heidelberg (2011)
19. Ślęzak, D., Wróblewski, J., Eastwood, V., Synak, P.: Brighthouse: an analytic data warehouse for ad-hoc queries. PVLDB 1(2), 1337–1345 (2008)
20. Szczuka, M., Janusz, A., Herba, K.: Clustering of Rough Set Related Documents with Use of Knowledge from DBpedia. In: Yao, J., Ramanna, S., Wang, G., Suraj, Z. (eds.) RSKT 2011. LNCS, vol. 6954, pp. 394–403. Springer, Heidelberg (2011)
21. Szczuka, M., Janusz, A., Herba, K.: Semantic clustering of scientific articles with use of DBpedia knowledge base. In: Bembenik, et al. [2], ch. 5, pp. 61–76
22. Tan, P.-N., Steinbach, M., Kumar, V.: Introduction to Data Mining. Addison Wesley, Boston (2006), http://www-users.cs.umn.edu/~kumar/dmbook/index.php
23. The DBPedia Community: The DBPedia knowledge base (2011), http://DBpedia.org/
24. United States National Library of Medicine: Introduction to MeSH - 2011 (2011), http://www.nlm.nih.gov/mesh/introduction.html
25. Wikipedia Community: Wikipedia - the free Encyclopedia (2011), http://en.wikipedia.org/

Maximal Clique Enumeration
in Finding Near Neighbourhoods*

Christopher J. Henry and Sheela Ramanna

University of Winnipeg, Department of Computer Science,
Winnipeg, Manitoba R3B 2E9, Canada
{ch.henry,s.ramanna}@uwinnipeg.ca

Abstract. The problem considered in this article stems from the observation that practical applications of near set theory require efficient determination of all the tolerance classes containing objects from the union of two disjoints sets. Near set theory consists in extracting perceptually relevant information from groups of objects based on their descriptions. Tolerance classes are sets where all the pairs of objects within a set must satisfy the tolerance relation and the set is maximal with respect to inclusion. Finding such classes is a computationally complex problem, especially in the case of large data sets or sets of objects with similar features. The contributions of this article are the observation that the problem of finding tolerance classes is equivalent to the MCE problem, empirical evidence verifying the conjecture from [15] that the extra perceptual information obtained by finding all tolerance classes on a set of objects obtained from a pair of images improves the CBIR results when using the tolerance nearness measure, and a new application of MCE to CBIR.

Keywords: Near sets, maximal clique enumeration, tolerance near sets, tolerance space, tolerance relation, pre-class, nearness measure, CBIR.

1 Introduction

The problem considered in this article is one of finding all the tolerance classes on a set of objects. In the proposed application to content-based image retrieval (CBIR) [36], classes in image covers determined by a tolerance relation provide the content used in CBIR and a feature-based tolerance space solution to detecting and measuring similarities in digital images. Specifically, the tolerance classes represent the extracted perceptual information which is used in quantizing the nearness of sets. The notion of nearness in mathematics and the more general notion of resemblance that is a dominant part of CBIR can be traced back to J.H. Poincaré [32]. Our approach stems from a recent extension of J.H. Poincaré's

---

* This research has been supported by the Natural Sciences and Engineering Research Council of Canada (NSERC) grants 194376 and 418413. Also, special thanks to Tariq Alusaifeer for recognizing the problem of finding tolerance classes is equivalent to maximal clique enumeration.

J.F. Peters et al. (Eds.): Transactions on Rough Sets XVI, LNCS 7736, pp. 103–124, 2013.
© Springer-Verlag Berlin Heidelberg 2013

representative spaces, tolerance spaces [40,37,31] and near sets introduced by J.F. Peters in 2007 [26,27], and elaborated in [28,30,7,29,11,39,9,33].

Tolerance classes are sets where all the pairs of objects within a set must satisfy the tolerance relation and the set is maximal with respect to inclusion. Finding such classes is a computationally complex problem, especially in CBIR involving sets of objects with similar features [11,13,15,12,10,14]. Previous work into finding tolerance classes was based on the observation that all tolerance classes containing an object are a subset of the neighbourhood of that object [11,13]. Reported algorithms include a serial approach for finding most tolerance classes using the Fast Library for Approximate Nearest Neighbours (FLANN) [11,13], and a parallel computing approach for finding all tolerance classes using NVIDIA's Compute Unified Device Architecture (CUDA) Graphics Processing Unit (GPU) [15]. This article presents a new solution to the problem of finding tolerance classes by observing that this problem can be mapped to the Maximal Clique Enumeration (MCE) problem. Consequently, the classes can be found using an algorithm with reduced complexity based in graph theory. In addition, the MCE approach to performing CBIR introduced in this article is compared with the well known Earth Movers Distance and Integrated Region Matching in [16]. Finally, the parallel approach is not considered in this article since the runtimes of the serial and MCE algorithms are more than 10 times faster.

The article is organized as follows: First, Section 2 introduces tolerance classes by way of near set theory, providing the context in which tolerance classes are used in this article. Next, previously reported algorithms for find tolerance classes are given in Section 3. Section 4 provides a brief review of the problem of MCE. Then, a discussion on the multitreaded implementation of each algorithm is given in Section 5. Section 6 defines tolerance near sets and presents the nearness measure used to perform CBIR. Finally, Section 7 presents the results and discussion. The contributions of this article are the observation that the problem of finding tolerance classes is equivalent to the MCE problem, empirical evidence verifying the conjecture from [15] that the extra perceptual information obtained by finding all tolerance classes on a set of objects obtained from a pair of images improves the CBIR results when using the tolerance nearness measure, and a new application of MCE to CBIR.

2   Tolerance Classes

Disjoint sets containing objects with similar descriptions are near sets. Similarity is determined quantitatively via some description of the objects. Near set theory provides a formal basis for identifying, comparing, and measuring resemblance of objects based on their descriptions, *i.e.* based on the features that describe the objects. The discovery of near sets begins with identifying feature vectors for describing and discerning affinities between sample objects. Objects that have, in some degree, affinities in their features are considered *perceptually near* each other. Groups of these objects, extracted from the disjoint sets, provide information and reveal patterns of interest.

Tolerance near sets are near sets defined by a description-based tolerance relation. Tolerance relations provide a view of the world without transitivity [37]. Consequently, tolerance near sets provide a formal foundation for *almost solutions*, solutions that are valid within some approximation, which is required for real world problems and applications [37]. In other words, tolerance near sets provide a basis for a quantitative approach for evaluating the similarity of objects without requiring object descriptions to be exact.

Let us begin with defining the content of the sets. All sets in near set theory consist of perceptual objects, which is anything in the physical world with characteristics observable to the senses such that they can be measured and are knowable to the mind. A feature characterizes some aspect of the makeup of a perceptual object. A probe function is a real-valued function representing a feature of a perceptual object [26]. In the context of near set theory, objects in our visual field are always presented with respect to the selected probe functions, which is in keeping with the approach to pattern recognition suggested by M. Pavel [23] where the features of an object are quantified by probe functions. In other words, probe functions are used to measure characteristics of visual objects and similarities among perceptual objects.

A perceptual system is a set of perceptual objects, together with a set of probe functions, *i.e.* a perceptual system $\langle O, \mathbb{F} \rangle$ consists of a non-empty set O of sample perceptual objects and a non-empty set $\mathbb{F}$ of real-valued functions $\phi \in \mathbb{F}$ such that $\phi : O \to \mathbb{R}$ [30]. The notion of a perceptual system admits a wide variety of different interpretations that result from the selection of sample perceptual objects contained in a particular sample space O. Two examples of perceptual systems are: a set of images together with a set of image processing probe functions, or a set of results from a web query together with some measures (probe functions) indicating, *e.g.*, relevancy or distance (*i.e.* geographical or conceptual distance) between web sources. The description of a perceptual object within a perceptual system can be defined as follows. Let $\langle O, \mathbb{F} \rangle$ be a perceptual system, and let $\mathcal{B} \subseteq \mathbb{F}$ be a set of probe functions. Then, the description of a perceptual object $x \in O$ is a feature vector given by

$$\phi_{\mathcal{B}}(x) = (\phi_1(x), \phi_2(x), \ldots, \phi_i(x), \ldots, \phi_l(x)),$$

where l is the length of the vector $\phi_{\mathcal{B}}$, and each $\phi_i(x)$ in $\phi_{\mathcal{B}}(x)$ is a probe function value that is part of the description of the object $x \in O$. Note, the idea of a feature space is implicitly introduced along with the definition of object description. An object description is the same as a feature vector as described in traditional pattern classification [5], yet different from the signature of an object defined in [24] (due to the use of features instead of attributes[1]). The description of an object can be considered a point in an l-dimensional Euclidean space $\mathbb{R}^l$ called a feature space. Thus, the relationship between objects is discovered in a feature space that is determined by the probe functions in $\mathcal{B}$.

Formally, a tolerance space can be defined as follows [40,37,31]. Let O be a set of sample perceptual objects, and let ξ be a binary relation (called a tolerance

---

[1] See, [25,27,39] for a discussion on the difference between features and attributes.

relation) on X ($\xi \subset X \times X$) that is reflexive (for all $x \in X$, $x\xi x$) and symmetric (for all $x, y \in X$, if $x\xi y$, then $y\xi x$) but transitivity of ξ is not required. Then a tolerance space is defined as $\langle X, \xi \rangle$. Considering the tolerance space definition, a specific tolerance relation [28,29] (see [7,8] for applications in image analysis) is given as follows. Let $\langle O, \mathbb{F} \rangle$ be a perceptual system and let $\varepsilon \in \mathbb{R}_0^+$. For every $\mathcal{B} \subseteq \mathbb{F}$, the perceptual tolerance relation $\cong_{\mathcal{B}, \varepsilon}$ is defined by:

$$\cong_{\mathcal{B}, \varepsilon} = \{(x, y) \in O \times O : \parallel \phi(x) - \phi(y) \parallel_2 \leq \varepsilon\},$$

where $\parallel \cdot \parallel_2$ is the L^2 norm.

Finally, the algorithms presented in Section 3 are based on the propositions involving neighbourhoods and tolerance classes. Formally, these concepts are defined as follows. Let $\langle O, \mathbb{F} \rangle$ be a perceptual system and let $x \in O$. For a set $\mathcal{B} \subseteq \mathbb{F}$ and $\varepsilon \in \mathbb{R}_0^+$, a neighbourhood is defined as

$$N(x) = \{y \in O : x \cong_{\mathcal{B}, \varepsilon} y\}.$$

Note, all objects satisfy the tolerance relation with a single object in a neighbourhood. In contrast, all the pairs of objects within a pre-class must satisfy the tolerance relation. Thus, let $\langle O, \mathbb{F} \rangle$ be a perceptual system. For $\mathcal{B} \subseteq \mathbb{F}$ and $\varepsilon \in \mathbb{R}_0^+$, a set $X \subseteq O$ is a pre-class iff $x \cong_{\mathcal{B}, \varepsilon} y$ for any pair $x, y \in X$. Similarly, a maximal pre-class with respect to inclusion is called a tolerance class.

3   Neighbourhood-Based Algorithms

The serial approach [11,13] and the parallel approach [15] to finding tolerance classes are both based on the propositions (proved in [11]) that all tolerance classes containing $x \in O$ are subsets of the neighbourhood of x, $N(x)$, and that tolerance classes are formed from the query points of successive neighbourhoods, *i.e.* from finding neighbourhoods within neighbourhoods. An illustrative example of the propositions central to these algorithms is given in Fig. 1, Fig. 1(a) gives a tolerance class from within a neighbourhood, Fig. 1(b) shows $N(20)$ obtained using only objects from $N(1)$, and Fig. 1(c) shows successive neighbourhoods using the objects within grey region as query points.

The serial approach attempted to mitigate the computational complexity of finding tolerance classes by using FLANN searches to find neighbourhoods, as well as a simple heuristic to reduce runtime. As a result, the serial approach produced found most (but not all) tolerance classes (see, *e.g.* [11]). Algorithm 1 gives the serial approach to finding tolerance classes, where *compsub* is list of the objects along the search path (*i.e.*, the objects in the grey region of Fig. 1(c)), and *cand* is a list objects that are not in *compsub* but satisfy the tolerance relation with every object in *compsub*. Notice the similarity of this approach to Algorithm 2, a similarity that was discovered independently of the body of literature devoted to the MCE problem [2,4]. Note, the variable names in Algorithm 1 were introduced here to maintain notational consistency with the algorithm reported in [34]. Lastly, Algorithm 1 produces duplicate classes. Consequently, at

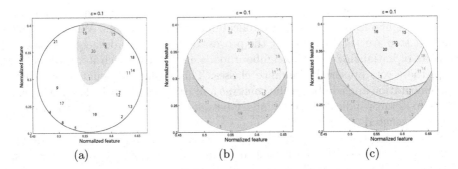

Fig. 1. Algorithm foundational ideas: a) Neighbourhood $N(1)$ in 2D feature space and tolerance class shown in orange, b) $N(20)$ found using only objects from $N(1)$, and c) series of successive neighbourhoods leading to the tolerance class depicted in (a), *i.e.* $N(3) \subset N(16) \subset N(15) \subset N(6) \subset N(10) \subset N(20)$.

Algorithm 1: Serial algorithm for finding tolerance classes

    Input   : Set of objects O
    Output: Set of tolerance classes $H_{\cong_{B,\varepsilon}}(O)$
1 **for** $x \in O$ **do**
2   **for** $y \in N(x)$ **do**
3         $compsub \leftarrow \{x, y\}$;
4         $cand \leftarrow$ All objects in $N(x)$ that satisfy $\cong_{B,\varepsilon}$ with y;
5   GenerateRemaining(y, *compsub*, *cand*);

Procedure GenerateRemaining(y, *compsub*, *cand*)

---

1 **if** *cand* $= \{\}$ **then**
2   **Output** *compsub*
3 **else**
4     $new\_y \leftarrow$ Object in $N(y)$ that is closest to y;
5     $new\_cand \leftarrow$ All objects in $N(y)$ that satisfy $\cong_{B,\varepsilon}$ with new_y;
6     $new\_cs \leftarrow compsub \cup new\_y$;
7   GenerateRemaining(new_y, new_cs, new_cand);

---

some point in the algorithm, the duplicate classes must be removed. In the case of Algorithm 1, this step is performed after the for loop in line 1 has completed.

Next, the parallel approach depends on CUDA GPU programming [21,22]. Briefly, a GPU consists of hundreds of cores (processors) and these cores are organized into groups call streaming multiprocessors that are capable of running code that is called the kernel. The abstraction used to execute this code is called a thread. To make full use of the GPU's stream processors one must generate 1000s of threads for execution. The parallel algorithm reported in [15] consists of the following three stages: Finding object neighbourhoods, finding pseudo neighbourhoods, deleting duplicate classes and subsets. While the first and last step are self explanatory, the process of finding pseudo neighbourhoods consists of initializing sets as neighbourhoods and, during each iteration of the loop, these sets approach tolerance classes. This structure facilated the GPU implementation and combined the *compsub* and *cand* sets described above, where the loop iterator indicates the boundary between the two. Note, only the second stage is executed on the GPU.

While this algorithm finds all the classes, it suffers from prohibitive runtimes. As reported, using $\varepsilon = 0.15$, the runtime for a single pair of images is 10 seconds. To generate the results in this article, tolerance classes need to be generated for 405,450 image pairs, giving a runtime of almost 47 days. As a result, the parallel approach is not considered in this article since the runtimes of the serial and MCE algorithms are 500 and 100 ms, respectively. Finally, note, a GPU approach to MCE is reported in [17], however, Jenkins, *et al.* report an inability to provide good performance against MCE algorithms that are non-GPU based.

4   Maximal Clique Enumeration Algorithm

The Maximal Clique Enumeration (MCE) problem consists of finding all maximal cliques among an undirected graph. Briefly, let $G = (V, E)$ denote an undirected graph, where V is a set of vertices and E is set of edges that connect pairs of distinct vertices from V. A clique is a set of vertices where each pair of vertices in the clique is connected by an edge in E. A maximal clique in G is a clique whose vertices are not all contained in some larger clique, *i.e.* there is no other vertex that is connected to all the vertices in the clique by edges in E.

The first serial algorithm for MCE was developed by Harary and Ross [6,2]. Since then, two main approaches have been established to solve the MCE problem [4], namely the greedy approach reported by Bron-Kerbosh [3] (and concurrent discovery by E. Akkoyunlu [1]), and output-sensitive approaches such as those in [38,19]. The implementation of the MCE algorithm used to generate the results in this paper is a modification of the one reported in [34], which is scalable and parallel version of the the the Bron-Kerbosh approach. The Bron-Kerbosh algorithm is given in Algorithm 2 (again, we are using the same notation as in [34]). The general idea is to use a tree structure to find all maximal cliques, where each call to *CliqueEnumerate* creates a new child node. Each node in the tree consists of four items: the current vertex used to make decisions (*cur_v*), a list of vertices consisting of the (non-maximal) clique formed up to the current node in the tree (*compsub*), a list of potential vertices that are connected to every vertex in *compsub* (*cand*), and a list of vertices that are connected to every vertex in *compsub*, but, if followed, constitute a redundant path in the search tree. Notice, in terms of the neighbourhood-based algorithms, that, at any given level in the tree, *new_cand* is the neighbourhood of *cur_v* using only objects in the list *cand*. Also, similar to the neighbourhood-based approach, both algorithms stop when there are no candidates left to process. Finally, a connection predicate is necessary for the repeated decisions on whether edges exist between vertices [34].

Options include: A linear search of a linked list of adjacent vertices, a lookup using an adjacency bit matrix, and a lookup using a hash table of edges. Since the number of objects generated from each image pair is small (456), the adjacency matrix was used since it is the fastest. The adjacency matrix was constructed by creating a $|V| \times |V|$ matrix, where a 1 (resp. 0) at position i, j represents the existence (non-existence) of an edge between the vertices v_i and v_j.

Algorithm 2: The BK algorithm

Input   : A graph G with vertex V and edge set E
Output: MCE for graph G

1 $compsub \leftarrow \{\}$;
2 $cand \leftarrow V$;
3 $not \leftarrow \{\}$;
4 CliqueEnumerate($compsub, cand, not$);

---

5   Multithreading Approach

As was mentioned, [34] presents a scalable and parallel, multi-threaded approach to solving the MCE problem. The algorithm is parallel in two different aspects. First, their algorithm generates multiple processes that communicate using the

---

Procedure CliqueEnumerate($compsub, cand, not$)

---

1 **if** $cand = \{\}$ **then**
2   **if** $not = \{\}$ **then**
3          $\lfloor$ **Output** $compsub$

4 **else**
5      $fixp \leftarrow$ The vertex in $cand$ that is connected to the greatest number of other vertices in $cand$;
6      $cur\_v \leftarrow fixp$;
7   **while** $cur\_v \neq NULL$ **do**
8          $new\_not \leftarrow$ All vertices in not that are connected to cur_v;
9          $new\_cand \leftarrow$ All vertices in $cand$ that are connected to cur_v;
10         $new\_cs \leftarrow compsub \cup cur\_v$;
11   CliqueEnumerate($new\_cs, new\_cand, new\_not$);
12         $not \leftarrow not \cup cur\_v$;
13         $cand \leftarrow cand \setminus cur\_v$;
14   **if** there is a vertex v in $cand$ that is not connected to $fixp$ **then**
15             $\lfloor$ $cur\_v \leftarrow v$;

16   **else**
17             $\lfloor$ $cur\_v \leftarrow NULL$;

Message Passing Interface (MPI), allowing their algorithm to run on a wide variety of parallel and networked computers. Second, each process generates multiple threads. To simplify the implementation, our results were generated using a single process with multiple threads since the amount of objects obtained from a pair of images in our experiments is 456, compared to the test sets used by Schmidt *et. al* in which the number of objects (vertices) range from 3,472 to 193,568. In fact, both the MCE and neighbourhood-based algorithms used a multi-thread approach to obtain results. The neighbourhood-based approach consisted of creating a stack of object to be processed. Then, each thread pops an object from the stack and finds all the tolerance classes containing that object. Thus, in the multi-threaded algorithm each thread runs an instance of Algorithm 1, except x is obtained from the stack in line 1 (rather than looping through all the objects in O). The MCE algorithm also uses a stack of structure. In this case, it contains the nodes in the tree and each thread process a single node at a time. The modified version of the algorithm in [34] is given in Algorithm 3.

Algorithm 3: The Multi-threaded BK algorithm

> **Input** : A graph G with vertex V and edge set E
> **Output**: MCE for graph G
> 1 **for** $i = 0;$ $i <$ num_threads; $i++$ **do**
> 2 $\quad$ Spawn thread T_i;
> 3 $\quad$ Have T_i run MCliqueEnumerate();
>
> 4 Wait for threads to finish processing;

---

6   Quantifying Nearness

The following two definitions enunciate the fundamental notion of nearness between two sets and provide the foundation for applying near set theory to the problem of CBIR.

Definition 1 Tolerance Nearness Relation [28,29]. *Let $\langle O, \mathbb{F} \rangle$ be a perceptual system and let $X, Y \subseteq O, \varepsilon \in \mathbb{R}_0^+$. A set X is near to a set Y within the perceptual system $\langle O, \mathbb{F} \rangle$ $(X \bowtie_{\mathbb{F}} Y)$ iff there exists $x \in X$ and $y \in Y$ and there is $\mathcal{B} \subseteq \mathbb{F}$ such that $x \cong_{\mathcal{B}, \varepsilon} y$.*

Definition 2 Tolerance Near Sets [28,29]. *Let $\langle O, \mathbb{F} \rangle$ be a perceptual system and let $\varepsilon \in \mathbb{R}_0^+, \mathcal{B} \subseteq \mathbb{F}$. Further, let $X, Y \subseteq O$, denote disjoint sets with coverings determined by the tolerance relation $\cong_{\mathcal{B}, \varepsilon}$, and let $H_{\cong_{\mathcal{B}, \varepsilon}}(X), H_{\cong_{\mathcal{B}, \varepsilon}}(Y)$ denote the set of tolerance classes for X, Y, respectively. Sets X, Y are tolerance near sets iff there are tolerance classes $A \in H_{\cong_{\mathcal{B}, \varepsilon}}(X), B \in H_{\cong_{\mathcal{B}, \varepsilon}}(Y)$ such that $A \bowtie_{\mathbb{F}} B$.*

Procedure MCliqueEnumerate

---

1 **foreach** vertex v_i assigned to the thread **do**
2   $cp \leftarrow$ New candidate path node structure for v_i;
3   **for** $v_j \in V$ **do**
4   **if** connected(v_i, v_j) **then**
5   **if** $i < j$ **then**
6   Vertex v_j is in cp's *cand* list;
7   **else**
8   Vertex v_j is in cp's *not* list;

9   Push cp onto shared stack;

10 **while** shared stack is not empty **do**
11   $cur \leftarrow$ Pop a candidate path node structure from stack;
12   **if** cur's *cand* and *not* lists are empty **then**
13   **Output** cur's *compsub*

14   **else**
15   Generate all cur's children (create child nodes and push onto stack);

Observe that two sets $X, Y \subseteq O$ are tolerance near sets, if they satisfy the tolerance nearness relation.

The tolerance nearness measure was created out of a need to determine the degree that near sets resemble each other, a need which arose during the application of near set theory to the practical applications of image correspondence (see, *e.g.* [7,11]). The tolerance nearness measure between two sets X, Y is based on the idea that tolerance classes formed from objects in the union $Z = X \cup Y$ should be evenly divided among X and Y if these sets are similar, where similarity is always determined with respect to the selected probe functions. The tolerance nearness measure is defined as follows. Let $\langle O, \mathbb{F} \rangle$ be a perceptual system, with $\varepsilon \in \mathbb{R}_0^+$, and $\mathcal{B} \subseteq \mathbb{F}$. Furthermore, let X and Y be two disjoint sets and let $Z = X \cup Y$. Then a tolerance nearness measure between two sets is given by

$$tNM_{\cong_{\mathcal{B},\varepsilon}}(X,Y) =$$

$$1 - \left(\sum_{C \in H_{\cong_{\mathcal{B},\varepsilon}}(Z)} |C| \right)^{-1} \cdot \sum_{C \in H_{\cong_{\mathcal{B},\varepsilon}}(Z)} |C| \frac{\min(|C \cap X|, |[C \cap Y|)}{\max(|C \cap X|, |C \cap Y|)}. \quad (1)$$

Finally, new measures inspired by the tNM have been reported in [35,20]. A systematic comparison of the tNM and these measures is outside the scope of this paper and is left for future work.

7   Results and Discussion

The algorithms presented here are compared using CBIR, where the goal is to retrieve images from databases based on the content of an image rather than on some semantic string or keywords associated with the image. The content of the image is determined by functions that characterize features such as colour, texture, shape of objects, and edges. In our approach to CBIR, a search entails analysis of content, based on the tNM nearness measure (see, *e.g.* [11]) between a query image and test image. Moreover, the nearness measure on tolerance classes of objects derived from two perspective images provides a quantitative approach for accessing the similarity of images. To generate our results, the SIMPLIcity image database [18], a database of images containing 10 categories with 100 images in each category was used (shown in Fig. 2).

The results were generated by partitioning the images into subimages, where each subimage was considered as an object in the near set sense, *i.e.* each subimage is a perceptual object, and each object description consists of the values obtained from image processing techniques on the subimage. This technique of partitioning an image, and assigning feature vectors to each subimage is an approach that has also been traditionally used in CBIR. Formally, an RGB

Fig. 2. Examples of each category of images. (a) - (d) Categories 0 - 3, and (e) - (i) categories 5 - 9.

image is defined as $f = \{\mathbf{p}_1, \mathbf{p}_2, \ldots, \mathbf{p}_T\}$, where $\mathbf{p}_i = (c, r, R, G, B)^{\mathrm{T}}$, $c \in [1, M]$, $r \in [1, N]$, $R, G, B \in [0, 255]$, and M, N respectively denote the width and height of the image and $M \times N = T$. Further, define a square subimage as $f_i \subset f$ such that $f_i \cap f_j = \{\}$ for $i \neq j$ and $f_1 \cup f_2 \ldots \cup f_s = f$, where s is the number of subimages in f. Next, O can be defined as the set of all subimages, $i.e.$, $O = \{f_1, \ldots, f_s\}$, and $\mathbb{F}$ is a set of image processing descriptors or functions that operate on images. Then, the nearness of two images can be discovered by partitioning each of the images into subimages and letting these represent objects in a perceptual system, $i.e$, let the sets X and Y represent the two images to be compared where each set consists of the subimages obtained by partitioning the images. Then, the set of all objects in this perceptual system is given by $Z = X \cup Y$.

Fig. 3. Example demonstrating the application of near set theory to images, namely the image is partitioned into subimages where each subimage is considered a perceptual object, and object descriptions are the results of image processing techniques on the subimage

The results in this article were obtained using a subimage size of 20×20 (resulting in 456 objects per image pair) and the 18 features used in [11], namely 4 texture features obtained from the grey-level co-occurrence matrix of a subimage, the first and second moments of u and v in the CIELUV colour space, an edge based feature, and the Zernike moments of order 4, excluding $\bar{A}_{00}$. Moreover, the results are presented using precision vs. recall plots, where the idea is to find tNM values between each pair of images in the database. Then, the measure values are sorted in ascending order, and the smallest value represents the results of the first query, the second value the results of the second query, $etc.$ Precision/recall plots are the common metric for evaluating CBIR systems where precision and recall are defined as

$$\text{precision} = \frac{|\{\text{relevant images}\} \cap \{\text{retrieved images}\}|}{|\{\text{retrieved images}\}}},$$

and

$$recall = \frac{|\{\text{relevant images}\} \cap \{\text{retrieved images}\}|}{|\{\text{relevant images}\}}.$$

In the ideal case, all images from the same category would be retrieved before any images from other categories. In this case, precision would be 100% until recall reached 100%, at which point precision would drop to # of images in query category / # of images in the database. As a result, our final value of precision will be ~11% since we used 9 categories each containing 100 images. Note, only 9 categories were used since the category of images shown in Fig. 4 are easy to retrieve and their inclusion in the test would only increase the runtime of the experiment.

(a)   (b)   (c)

Fig. 4. Examples of images from category 4

The results are presented in Fig. 5 - 17, where the average precision vs. recall plots are given in Fig. 5 & 7, the precision vs. recall results of the best query image are given in Fig. 6 & 8, and Fig. 9 - 17 are the top 40 retrieved images from the best search in each category[2]. These plots present a comparison of the two approaches described in Sections 3 & 4 respectively: neighbourhood-based (most tolerance classes) vs. MCE (all tolerance classes.) In the case of the neighbourhood-based algorithm, only results for $\varepsilon = 0.2$ are given since it was reported in [11] that this value produces the best results that are achievable with reasonable runtime. In other words, the optimal value of ε for the neighbourhood based algorithm on this test and feature set may be greater than $\varepsilon = 0.2$, but, due to prohibitive runtimes, these experiments were not performed. Recall, in any given application (regardless of the distance metric), there is always an optimal ε when performing experiments using the perceptual tolerance relation [11]. For instance, a value of $\varepsilon = 0$ produces little or no pairs of objects that satisfy the perceptual tolerance relation, and a value of $\varepsilon = \sqrt{l}$, means that all pairs of objects satisfy the tolerance relation[3]. Consequently, ε should be selected such that the objects that are relatively[4] close in feature space satisfy the tolerance

---

[2] The query image is in the top left position, where the images are ranked from the top down, then left to right.

[3] For normalized feature values, the largest distance between two objects occurs in the interval $[0, \sqrt{l}]$, where l is the length of the feature vectors.

[4] Here, distance of "objects that are relatively close" will be determined by the application.

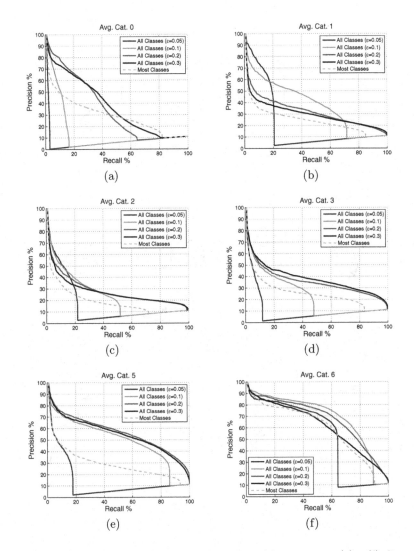

Fig. 5. Average precision versus recall plots grouped by category. (a) - (f) Categories 0 - 6 (excluding category 4).

relation, and the rest of the pairs of objects do not. The selection of ε is straightforward when a metric is available for measuring the success of the experiment. Thus, if runtime were not an issue, the value of ε should be selected based on the best result of the evaluation metric, which, in the context of CBIR, is the best results in terms of precision vs. recall.

Next, the following presents some observations of the reported results. First, notice that some of the curves have a sharp point of inflection (see, *e.g.*, $\varepsilon = 0.05$ at 20% recall in Fig. 5(b)). These points represents the location at which the remaining tNM values for a particular query become zero. In the case of Fig. 5 & 7,

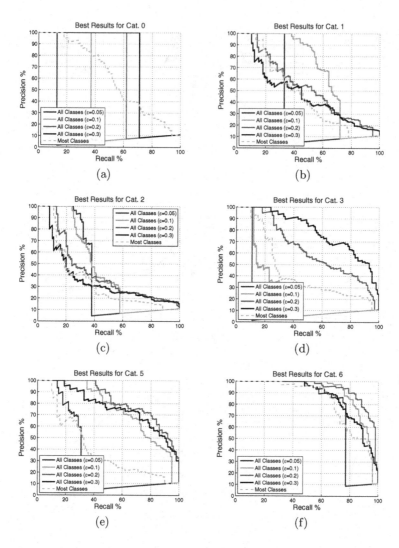

Fig. 6. Best precision versus recall plots grouped by category. (a) - (f) Categories 0 - 6 (excluding category 4).

these points represent the location at which all query images in the category produce a tNM value of zero. In order to provide this clear demarcation, any images from the same category as the query image that produced a tNM value of zero were ranked last in the search[5]. Next, results are not reported for $\varepsilon = 0.3$ for images from category 7 (see, e.g. Fig. 2(g)), since the runtime was too large for some of the images in this category. For instance, some image pairs produced

---

[5] This was not the case in [11], which accounts for some small discrepencies in the plots of this article near the end of the curve.

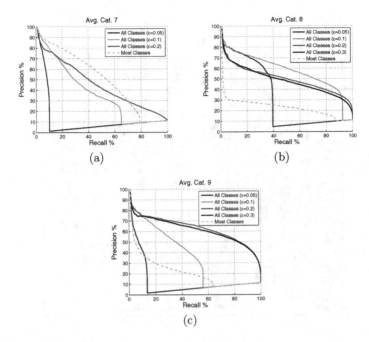

Fig. 7. Average precision versus recall plots grouped by category. (a) - (c) Categories 7 - 9 (excluding category 4).

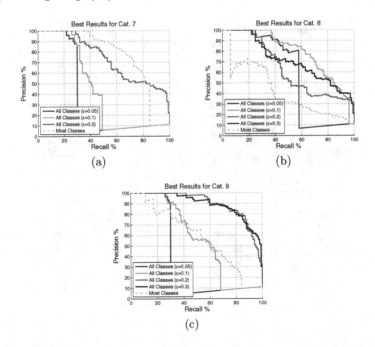

Fig. 8. Best precision versus recall plots grouped by category. (a) - (c) Categories 7 - 9 (excluding category 4).

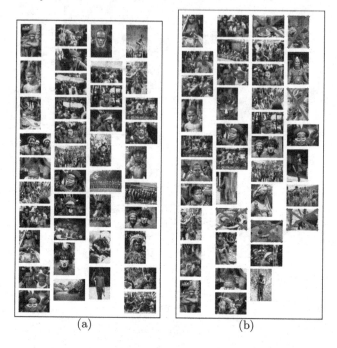

Fig. 9. Top 40 retrieved images using $\varepsilon = 0.2$ for category 0. (a) Results obtained using all classes, and (b) results from using most classes.

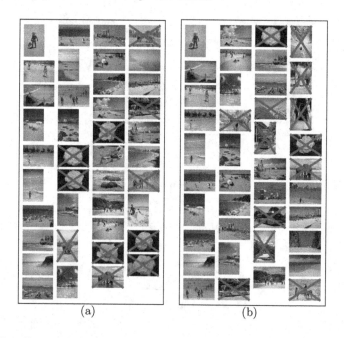

Fig. 10. Top 40 retrieved images using $\varepsilon = 0.2$ for category 1. (a) Results obtained using all classes, and (b) results from using most classes.

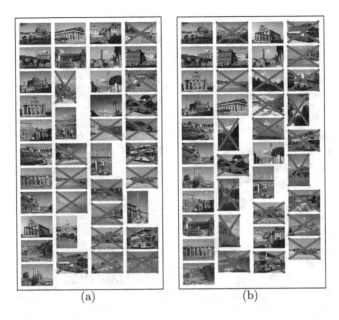

Fig. 11. Top 40 retrieved images using $\varepsilon = 0.2$ for category 2. (a) Results obtained using all classes, and (b) results from using most classes.

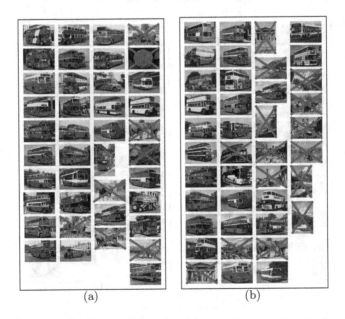

Fig. 12. Top 40 retrieved images using $\varepsilon = 0.2$ for category 3. (a) Results obtained using all classes, and (b) results from using most classes.

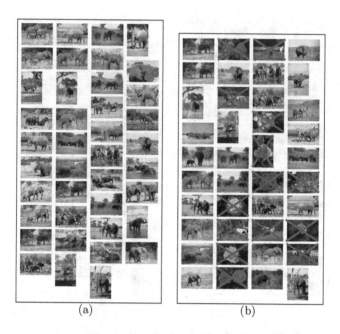

(a)   (b)

Fig. 13. Top 40 retrieved images using $\varepsilon = 0.2$ for category 5. (a) Results obtained using all classes, and (b) results from using most classes.

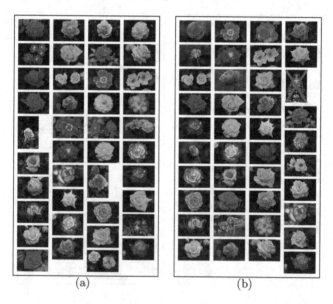

(a)   (b)

Fig. 14. Top 40 retrieved images using $\varepsilon = 0.2$ for category 6. (a) Results obtained using all classes, and (b) results from using most classes.

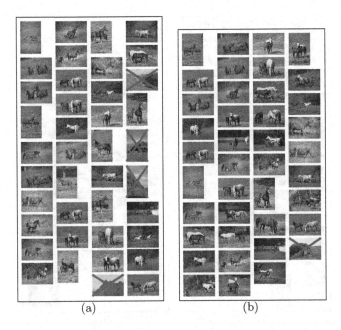

Fig. 15. Top 40 retrieved images using $\varepsilon = 0.2$ for category 7. (a) Results obtained using all classes, and (b) results from using most classes.

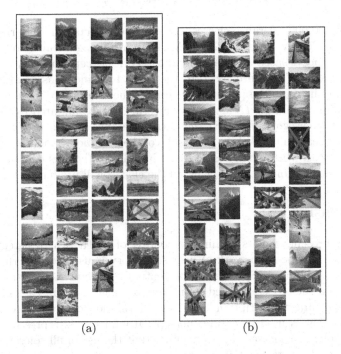

Fig. 16. Top 40 retrieved images using $\varepsilon = 0.2$ for category 8. (a) Results obtained using all classes, and (b) results from using most classes.

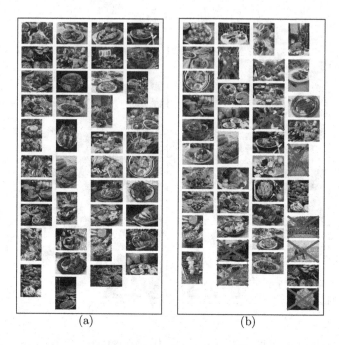

Fig. 17. Top 40 retrieved images using $\varepsilon = 0.2$ for category 9. (a) Results obtained using all classes, and (b) results from using most classes.

in excess of 700,000 tolerance classes (on only 456 objects) and had runtimes of over 2 hours. Finally, for $\varepsilon = 0.2$, the MCE approach significantly outperforms the neighbourhood-based approach in all categories except category 7, which is due to the value of ε. We conjecture the extra tolerance classes produced at $\varepsilon \geq 0.3$ would increase the performance due to the addition of more perceptual information in calculating tNM. This conjecture is substantiated by the results of every other category in which the extra tolerance classes produced better precision vs. recall values. While, the results of the neighbourhood-based approah may also increase with ε, the result from the other categories demonstrate the additional information obtained using all the classes will produce better results.

8   Conclusion

This article presents results within the context of CBIR, where perceptual information within the framework of near set theory is used to discern affinities between pairs of images. Specifically, perceptually relevant information was extracted from a set objects formed from pairs of images, where each object has an associated object description. It is the information contained in these feature vectors that is used to extract perceptual information represented by the discovered tolerance classes. The conjecture that the use of all tolerance classes in a covering of image pairs *increases* the perceptual information available to make decisions on nearness leading to an *improvement* of precision and recall was substantiated by the results presented here. This article also demonstrates that discovery of all tolerance classes is equivalent to the MCE problem. Finally, this article presents a new application of MCE.

References

1. Akkoyunlu, E.A.: The enumeration of maximal cliques of large graphs. SIAM Journal on Computing 2(1), 1–6 (1973)
2. Bomze, I., Budinich, M., Pardalos, P., Pelillo, M.: The maximum clique problem. In: Du, D.Z., Pardalos, P.M. (eds.) Handbook of Combinatorial Optimization, vol. 4. Kluwer (1999)
3. Bron, C., Kerbosch, J.: Algorithm 457: finding all cliques of an undirected graph. Communications of the ACM 16(9), 575–577 (1973)
4. Cazals, F., Karande, C.: A note on the problem of reporting maximal cliques. Theoretical Computer Science 407(1), 564–568 (2008)
5. Duda, R., Hart, P., Stork, D.: Pattern Classification, 2nd edn. Wiley (2001)
6. Harary, F., Ross, I.C.: A procedure for clique detection using the group matrix. Sociometry 20(3), 205–215 (1957)
7. Hassanien, A.E., Abraham, A., Peters, J.F., Schaefer, G., Henry, C.: Rough sets and near sets in medical imaging: A review. IEEE Transactions on Information Technology in Biomedicine 13(6), 955–968 (2009)
8. Henry, C.: Near set Evaluation And Recognition (NEAR) system. In: Pal, S.K., Peters, J.F. (eds.) Rough Fuzzy Analysis Foundations and Applications, pp. 7-1–7-22. CRC Press, Taylor & Francis Group (2010), http://wren.ee.umanitoba.ca
9. Henry, C., Peters, J.F.: Perception-based image classification. International Journal of Intelligent Computing and Cybernetics 3(3), 410–430 (2010), Emerald Literati Network 2011 Award for Excellence
10. Henry, C., Peters, J.F.: Arthritic hand-finger movement similarity measurements: Tolerance near set approach. Computational and Mathematical Methods in Medicine, article ID 569898, 14 pp (2011)
11. Henry, C.J.: Near Sets: Theory and Applications. Ph.D. thesis (2010), https://mspace.lib.umanitoba.ca/handle/1993/4267
12. Henry, C.J.: Neighbourhoods, classes and near sets. Applied Mathematical Sciences 5(35), 1727–1732 (2011)
13. Henry, C.J.: Perceptual Indiscernibility, Rough Sets, Descriptively Near Sets, and Image Analysis. In: Peters, J.F., Skowron, A. (eds.) Transactions on Rough Sets XV. LNCS, vol. 7255, pp. 41–121. Springer, Heidelberg (2012)
14. Henry, C.J., Peters, J.F.: Neighbourhood-based vision systems. Cybernetics and Systems 42(1), 33–44 (2011)
15. Henry, C.J., Ramanna, S.: Parallel Computation in Finding Near Neighbourhoods. In: Yao, J., Ramanna, S., Wang, G., Suraj, Z. (eds.) RSKT 2011. LNCS, vol. 6954, pp. 523–532. Springer, Heidelberg (2011)
16. Henry, C.J., Ramanna, S.: Signature-based perceptual nearness. Application of near sets to image retrieval. Mathematics in Computer Science p. 21 (submitted, 2012)
17. Jenkins, J., Arkatkar, I., Owens, J.D., Choudhary, A., Samatova, N.F.: Lessons learned from exploring the backtracking paradigm on the GPU. In: Proceedings of the 17th International Conference on Parallel Processing, vol. II, pp. 425–437 (2011)
18. Li, J., Wang, J.Z.: Automatic linguistic indexing of pictures by a statistical modeling approach. IEEE Transactions on Pattern Analysis and Machine Intelligence 25(9), 1075–1088 (2003)

19. Makino, K., Uno, T.: New Algorithms for Enumerating All Maximal Cliques. In: Hagerup, T., Katajainen, J. (eds.) SWAT 2004. LNCS, vol. 3111, pp. 260–272. Springer, Heidelberg (2004)
20. Meghdadi, A.H.: Fuzzy Tolerance Neighborhood Approach to Image Similarity in Content-based Image Retrieval. Ph.D. thesis (2012)
21. NVIDIA: NVIDIA CUDA programming guide v3.0 (2010), http://docs.nvidia.com/cuda/index.html
22. Patel, S.J.: Applied parallel programming (2010), http://courses.engr.illinois.edu/ece498/al/
23. Pavel, M.: Fundamentals of Pattern Recognition. Marcel Dekker, Inc., NY (1993)
24. Pawlak, Z., Skowron, A.: Rudiments of rough sets. Information Sciences 177, 3–27 (2007)
25. Peters, J.F.: Classification of objects by means of features. In: Proceedings of the IEEE Symposium Series on Foundations of Computational Intelligence (IEEE SCCI 2007), pp. 1–8 (2007)
26. Peters, J.F.: Near sets. General theory about nearness of objects. Applied Mathematical Sciences 1(53), 2609–2629 (2007)
27. Peters, J.F.: Near sets. Special theory about nearness of objects. Fundamenta Informaticae 75(1-4), 407–433 (2007)
28. Peters, J.F.: Tolerance near sets and image correspondence. International Journal of Bio-Inspired Computation 1(4), 239–245 (2009)
29. Peters, J.F.: Corrigenda and addenda: Tolerance near sets and image correspondence. International Journal of Bio-Inspired Computation 2(5), 310–318 (2010)
30. Peters, J.F., Wasilewski, P.: Foundations of near sets. Info. Sci. 179(18), 3091–3109 (2009)
31. Peters, J.F., Wasilewski, P.: Tolerance spaces: Origins, theoretical aspects and applications. Information Sciences 195, 211–225 (2012)
32. Poincaré, H.: L'espace et la géomètrie. Revue de métaphysique et de morale 3, 631–646 (1895)
33. Ramanna, S., Meghdadi, A.H., Peters, J.F.: Nature-inspired framework for measuring image resemblance: A near rough set approach. Theoretical Computer Science 412(42), 5926–5938 (2011), doi:10.1016/j.tcs.2011.05.044
34. Schmidt, M.C., Samatova, N.F., Thomas, K., Byung-Hoon, P.: A scalable, parallel algorithm for maximal clique enumeration. Journal of Parallel and Distributed Computing 69, 417–428 (2009)
35. Shahfar, S.: Near Images: A Tolerance Based Approach to Image Similarity and Its Robustness to Noise and Lightening. M.Sc. thesis (2011)
36. Smeulders, A.W.M., Worring, M., Santini, S., Gupta, A., Jain, R.: Content-based image retrieval at the end of the early years. IEEE Transactions on Pattern Analysis and Machine Intelligence 22(12), 1349–1380 (2000)
37. Sossinsky, A.B.: Tolerance space theory and some applications. Acta Applicandae Mathematicae: An International Survey Journal on Applying Mathematics and Mathematical Applications 5(2), 137–167 (1986)
38. Tsukiyama, S., Ide, M., Ariyoshi, H., Shirakawa, I.: A new algorithm for generating all the maximal independent sets. SIAM Journal on Computing 6, 505–517 (1977)
39. Wolski, M.: Perception and classification. A Note on near sets and rough sets. Fundamenta Informaticae 101, 143–155 (2010)
40. Zeeman, E.C.: The topology of the brain and the visual perception. In: Fort, K.M. (ed.) Topoloy of 3-manifolds and Selected Topices, pp. 240–256. Prentice Hall, New Jersey (1965)

On Fuzzy Topological Structures
of Rough Fuzzy Sets

.

Wei-Zhi Wu and You-Hong Xu

School of Mathematics, Physics and Information Science
Zhejiang Ocean University, Zhoushan, Zhejiang, 316004, P.R. China
{wuwz,xyh}@zjou.edu.cn

Abstract. A rough fuzzy set is the result of approximation of a fuzzy set with respect to a crisp approximation space. In this paper, we investigate topological structures of rough fuzzy sets. We first show that a reflexive crisp rough approximation space can induce a fuzzy Alexandrov space. We then prove that the lower and upper rough fuzzy approximation operators are, respectively, the fuzzy interior operator and fuzzy closure operator if and only if the binary relation in the crisp approximation space is reflexive and transitive. We also examine that a similarity crisp approximation space can produce a fuzzy clopen topological space. Finally, we present the sufficient and necessary conditions that a fuzzy interior (closure, respectively) operator derived from a fuzzy topological space can associate with a reflexive and transitive crisp relation such that the induced lower (upper, respectively) rough fuzzy approximation operator is exactly the fuzzy interior (closure, respectively) operator.

Keywords: approximation operators, binary relations, fuzzy topologies, rough fuzzy sets, rough sets.

1 Introduction

The basic structure of the rough set theory [23] is an approximation space consisting of a universe of discourse and a binary relation imposed on it. Based on the approximation space, the notions of lower and upper approximation operators can be induced. Using the concepts of lower and upper approximations in rough set theory, knowledge hidden in information systems may be unraveled and expressed in the form of decision rules. Rough set theory can be viewed as a set-based granular computing method that advances research in this area.

The Pawlak's rough approximations, originally introduced with reference to an indiscernibility (equivalence) relation, are useful in the analysis of data presented in terms of complete information. The equivalence relation in Pawlak's rough set model provides the basis of "information granules" for database analysis. However, the requirement of an equivalence relation in Pawlak's rough set model seems to be a very restrictive condition that may limit the applications of the rough set model. Thus, Pawlak's rough set approximations may be generalized by using non-equivalence relations. The extensive models have been used

J.F. Peters et al. (Eds.): Transactions on Rough Sets XVI, LNCS 7736, pp. 125–143, 2013.

in reasoning and knowledge acquisition in incomplete information. On the other hand, data with fuzzy set values are commonly seen in real-world applications. Fuzzy-set concepts are often used to represent quantitative data expressed in linguistic terms and membership functions in intelligent systems. Based on this observation, many authors have generalized rough set model to the fuzzy environment. The results of these studies lead to the introduction of notions of rough fuzzy sets (fuzzy sets approximated by a crisp approximation space) [9,34,35] and fuzzy rough sets (fuzzy or crisp sets approximated by a fuzzy approximation space) [9,20,21,22,27,33,34,35]. The rough fuzzy set model may be used to handle knowledge acquisition in information systems with fuzzy decisions while the fuzzy rough set model may be employed to unravel knowledge hidden in fuzzy decision systems.

An interesting and important research for rough approximation operators is to compare them with the topological properties and structures. Topology is a branch of mathematics, whose concepts exist not only in almost all branches of mathematics, but also in many real life applications. Topological structure is an important base for knowledge extraction and processing (see e.g. [1,2,6,13,14,17,30]). For example, Koretelainen [13,14] used topologies to detect dependencies of attributes in information systems with respect to gradual rules. Choudhury and Zaman [6] applied the mathematical theory of topology to study the evolutionary impact of learning on social problems. Wu *et al.* [30] investigated the topological space on rough sets and the corresponding topological properties, and provided some applications on image processing and some topological diagram as well as the application on knowledge and attribute reduction. Li and Zhang [17] proposed reduction of subbases of topological spaces in rough set data analysis. The concept of topological structures and their generalizations are the most powerful notions and are important bases in data and system analysis.

To improve the applications of topology and rough set theory on uncertain and incomplete information and to study further the cosmical structure, origin, and its evolvement, the topological spaces and the topological properties of rough sets need to be studied. In fact, many authors studied rough set approximations by comparing them with the topological properties and structures. For example, Chuchro [7,8], Kondo [11], Lashin *et al.* [16], Pei *et al.* [24], Qin *et al.* [26], Wiweger [29], Wu *et al.* [30], Yang and Xu [37], and Zhu [41] studied the topological structures for crisp rough sets. Boixader *et al.* [4], Hao and Li [10], Kortelainen [12], Qin and Pei [25], Thiele [28], Wu [31,32], Wu and Zhou [36], Zhou and Wu [40], respectively, discussed topological structures of rough sets in the fuzzy environment. One of the main results is that a reflexive and transitive approximation space can yield a topology on the same universe, and conversely, under some conditions, a topology can be associated with a reflexive and transitive approximation space which produces the same topology. There exists a one-to-one correspondence between the set of all reflexive, transitive relations and the set of Alexandrov topologies on an arbitrary universe [10,15,36,39].

In the present paper, we further investigate topological structures of rough fuzzy sets. In the next section, we review basic concepts related to rough fuzzy sets and present some properties of rough fuzzy approximation operators. Section 3 provides the axiomatic characterization of rough fuzzy approximation operators. In Section 4, we introduce some notions and theoretical results of fuzzy topological spaces. In Section 5, we examine fuzzy topological structures of rough fuzzy sets. In Section 6, we investigate under which conditions that a fuzzy topology can be associated with a crisp approximation space which produces the same fuzzy topology. We then conclude the paper with a summary in Section 7.

2   Rough Fuzzy Approximation Operators

Throughout this paper, U will be a nonempty set called the universe of discourse which may be infinite. By a fuzzy set in U we mean a mapping $F : U \to [0,1]$. The class of all subsets (fuzzy subsets, respectively) of U will be denoted by $\mathcal{P}(U)$ (by $\mathcal{F}(U)$, respectively). Zadeh's fuzzy union and fuzzy intersection will be denoted by $\cup$ and $\cap$, respectively. For $A \in \mathcal{F}(U)$, $\sim A$ will be used to denote the fuzzy complement of the fuzzy set A in U, i.e. for every $x \in U$, $(\sim A)(x) = 1 - A(x)$. For $y \in U$, 1_y will denote the fuzzy singleton with value 1 at y and 0 elsewhere; 1_M will denote the characteristic function of a crisp set $M \subseteq U$, and, for $\alpha \in [0,1]$, $\widehat{\alpha}$ will denote the constant fuzzy set: $\widehat{\alpha}(x) = \alpha$ for all $x \in U$, obviously, $\widehat{1} = 1_U$ and $\widehat{0} = 1_\emptyset$. We will use the symbols $\vee$ and $\wedge$ to denote the supremum and the infimum, respectively.

Definition 1. *Let U and W be two nonempty universes of discourse which may be infinite. A subset $R \in \mathcal{P}(U \times W)$ is referred to as a (crisp) binary relation from U to W. The relation R is referred to as serial if for each $x \in U$ there exists $y \in W$ such that $(x,y) \in R$; If $U = W$, R is referred to as a binary relation on U. R is referred to as reflexive if for all $x \in U$, $(x,x) \in R$; R is referred to as symmetric if for all $x,y \in U$, $(x,y) \in R$ implies $(y,x) \in R$; R is referred to as transitive if for all $x,y,z \in U$, $(x,y) \in R$ and $(y,z) \in R$ imply $(x,z) \in R$; R is referred to as a similarity relation if R is reflexive and symmetric; R is referred to as a preorder if R is reflexive and transitive; and R is referred to as an equivalence relation if R is reflexive, symmetric and transitive.*

For an arbitrary crisp relation R from U to W, we can define a set-valued mapping $R_s : U \to \mathcal{P}(W)$ by:

$$R_s(x) = \{y \in W : (x,y) \in R\}, \quad x \in U. \tag{1}$$

$R_s(x)$ is referred to as the successor neighborhood of x with respect to (w.r.t.) R.

A rough fuzzy set is the approximation of a fuzzy set w.r.t. a crisp approximation space [9,35].

Definition 2. *Let U and W be two non-empty universes of discourse and R a crisp binary relation from U to W, then the triple (U, W, R) is called an approximation space. For any fuzzy set $A \in \mathcal{F}(W)$, the lower and upper approximations* of A, $\underline{R}(A)$ and $\overline{R}(A)$, w.r.t. the approximation space (U, W, R) are fuzzy sets of U whose membership functions, for each $x \in U$, are, respectively, defined by

$$\underline{R}(A)(x) = \bigwedge_{y \in R_s(x)} A(y), \quad \overline{R}(A)(x) = \bigvee_{y \in R_s(x)} A(y). \tag{2}$$

The pair $(\underline{R}(A), \overline{R}(A))$ is called a generalized rough fuzzy set, and $\underline{R}$ and $\overline{R}$: $\mathcal{F}(W) \to \mathcal{F}(U)$ are referred to as the lower and upper rough fuzzy approximation operators, respectively.

By Definition 2, we can obtain properties of rough fuzzy approximation operators [33,34,35].

Theorem 1. *The lower and upper rough fuzzy approximation operators, $\underline{R}$ and $\overline{R}$, defined by Eq. (2) satisfy the following properties: For all $A, B \in \mathcal{F}(W)$, $A_j \in \mathcal{F}(W)(\forall j \in J)$, J is an index set, and $\alpha \in [0, 1]$,*

(FL1) $\underline{R}(A) =\sim \overline{R}(\sim A)$,   (FU1) $\overline{R}(A) =\sim \underline{R}(\sim A)$,

(FL2) $\underline{R}(A \cup \widehat{\alpha}) = \underline{R}(A) \cup \widehat{\alpha}$,   (FU2) $\overline{R}(A \cap \widehat{\alpha}) = \overline{R}(A) \cap \widehat{\alpha}$;

(FL3) $\underline{R}(\bigcap_{j \in J} A_j) = \bigcap_{j \in J} \underline{R}(A_j)$,   (FU3) $\overline{R}(\bigcup_{j \in J} A_j) = \bigcup_{j \in J} \overline{R}(A_j)$,

(FL4) $A \subseteq B \Longrightarrow \underline{R}(A) \subseteq \underline{R}(B)$, (FU4) $A \subseteq B \Longrightarrow \overline{R}(A) \subseteq \overline{R}(B)$,

(FL5) $\underline{R}(\bigcup_{j \in J} A_j) \supseteq \bigcup_{j \in J} \underline{R}(A_j)$,   (FU5) $\overline{R}(\bigcap_{j \in J} A_j) \subseteq \bigcap_{j \in J} \overline{R}(A_j)$.

Properties (FL1) and (FU1) show that the rough fuzzy approximation operators $\underline{R}$ and $\overline{R}$ are dual to each other. Properties with the same number may be regarded as dual properties. Properties (FL3) and (FU3) state that the lower rough fuzzy approximation operator $\underline{R}$ is multiplicative, and the upper rough fuzzy approximation operator $\overline{R}$ is additive. One may also say that $\underline{R}$ is distributive w.r.t. the intersection of fuzzy sets, and $\overline{R}$ is distributive w.r.t. the union of fuzzy sets. Properties (FL5) and (FU5) imply that $\underline{R}$ is not distributive w.r.t. set union, and $\overline{R}$ is not distributive w.r.t. set intersection. However, properties (FL2) and (FU2) show that $\underline{R}$ is distributive w.r.t. the union of a fuzzy set and a fuzzy constant set, and $\overline{R}$ is distributive w.r.t. the intersection of a fuzzy set and a constant fuzzy set. Evidently, properties (FL2) and (FU2) imply the following properties:

$$(FL2)' \quad \underline{R}(1_W) = 1_U, \quad (FU2)' \quad \overline{R}(1_\emptyset) = 1_\emptyset.$$

Analogous to Yao's study in [38], a serial rough fuzzy set model is obtained from a serial binary relation. The property of a serial relation can be characterized by the properties of its induced rough fuzzy approximation operators [33,34,35].

Theorem 2. *If R is an arbitrary crisp relation from U to W, and $\underline{R}$ and $\overline{R}$ are the rough fuzzy approximation operators defined by Eq. (2), then*

$$
\begin{aligned}
R \text{ is serial} &\Longleftrightarrow \text{(FL0)} \quad \underline{R}(1_\emptyset) = 1_\emptyset, \\
&\Longleftrightarrow \text{(FU0)} \quad \overline{R}(1_W) = 1_U, \\
&\Longleftrightarrow \text{(FL0)}' \quad \underline{R}(\widehat{\alpha}) = \widehat{\alpha}, \qquad \forall \alpha \in [0,1], \\
&\Longleftrightarrow \text{(FU0)}' \quad \overline{R}(\widehat{\alpha}) = \widehat{\alpha}, \qquad \forall \alpha \in [0,1], \\
&\Longleftrightarrow \text{(FLU0)} \; \underline{R}(A) \subseteq \overline{R}(A), \forall A \in \mathcal{F}(W).
\end{aligned}
$$

In the case of connections between other special crisp relations and rough fuzzy approximation operators, we have the following [34,35]

Theorem 3. *Let R be an arbitrary crisp relation on U, and $\underline{R}$ and $\overline{R}$ the lower and upper rough fuzzy approximation operators defined by Eq. (2). Then*

$$
\begin{aligned}
R \text{ is reflexive} &\Longleftrightarrow \text{(FL6)} \quad \underline{R}(A) \subseteq A, \quad \forall A \in \mathcal{F}(U), \\
&\Longleftrightarrow \text{(FU6)} \quad A \subseteq \overline{R}(A), \quad \forall A \in \mathcal{F}(U). \\
R \text{ is symmetric} &\Longleftrightarrow \text{(FL7)} \quad \overline{R}(\underline{R}(A)) \subseteq A, \quad \forall A \in \mathcal{F}(U), \\
&\Longleftrightarrow \text{(FU7)} \quad A \subseteq \underline{R}(\overline{R}(A)), \quad \forall A \in \mathcal{F}(U), \\
&\Longleftrightarrow \text{(FL7)}' \; \underline{R}(1_{U-\{x\}})(y) = \underline{R}(1_{U-\{y\}})(x), \quad \forall (x,y) \in U \times U, \\
&\Longleftrightarrow \text{(FU7)}' \; \overline{R}(1_x)(y) = \overline{R}(1_y)(x), \quad \forall (x,y) \in U \times U. \\
R \text{ is transitive} &\Longleftrightarrow \text{(FL8)} \quad \underline{R}(A) \subseteq \underline{R}(\underline{R}(A)), \quad \forall A \in \mathcal{F}(U), \\
&\Longleftrightarrow \text{(FU8)} \quad \overline{R}(\overline{R}(A)) \subseteq \overline{R}(A), \quad \forall A \in \mathcal{F}(U).
\end{aligned}
$$

3 Axiomatic Characterization of Rough Fuzzy Approximation Operators

In the axiomatic approach, rough sets are characterized by abstract operators. For the case of rough fuzzy sets, the primitive notion is a system $(\mathcal{F}(U), \mathcal{F}(W), \cap, \cup, \sim, L, H)$, where L and H are unary operators from $\mathcal{F}(W)$ to $\mathcal{F}(U)$. In this section, we review the axiomatic characterization of rough fuzzy approximation operators [34,35].

Definition 3. *Let $L, H : \mathcal{F}(W) \to \mathcal{F}(U)$ be two operators. They are referred to as dual operators if for all $A \in \mathcal{F}(W)$,*

$$
\begin{aligned}
&\text{(fl1)} \quad L(A) =\sim H(\sim A), \\
&\text{(fu1)} \quad H(A) =\sim L(\sim A).
\end{aligned}
$$

By the dual properties of the operators, we only need to define one operator. For example, one may define the operator H and regard L as an abbreviation of $\sim H \sim$. From an operator $H : \mathcal{F}(W) \to \mathcal{F}(U)$, we define a binary relation R_H from U to W as follows:

$$
R_H(x,y) = H(1_y)(x), \quad (x,y) \in U \times W. \tag{3}
$$

By employing Eq. (3), we can conclude following theorem via the discussion on the constructive approach in [32,34,35]:

Theorem 4. *Suppose that* $L, H : \mathcal{F}(W) \to \mathcal{F}(U)$ *are dual operators. Then there exists a crisp binary relation* R_H *from* U *to* W *such that for all* $A \in \mathcal{F}(W)$

$$L(A) = \underline{R_H}(A), \qquad and \qquad H(A) = \overline{R_H}(A)$$

iff L *satisfies axioms* (flc), (fl2), (fl3), *or equivalently* H *satisfies axioms* (fuc), (fu2), (fu3):

$\qquad$ (flc) $L(1_{W-\{y\}}) \in \mathcal{P}(U), \qquad \forall y \in W,$

$\qquad$ (fl2) $L(A \cup \widehat{\alpha}) = L(A) \cup \widehat{\alpha}, \quad \forall A \in \mathcal{F}(W), \ \forall \alpha \in [0,1];$

$\qquad$ (fl3) $L(\bigcap_{j \in J} A_j) = \bigcap_{j \in J} L(A_j), \ \forall A_j \in \mathcal{F}(W), j \in J;$

$\qquad$ (fuc) $H(1_y) \in \mathcal{P}(U), \qquad \forall y \in W;$

$\qquad$ (fu2) $H(A \cap \widehat{\alpha}) = H(A) \cap \widehat{\alpha}, \ \forall A \in \mathcal{F}(W), \ \forall \alpha \in [0,1];$

$\qquad$ (fu3) $H(\bigcup_{j \in J} A_j) = \bigcup_{j \in J} H(A_j), \forall A_j \in \mathcal{F}(W), j \in J.$

According to Theorem 4, axioms (flc),(fl1),(fl2), (fl3), or equivalently, axioms (fuc), (fu1), (fu2), (fu3) are considered to be basic axioms of rough fuzzy approximation operators. These lead to the following definitions of rough fuzzy set algebras:

Definition 4. *Let* $L, H : \mathcal{F}(W) \to \mathcal{F}(U)$ *be a pair of dual operators. If* L *satisfies axioms* (flc), (fl2), *and* (fl3), *or equivalently* H *satisfies axioms* (fuc), (fu2), *and* (fu3), *then the system* $(\mathcal{F}(U), \mathcal{F}(W), \cap, \cup, \sim, L, H)$ *is referred to as a rough fuzzy set algebra, and* L *and* H *are referred to as rough fuzzy approximation operators. When* $U = W$, $(\mathcal{F}(U), \cap, \cup, \sim, L, H)$ *is also called a rough fuzzy set algebra, in such a case, if there exists a serial (a reflexive, a symmetric, a transitive, an equivalence) crisp relation* R *on* U *such that* $L(A) = \underline{R}(A)$ *and* $H(A) = \overline{R}(A)$ *for all* $A \in \mathcal{F}(U)$, *then* $(\mathcal{F}(U), \cap, \cup, \sim, L, H)$ *is referred to as a serial (a reflexive, a symmetric, a transitive, a Pawlak) rough fuzzy set algebra.*

Axiomatic characterization of serial rough fuzzy set algebra is summarized as the following [32,34,35]

Theorem 5. *Suppose that* $(\mathcal{F}(U), \mathcal{F}(W), \cap, \cup, \sim, L, H)$ *is a rough fuzzy set algebra, i.e.,* L *satisfies axioms* (flc), (fl1), (fl2) *and* (fl3), *and* H *satisfies* (fuc), (fu1), (fu2) *and* (fu3). *Then it is a serial rough fuzzy set algebra iff one of following equivalent axioms holds:*

$\qquad$ (fl0) $\quad L(\widehat{\alpha}) = \widehat{\alpha}, \quad \forall \alpha \in [0,1],$

$\qquad$ (fu0) $\quad H(\widehat{\alpha}) = \widehat{\alpha}, \quad \forall \alpha \in [0,1],$

$\qquad$ (fl0)$'$ $\quad L(1_{\emptyset}) = 1_{\emptyset},$

$\qquad$ (fu0)$'$ $\quad H(1_W) = 1_U,$

$\qquad$ (flu0)$'$ $\quad L(A) \subseteq H(A), \quad \forall A \in \mathcal{F}(W).$

Axiom (flu0)$'$ states that $L(A)$ is a fuzzy subset of $H(A)$. In such a case, $L, H : \mathcal{F}(W) \to \mathcal{F}(U)$ are called the lower and upper rough fuzzy approximation operators and the system $(\mathcal{F}(U), \mathcal{F}(W), \cap, \cup, \sim, L, H)$ is an interval structure. Axiomatic characterizations of other special rough fuzzy operators are summarized in the following Theorems 6 and 7 [35]:

Theorem 6. *Suppose that* $(\mathcal{F}(U), \cap, \cup, \sim, L, H)$ *is a rough fuzzy set algebra. Then*

(1) *it is a reflexive rough fuzzy set algebra iff one of following equivalent axioms holds:*

$$(\text{fl6}) \quad L(A) \subseteq A, \quad \forall A \in \mathcal{F}(U),$$
$$(\text{fu6}) \quad A \subseteq H(A), \quad \forall A \in \mathcal{F}(U).$$

(2) *it is a symmetric rough fuzzy set algebra iff one of the following equivalent axioms holds:*

$$(\text{fl7})' \quad L(1_{U-\{x\}})(y) = L(1_{U-\{y\}})(x), \forall (x, y) \in U \times U,$$
$$(\text{fu7})' \quad H(1_x)(y) = H(1_y)(x), \qquad \forall (x, y) \in U \times U,$$
$$(\text{fl7}) \quad \sim A \subseteq L(\sim L(A)), \qquad \forall A \in \mathcal{F}(U),$$
$$(\text{fu7}) \quad H(\sim H(A)) \subseteq \sim A, \qquad \forall A \in \mathcal{F}(U).$$

(3) *it is a transitive rough fuzzy set algebra iff one of following equivalent axioms holds:*

$$(\text{fl8}) \quad L(A) \subseteq L(L(A)), \quad \forall A \in \mathcal{F}(U),$$
$$(\text{fu8}) \quad H(H(A)) \subseteq H(A), \forall A \in \mathcal{F}(U).$$

Theorem 6 implies that a rough fuzzy algebra $(\mathcal{F}(U), \cap, \cup, \sim, L, H)$ is a reflexive rough fuzzy algebra iff H is an embedding on $\mathcal{F}(U)$ [21,28] and it is a transitive rough fuzzy algebra iff H is closed on $\mathcal{F}(U)$ [21].

Theorem 7. *Suppose that* $(\mathcal{F}(U), \cap, \cup, \sim, L, H)$ *is a rough fuzzy set algebra. Then it is a Pawlak rough fuzzy set algebra iff* L *satisfies axioms* (fl6), (fl7) *and* (fl8) *or equivalently,* H *satisfies axioms* (fu6), (fu7) *and* (fu8).

Theorem 7 implies that a rough fuzzy algebra $(\mathcal{F}(U), \cap, \cup, \sim, L, H)$ is a Pawlak rough fuzzy algebra iff H is a symmetric closure operator on $\mathcal{F}(U)$ [21].

Theorem 8. *Suppose that* $L, H : \mathcal{F}(U) \to \mathcal{F}(U)$ *are dual operators. Then*

$$H(L(A)) \subseteq A, \quad \forall A \in \mathcal{F}(U) \Longrightarrow H(1_y) \in \mathcal{P}(U), \forall y \in U.$$

Proof. For any $A \in \mathcal{F}(U)$, notice that

$$A = \bigcup_{y \in U} [1_y \cap \widehat{A(y)}], \tag{4}$$

where $\widehat{A(y)}$ denotes the constant fuzzy set of $A(y)$, i.e. $\widehat{A(y)}(x) = A(y)$ for all $x \in U$. Then, by axioms (fu2) and (fu3), we have

$$H(A) = \bigcup_{y \in U} [H(1_y) \cap \widehat{A(y)}]. \tag{5}$$

According to the dual properties of L and H, we can easily conclude

$$L(A) = \bigcap_{y \in U} \left[(\sim H(1_y)) \cup \widehat{A(y)} \right]. \tag{6}$$

For any $z \in U$, let $A = 1_{U-\{z\}}$, by Eq. (6), we have

$$L(A) = \bigcap_{y \in U} \left[(\sim H(1_y)) \cup \widehat{A(y)} \right]$$
$$= \bigcap_{y \in U} \left[(\sim H(1_y)) \cup \widehat{1_{U-\{z\}}(y)} \right]$$
$$= \sim H(\sim A) = \sim H(1_z).$$

Thence, by employing Eq. (5), we have

$$HL(A) = \bigcup_{y \in U} \left[H(1_y) \cap \widehat{(1 - H(1_z)(y))} \right]. \tag{7}$$

By the condition, we can obtain that

$$HL(A)(x) = \bigvee_{y \in U} \left[H(1_y)(x) \wedge (1 - H(1_z)(y)) \right] \leq A(x) = 1_{U-\{z\}}(x), \; \forall x \in U. \tag{8}$$

By setting $x = z$ in Eq. (8), we then conclude

$$\bigvee_{y \in U} \left[H(1_y)(z) \wedge (1 - H(1_z)(y)) \right] = 0. \tag{9}$$

It follows that

$$H(1_y)(z) \wedge (1 - H(1_z)(y)) = 0, \; \forall (y, z) \in U \times U. \tag{10}$$

Thus $H(1_y)(z) = 0$ or $H(1_z)(y) = 1$. If $H(1_y)(z) \neq 0$, then $H(1_z)(y) = 1$, by Eq. (10), we have

$$H(1_z)(y) \wedge (1 - H(1_y)(z)) = 0, \tag{11}$$

which implies that $H(1_y)(z) = 1$.

Thus we have proved that, for each $(y, z) \in U \times U$, $H(1_y)(z) = 0$ or $H(1_y)(z) = 1$, that is, $H(1_y) \in \mathcal{P}(U)$.

4 Basic Concepts and Properties of Fuzzy Topological Spaces

In this section, we recall some basic notions and theoretical results of fuzzy topologies.

Definition 5. [19] *A fuzzy topology on a nonempty set U is a family τ of fuzzy subsets in U satisfying the following axioms:*

(T_1) $\widehat{\alpha} \in \tau$ for all $\alpha \in [0,1]$,

(T_2) $G_1 \cap G_2 \in \tau$ for any $G_1, G_2 \in \tau$,

(T_3) $\bigcup_{i \in J} G_i \in \tau$ for an arbitrary family $\{G_i : i \in J\} \subseteq \tau$, where J is an index set.

In this case the pair (U, τ) is called a fuzzy topological space and each fuzzy set in τ is referred to as a fuzzy open set in (U, τ). The complement of a fuzzy open set in the fuzzy topological space (U, τ) is called a fuzzy closed set in (U, τ).

It should be pointed out that if axiom (T_1) in Definition 5 is replaced by axiom

(T_1') $\emptyset \in \tau$ and $U \in \tau$,

then τ is a fuzzy topology in the sense of Chang [5]. We can see that a fuzzy topology in the sense of Lowen must be a fuzzy topology in the sense of Chang. Since a fuzzy topology induced from an approximation space must be in the sense of Lowen, in the present paper, we only use the Lowen's fuzzy topology. For more detail about fuzzy topology, we refer the reader to [18].

Now we define fuzzy closure and interior operations in a fuzzy topological space.

Definition 6. *Let (U, τ) be a fuzzy topological space and $A \in \mathcal{F}(U)$. The fuzzy interior $int(A)$ and fuzzy closure $cl(A)$ of A are, respectively, defined as follows:*

$$int(A) = \cup\{G : G \text{ is a fuzzy open set and } G \subseteq A\}, \tag{12}$$

$$cl(A) = \cap\{K : K \text{ is a fuzzy closed set and } A \subseteq K\}, \tag{13}$$

and $int : \mathcal{F}(U) \to \mathcal{F}(U)$ and $cl : \mathcal{F}(U) \to \mathcal{F}(U)$ are, respectively, called the fuzzy interior operator and the fuzzy closure operator of τ. And sometimes in order to distinguish, we denote them by int_τ and cl_τ, respectively.

It can be shown that $cl(A)$ is a fuzzy closed set and $int(A)$ is a fuzzy open set in (U, τ). A is a fuzzy open set in (U, τ) if and only if $int(A) = A$, and A is a fuzzy closed set in (U, τ) if and only if $cl(A) = A$. Moreover, the fuzzy interior operator and the fuzzy closure operator derived from the same topological space (U, τ) are dual with each other, i.e.,

$$cl(\sim A) = \sim int(A), \quad \forall A \in \mathcal{F}(U), \tag{14}$$

$$int(\sim A) = \sim cl(A), \quad \forall A \in \mathcal{F}(U). \tag{15}$$

The fuzzy closure operator can also be defined by axioms which are called the closure axioms.

Definition 7. *A mapping $cl : \mathcal{F}(U) \to \mathcal{F}(U)$ is referred to as a fuzzy closure operator on U if it satisfies following axioms:*

(Cl1) $A \subseteq cl(A)$, $\forall A \in \mathcal{F}(U)$,

(Cl2) $cl(A \cup B) = cl(A) \cup cl(B)$, $\forall A, B \in \mathcal{F}(U)$,

(Cl3) $cl(cl(A)) = cl(A)$, $\forall A \in \mathcal{F}(U)$,

(Cl4) $cl(\widehat{\alpha}) = \widehat{\alpha}$, $\forall \alpha \in [0,1]$.

Similarly, the fuzzy interior operator can be defined by corresponding axioms.

Definition 8. *A mapping* $int : \mathcal{F}(U) \to \mathcal{F}(U)$ *is referred to as a fuzzy interior operator on U if it satisfies following axioms:*

(Int1) $int(A) \subseteq A, \quad \forall A \in \mathcal{F}(U)$,

(Int2) $int(A \cap B) = int(A) \cap int(B), \quad \forall A, B \in \mathcal{F}(U)$,

(Int3) $int(int(A)) = int(A), \quad \forall A \in \mathcal{F}(U)$,

(Int4) $int(\widehat{\alpha}) = \widehat{\alpha}, \quad \forall \alpha \in [0, 1]$.

It is easy to show that a fuzzy interior operator int defined in Definition 8 determines a fuzzy topology

$$\tau_{int} = \{A \in \mathcal{F}(U) : int(A) = A\}. \tag{16}$$

So, the fuzzy open sets are the fixed points of int. Dually, from a fuzzy closure operator defined in Definition 7, we can obtain a fuzzy topology on U by setting

$$\tau_{cl} = \{A \in \mathcal{F}(U) : cl(\sim A) = \sim A\}. \tag{17}$$

The results are summarized as the following

Theorem 9. (1) *If an operator* $int : \mathcal{F}(U) \to \mathcal{F}(U)$ *satisfies axioms* (Int1)-(Int4), *then* τ_{int} *defined in Eq.* (16) *is a fuzzy topology on U and*

$$int_{\tau_{int}} = int. \tag{18}$$

(2) *If an operator* $cl : \mathcal{F}(U) \to \mathcal{F}(U)$ *satisfies axioms* (Cl1)-(Cl4), *then* τ_{cl} *defined in Eq.* (17) *is a fuzzy topology on U and*

$$cl_{\tau_{cl}} = cl. \tag{19}$$

Similar to the crisp Alexandrov topology [3] and crisp clopen topology [11], we now introduce the concepts of a fuzzy Alexandrov topology and a fuzzy clopen topology.

Definition 9. *A fuzzy topology* τ *on U is called a fuzzy Alexandrov topology* [15] *if the intersection of arbitrarily many fuzzy open sets is still open, or equivalently, the union of arbitrarily many fuzzy closed sets is still closed. A fuzzy topological space* (U, τ) *is said to be a fuzzy Alexandrov space if* τ *is a fuzzy Alexandrov topology on U. A fuzzy topology* τ *on U is called a fuzzy clopen topology if, for every* $A \in \mathcal{F}(U)$, *A is fuzzy open in* (U, τ) *if and only if A is fuzzy closed in* (U, τ). *A fuzzy topological space* (U, τ) *is said to be a fuzzy clopen space if* τ *is a fuzzy clopen topology on U.*

Theorem 10. *Let* $int : \mathcal{F}(U) \to \mathcal{F}(U)$ *be a fuzzy interior operator. The the following two conditions are equivalent:*

(1) *int satisfies axiom* (fI7), *i.e.,* $\sim A \subseteq int(\sim int(A))$ *for all* $A \in \mathcal{F}(U)$;

(2) τ_{int} *is fuzzy clopen topology, i.e., for every* $A \in \mathcal{F}(U)$, *A is fuzzy open in* (U, τ) *if and only if A is fuzzy closed in* (U, τ).

Proof. "(1) $\Rightarrow$ (2)." Assume that *int* satisfies axiom (fl7). If $A \in \mathcal{F}(U)$ is fuzzy open, that is,

$$int(A) = A, \tag{20}$$

then

$$\sim A =\sim int(A) \subseteq int(\sim int(A)). \tag{21}$$

Since *int* is a fuzzy interior operator, by (Int1), we have

$$int(\sim int(A)) \subseteq\sim int(A). \tag{22}$$

Combining Eqs. (21) and (22), we obtain

$$int(\sim int(A)) =\sim int(A). \tag{23}$$

Thus, $\sim int(A)$ is fuzzy open. By using Eq. (20), we see that $\sim A$ is fuzzy open, and hence A is fuzzy closed.

On the other hand, if $A \in \mathcal{F}(U)$ is fuzzy closed, then $\sim A$ is fuzzy open. By employing the above proof, we can observe that $\sim A$ is fuzzy closed. Hence $A =\sim (\sim A)$ is fuzzy open.

"(2) $\Rightarrow$ (1)." For any $A \in \mathcal{F}(U)$, since $int : \mathcal{F}(U) \to \mathcal{F}(U)$ is a fuzzy interior operator, $int(A)$ is open. Hence $\sim int(A)$ is fuzzy closed. Since (U, τ_{int}) is a fuzzy clopen space, therefore $\sim int(A)$ is fuzzy open. Hence

$$int(\sim int(A)) =\sim intA. \tag{24}$$

Since *int* is a fuzzy interior operator, by (Int1), we have

$$int(A) \subseteq A. \tag{25}$$

Therefore, by employing Eq. (24), we conclude

$$\sim A \subseteq\sim int(A) = int(\sim int(A)). \tag{26}$$

Thus, we have proved that *int* satisfies axiom (fl7). $\quad\blacksquare$

5   From Rough Fuzzy Sets to Fuzzy Topologies

In this section we discuss the relationship between fuzzy topological spaces and rough fuzzy sets. Throughout this section we always assume that U is a nonempty universe of discourse, R a crisp binary relation on U, and $\underline{R}$ and $\overline{R}$ the rough fuzzy approximation operators defined in Definition 2.

Denote

$$\tau_R = \{A \in \mathcal{F}(U) : \underline{R}(A) = A\}. \tag{27}$$

The next theorem shows that any reflexive binary relation determines a fuzzy Alexandrov topology.

Theorem 11. *If R is a reflexive crisp binary relation on U, then τ_R defined by Eq. (27) is a fuzzy Alexandrov topology on U.*

Proof. (T_1) For any $\alpha \in [0,1]$, since a reflexive binary relation must be serial, in terms of Theorem 2, we have $\underline{R}(\widehat{\alpha}) = \widehat{\alpha}$, thus $\widehat{\alpha} \in \tau_R$.

(T_2) For any $A, B \in \tau_R$, that is, $\underline{R}(A) = A$ and $\underline{R}(B) = B$, by Theorem 1, we have $\underline{R}(A \cap B) = \underline{R}(A) \cap \underline{R}(B) = A \cap B$. Thus, $A \cap B \in \tau_R$.

(T_3) Assume that $A_i \in \tau_R, i \in J$, J is an index set. Since R is reflexive, by Theorem 3, we have

$$\underline{R}(\bigcup_{i \in J} A_i) \subseteq \bigcup_{i \in J} A_i. \tag{28}$$

For any $x \in U$, let

$$\alpha = (\bigcup_{i \in J} A_i)(x) = \sup_{i \in J} A_i(x). \tag{29}$$

Since $\alpha = \sup_{i \in J} A_i(x)$, we have $A_i(x) \le \alpha$ for all $i \in J$, and, on the other hand, for an arbitrary $\varepsilon > 0$, there exists an $i_0 \in J$ such that $\alpha < A_{i_0}(x) + \varepsilon$. Since $A_i \in \tau_R$ for all $i \in J$, that is, $\underline{R}(A_i) = A_i$ for all $i \in J$, we have $\alpha < A_{i_0}(x) + \varepsilon = \underline{R}(A_{i_0})(x) + \varepsilon$, then, by (FL5) in Theorem 1, we conclude

$$\alpha < \underline{R}(A_{i_0})(x) + \varepsilon \le \bigvee_{i \in J} \underline{R}(A_i)(x) + \varepsilon = (\bigcup_{i \in J} \underline{R}(A_i))(x) + \varepsilon \le \underline{R}(\bigcup_{i \in J} A_i)(x) + \varepsilon. \tag{30}$$

Since $\varepsilon > 0$ is arbitrary, it follows that

$$\alpha \le \underline{R}(\bigcup_{i \in J} A_i)(x), \tag{31}$$

that is,

$$(\bigcup_{i \in J} A_i)(x) \le \underline{R}(\bigcup_{i \in J} A_i)(x). \tag{32}$$

Hence

$$\bigcup_{i \in J} A_i \subseteq \underline{R}(\bigcup_{i \in J} A_i). \tag{33}$$

Combining Eqs. (28) and (33), we obtain

$$\bigcup_{i \in J} A_i = \underline{R}(\bigcup_{i \in J} A_i). \tag{34}$$

Thus, we conclude that $\bigcup_{i \in J} A_i \in \tau_R$.

Therefore, τ_R is a fuzzy topology on U.

Finally, by (FL3) and (FU3) in Theorem 1, we see that τ_R defined by Eq. (27) is a fuzzy Alexandrov topology on U.

Theorem 12. *Assume that R is a crisp binary relation on U. Then the following statements are equivalent:*

(1) R is a preorder, i.e., R is a reflexive and transitive relation;

(2) *the upper rough fuzzy approximation operator* $\overline{R} : \mathcal{F}(U) \to \mathcal{F}(U)$ *is a fuzzy closure operator;*

(3) *the lower rough fuzzy approximation operator* $\underline{R} : \mathcal{F}(U) \to \mathcal{F}(U)$ *is a fuzzy interior operator.*

Proof. By the dual properties of lower and upper rough fuzzy approximation operators, we can easily conclude that (2) and (3) are equivalent. We only need to prove that (1) and (2) are equivalent.

"(1)⇒(2)". Assume that R is a preorder on U. Firstly, by the reflexivity of R and property (FU6) in Theorem 3, we observe that $A \subseteq \overline{R}(A)$ for all $A \in \mathcal{F}(U)$. Thus $\overline{R}$ obeys axiom (Cl1). Secondly, according to property (FU3) in Theorem 1, we see that $\overline{R}$ satisfies axiom (Cl2). Thirdly, since a preorder is reflexive and transitive, $\overline{R}$ satisfies properties (FU6) and (FU8) in Theorem 3. On the other hand, properties (FU6) and (FU8) imply following property

$$\overline{R}(A) = \overline{R}(\overline{R}(A)), \quad \forall A \in \mathcal{F}(U). \tag{35}$$

Thus $\overline{R}$ obeys axiom (Cl3). Finally, notice that a preorder must be a serial relation, then by Theorem 2, we conclude that $\overline{R}$ obeys axiom (Cl4). Therefore, $\overline{R}$ is a fuzzy closure operator.

"(2)⇒(1)". Assume that $\overline{R} : \mathcal{F}(U) \to \mathcal{F}(U)$ is a fuzzy closure operator. By axiom (Cl1), we see that

$$A \subseteq \overline{R}(A), \quad \forall A \in \mathcal{F}(U). \tag{36}$$

Then, by Theorem 3, we conclude that R is a reflexive relation. Moreover, by axiom (Cl1) again, we have

$$\overline{R}(A) \subseteq \overline{R}(\overline{R}(A)), \quad \forall A \in \mathcal{F}(U). \tag{37}$$

On the other hand, by axiom (Cl3), we observe that

$$\overline{R}(\overline{R}(A)) = \overline{R}(A), \quad \forall A \in \mathcal{F}(U). \tag{38}$$

Hence, in terms of Eqs. (37) and (38), we must have

$$\overline{R}(\overline{R}(A)) \subseteq \overline{R}(A), \quad \forall A \in \mathcal{F}(U). \tag{39}$$

According to Theorem 3, we then conclude that R is a transitive relation. Thus we have proved that R is a preorder.

Remark 1. According to Theorem 11, an Alexandrov fuzzy topology can be obtained from a reflexive relation R by using Eq. (27), by Eq. (12) we see that any topology τ induces an interior operator int_τ, which in turn induces a topology τ_{int_τ}. Of course, int_τ is a fuzzy interior operator. It also holds that $\tau_{int_\tau} = \tau$. Now let us take τ_R, then its interior is int_{τ_R} which produces a fuzzy topology $\tau_{int_{\tau_R}}$. Since $\tau_{int_{\tau_R}} = \tau_R$, we have:

$$\tau_R = \{A \in \mathcal{F}(U) : \underline{R}(A) = A\} = \{A \in \mathcal{F}(U) : int_{\tau_R}(A) = A\}. \tag{40}$$

We should point out that though $\underline{R}$ and int_{τ_R} produce the same topology, in general, $\underline{R} \neq int_{\tau_R}$, the reason is that int_{τ_R} is a fuzzy interior operator whereas $\underline{R}$ is not. In fact, Theorem 12 tell us that $\underline{R} = int_{\tau_R}$ if and only if R is a preorder.

Lemma 1. *If R is a symmetric crisp binary relation on U, then for all $A, B \in \mathcal{F}(U)$,*

$$\overline{R}(A) \subseteq B \Longleftrightarrow A \subseteq \underline{R}(B). \tag{41}$$

Proof. "$\Rightarrow$" Let $A, B \in \mathcal{F}(U)$, if $\overline{R}(A) \subseteq B$, by (FU1) in Theorem 1, we have $\sim \underline{R}(\sim A) \subseteq B$, then, $\sim B \subseteq \underline{R}(\sim A)$. By (FU4) in Theorem 1 and (FL7) in Theorem 3, it follows that

$$\overline{R}(\sim B) \subseteq \overline{R}(\underline{R}(\sim A)) \subseteq \sim A. \tag{42}$$

Hence

$$A \subseteq \sim \overline{R}(\sim B) = \underline{R}(B). \tag{43}$$

"$\Leftarrow$" Assume that $A \subseteq \underline{R}(B)$, by (FL1) in Theorem 1, we have $A \subseteq \sim \overline{R}(\sim B)$, that is, $\overline{R}(\sim B) \subseteq \sim A$, according to (FL4) in Theorem 1, we then conclude

$$\underline{R}(\overline{R}(\sim B)) \subseteq \underline{R}(\sim A). \tag{44}$$

By (FL1) and (FU1) in Theorem 1, it is easy to obtain that

$$\sim \overline{R}(\underline{R}(B)) \subseteq \underline{R}(\sim A) = \sim \overline{R}(A). \tag{45}$$

Consequently, by (FL7) in Theorem 3, we conclude that

$$\overline{R}(A) \subseteq \overline{R}(\underline{R}(B)) \subseteq B. \tag{46}$$

Theorem 13. *Let R be a similarity crisp binary relation on U, and $\underline{R}$ and $\overline{R}$ the rough fuzzy approximation operators defined in Definition 2. Then $\underline{R}$ and $\overline{R}$ satisfy property* (Clop): *for $A \in \mathcal{F}(U)$,*

$$\text{(Clop)} \quad \underline{R}(A) = A \Longleftrightarrow A = \overline{R}(A) \Longleftrightarrow \underline{R}(\sim A) = \sim A \Longleftrightarrow \sim A = \overline{R}(\sim A). \tag{47}$$

Proof. For $A \in \mathcal{F}(U)$, assume that $\underline{R}(A) = A$. Since R is reflexive, by (FL6) in Theorem 3, we see that $\underline{R}(A) = A$ implies $A \subseteq \overline{R}(A)$. Then, by Lemma 1, we conclude $\overline{R}(A) \subseteq A$. Furthermore, since R is reflexive, in terms of (FU6) in Theorem 3 we obtain $A = \overline{R}(A)$. Similarly, we can prove that $A = \overline{R}(A)$ implies $\underline{R}(A) = A$. Moreover, by using (FL1) and (FU1) in Theorem 1, it is easy to verify that Eq. (47) holds.

Theorem 14. *Let R be a similarity crisp binary relation on U, and $\underline{R}$ and $\overline{R}$ the rough fuzzy approximation operators defined in Definition 2. Then τ_R defined in Eq. (27) is a fuzzy clopen topology on U.*

Proof. For $A \in \mathcal{F}(U)$, since R is a similarity crisp binary relation, by Theorem 13, we have

$$A \text{ is fuzzy open} \Longleftrightarrow A \in \tau_R$$
$$\Longleftrightarrow A = \underline{R}(A)$$
$$\Longleftrightarrow \sim A = \underline{R}(\sim A)$$
$$\Longleftrightarrow \sim A \in \tau_R$$
$$\Longleftrightarrow A \text{ is fuzzy closed.}$$

Thus, τ_R is a fuzzy clopen topology on U.

Corollary 1. *Let R be an equivalent crisp binary relation on U, $\underline{R}$ and $\overline{R}$: $\mathcal{F}(U) \to \mathcal{F}(U)$ are the rough fuzzy approximation operators of (U, R). Let $\tau_R = \{A \in \mathcal{F}(U) : \underline{R}(A) = A\}$, then*

(1) $\underline{R}$ and $\overline{R} : \mathcal{F}(U) \to \mathcal{F}(U)$ are the fuzzy interior operator and the fuzzy closure operator of the fuzzy topology τ_R

(2) τ_R is a fuzzy clopen topology on U.

6 From Fuzzy Topological Spaces to Rough Fuzzy Sets

As can be seen from Section 5, a preorder rough fuzzy algebra yields a fuzzy topological space such that its fuzzy interior and closure operators are, respectively, the lower and upper rough fuzzy approximation operators of the given crisp approximation space. In this section, we consider the reverse problem, that is, under which conditions can a fuzzy topological space be associated with a crisp approximation space which produces the given fuzzy topological space?

The following Theorem 15 gives the sufficient and necessary conditions that a fuzzy interior (respectively, a fuzzy closure) operator in a fuzzy topological space can be associated with a crisp preorder such that the induced lower (respectively, upper) rough fuzzy approximation operator is exactly the fuzzy interior (respectively, fuzzy closure) operator.

Theorem 15. *Let (U, τ) be a fuzzy topological space and $cl, int : \mathcal{F}(U) \to \mathcal{F}(U)$ its fuzzy closure and interior operators, respectively. Then there exists a preorder R_τ on U such that*

$$\overline{R_\tau}(A) = cl(A) \text{ and } \underline{R_\tau}(A) = int(A), \quad \forall A \in \mathcal{F}(U) \tag{48}$$

iff the operator cl satisfies axioms (fuc) (fu1) and (fu2), or equivalently, int obeys axioms, (flc), (fl1) and (fl2), that is,

(fuc) $cl(1_y) \in \mathcal{P}(U), \; \forall y \in U.$

(fu1) $cl(A \cap \widehat{\alpha}) = (cl(A)) \cap \widehat{\alpha}, \; \forall A \in \mathcal{F}(U), \forall \alpha \in [0, 1].$

(fu2) $cl(\bigcup_{i \in J} A_i) = \bigcup_{i \in J} cl(A_i), \; A_i \in \mathcal{F}(U), \; i \in J, \; J \text{ is any index set.}$

(flC) $int(1_{U-\{y\}}) \in \mathcal{P}(U), \; \forall y \in U.$

(fl1) $int(A \cup \widehat{\alpha}) = (int(A)) \cup \widehat{\alpha} \; \forall A \in \mathcal{F}(U), \forall \alpha \in [0, 1].$

(fl2) $int(\bigcap_{i \in J} A_i) = \bigcap_{i \in J} int(A_i), \; A_i \in \mathcal{F}(U), \; i \in J, \; J \text{ is any index set.}$

Proof. "⇒" Assume that there exists a preorder R_τ on U such that Eq. (48) holds, then, by Theorem 1, it can easily be observed that the operator cl satisfies axioms (fuc), (fu1), and (fu2), and int obeys axioms (flc), (fl1), and (fl2).

"⇐" If the fuzzy closure operator $cl : \mathcal{F}(U) \to \mathcal{F}(U)$ satisfies axioms (fuc), (fu1), and (fu2), and the fuzzy interior operator $int : \mathcal{F}(U) \to \mathcal{F}(U)$ obeys axioms (flc), (fl1), and (fl2), then, by Theorem 4, we can define a binary relation R_τ on U by setting

$$(x, y) \in R_\tau \iff cl(1_y)(x) = 1, \quad (x, y) \in U \times U \tag{49}$$

such that Eq. (48) holds. Moreover, by Theorem 12, we conclude that R_τ is a preorder.

Remark 2. Notice that a fuzzy topological space satisfying axioms (fl2) and (fu2) is a fuzzy Alexandrov space. Theorem 15 shows that a fuzzy topological space can be associated with a preorder such that the induced lower and upper rough fuzzy approximation operators are, respectively, the fuzzy interior and closure operators of the fuzzy topology if and only if the given topological space must be a fuzzy Alexandrov space and the fuzzy interior satisfies axioms (flc) and (fl1), and the closure operators obeys axioms (fuc) and (fu1).

Let $\mathcal{R}$ be the set of all crisp preorders on U and $\mathcal{T}$ the set of all fuzzy Alexandrov spaces on U in which the fuzzy interior operator satisfies axioms (flc) and (fl1), and the fuzzy closure operator obey axioms (fuc) and (fu1). Then, we can easily conclude following Theorems 16 and 17.

Theorem 16. (1) If $R \in \mathcal{R}$, τ_R is defined by Eq. (27) and R_{τ_R} is defined by Eq. (49), then $R_{\tau_R} = R$.

(2) If $\tau \in \mathcal{T}$, R_τ is defined by Eq. (49), and τ_{R_τ} is defined by Eq. (27), then $\tau_{R_\tau} = \tau$.

Remark 3. Result (1) in Theorem 16 shows that, for a given crisp preorder R, the binary relation R_{τ_R}, which is defined by the induced fuzzy topology τ_R of R, is no other than the given R. And similarly, results (2) in Theorem 16 implies that, for a given fuzzy topology τ, the topology τ_{R_τ}, which is defined by the induced binary relation R_τ of the given topology τ, is identified with the given topology τ.

Theorem 17. *There exists a one-to-one correspondence between $\mathcal{R}$ and $\mathcal{T}$.*

Proof. Define a mapping $f : \mathcal{R} \to \mathcal{T}$ as follows:

$$f(R) = \tau_R, \quad R \in \mathcal{R}.$$

Then, by Theorem 16, it is easy to verify that f is a one-to-one correspondences between $\mathcal{R}$ and $\mathcal{T}$.

Theorem 18. *Let (U, τ) be a fuzzy topological space and $cl_\tau, int_\tau : \mathcal{F}(U) \to \mathcal{F}(U)$ its fuzzy closure and interior operators, respectively. If there exists a crisp binary R_τ on U such that Eq. (48) holds, then (U, τ) is a fuzzy clopen space if and only R_τ is an equivalence relation on U.*

Proof. "$\Rightarrow$" If exists a crisp binary R_τ on U such that Eq. (48) holds, then, by Theorem 12, we conclude that R is a preorder. On the other hand, since (U, τ) is a fuzzy clopen topological space, according to Theorem 10, we see that the fuzzy interior operator $int_\tau : \mathcal{F}(U) \to \mathcal{F}(U)$ and fuzzy closure operator $cl_\tau : \mathcal{F}(U) \to \mathcal{F}(U)$ satisfy axioms (fl7) and (fu7), respectively. That is, $\underline{R} : \mathcal{F}(U) \to \mathcal{F}(U)$ and $\overline{R} : \mathcal{F}(U) \to \mathcal{F}(U)$ obey properties (FL7) and (FU7). Hence, by Theorem 3, we conclude that R is symmetric. Therefore, R is an equivalence relation on U.

"$\Leftarrow$" It follows immediately from Corollary 1.

7   Conclusion

In this paper we have studied the topological structure of rough fuzzy sets in infinite universes of discourse. We have shown that a reflexive crisp rough approximation space can induce a fuzzy Alexandrov space. We have also examined that the lower and upper rough fuzzy approximation operators are, respectively, a fuzzy interior operator and a fuzzy closure operator if and only if the binary relation in the crisp approximation space is reflexive and transitive. We have further proved that a fuzzy topological space induced from a reflexive and symmetric crisp approximation space is a fuzzy clopen topological space. Finally, we have explored the sufficient and necessary conditions that a fuzzy interior (closure, respectively) operator derived from a fuzzy topological space can associate with a reflexive and transitive crisp approximation space such that the induced lower (upper, respectively) rough fuzzy approximation operator is exactly the fuzzy interior (closure, respectively) operator.

Acknowledgement. The authors would like to thank the anonymous referees for their valuable comments and suggestions. This work was supported by grants from the National Natural Science Foundation of China (Nos. 61272021, 61075120, 11071284, and 61173181) and the Zhejiang Provincial Natural Science Foundation of China (No. LZ12F03002).

References

1. Acencio, M.L., Lemke, N.: Towards the Prediction of Essential Genes by Integration of Network Topology, Cellular Localization and Biological Process Information. BMC Bioinformatics 10, 290 (2009)
2. Albizuri, F.X., Danjou, A., Grana, M., et al.: The High-order Boltzmann Machine: Learned Distribution and Topology. IEEE Transactions on Neural Networks 6, 767–770 (1995)
3. Arenas, F.G.: Alexandroff Spaces. Acta Mathematica Universitatis Comenianae 68, 17–25 (1999)
4. Boixader, D., Jacas, J., Recasens, J.: Upper and Lower Approximations of Fuzzy Sets. International Journal of General Systems 29, 555–568 (2000)
5. Chang, C.L.: Fuzzy Topological Spaces. Journal of Mathematical Analysis and Applications 24, 182–189 (1968)

6. Choudhury, M.A., Zaman, S.I.: Learning Sets and Topologies. Kybernetes 35, 1567–1578 (2006)
7. Chuchro, M.: On Rough Sets in Topological Boolean Algebras. In: Ziarko, W. (ed.) Rough Sets, Fuzzy Sets and Knowledge Discovery, pp. 157–160. Springer, Berlin (1994)
8. Chuchro, M.: A Certain Conception of Rough Sets in Topological Boolean Algebras. Bulletin of the Section of Logic 22, 9–12 (1993)
9. Dubois, D., Prade, H.: Rough Fuzzy Sets and Fuzzy Rough Sets. International Journal of General Systems 17, 191–209 (1990)
10. Hao, J., Li, Q.G.: The Relationship between L-Fuzzy Rough Set and L-Topology. Fuzzy Sets and Systems 178, 74–83 (2011)
11. Kondo, M.: On the Structure of Generalized Rough Sets. Information Sciences 176, 589–600 (2006)
12. Kortelainen, J.: On Relationship between Modified Sets, Topological Space and Rough Sets. Fuzzy Sets and Systems 61, 91–95 (1994)
13. Kortelainen, J.: On the Evaluation of Compatibility with Gradual Rules in Information Systems: A Topological Approach. Control and Cybernetics 28, 121–131 (1999)
14. Kortelainen, J.: Applying Modifiers to Knowledge Acquisition. Information Sciences 134, 39–51 (2001)
15. Lai, H., Zhang, D.: Fuzzy Preorder and Fuzzy Topology. Fuzzy Sets and Systems 157, 1865–1885 (2006)
16. Lashin, E.F., Kozae, A.M., Khadra, A.A.A., Medhat, T.: Rough Set Theory for Topological Spaces. International Journal of Approximate Reasoning 40, 35–43 (2005)
17. Li, J.J., Zhang, Y.L.: Reduction of Subbases and Its Applications. Utilitas Mathematica 82, 179–192 (2010)
18. Liu, Y.-M., Luo, M.-K.: Fuzzy Topology. World Scientific Publishing Co. Pte. Ltd., Siingapore (1997)
19. Lowen, R.: Fuzzy Topological Spaces and Fuzzy Compactness. Journal of Mathematical Analysis and Applications 56, 621–633 (1976)
20. Mi, J.-S., Leung, Y., Zhao, H.-Y., Feng, T.: Generalized Fuzzy Rough Sets Determined by a Triangular Norm. Information Sciences 178, 3203–3213 (2008)
21. Mi, J.-S., Zhang, W.-X.: An Axiomatic Characterization of a Fuzzy Generalization of Rough Sets. Information Sciences 160, 235–249 (2004)
22. Morsi, N.N., Yakout, M.M.: Axiomatics for Fuzzy Rough Sets. Fuzzy Sets and Systems 100, 327–342 (1998)
23. Pawlak, Z.: Rough Sets: Theoretical Aspects of Reasoning about Data. Kluwer Academic Publishers, Boston (1991)
24. Pei, Z., Pei, D.W., Zheng, L.: Topology vs Generalized Rough Sets. International Journal of Approximate Reasoning 52, 231–239 (2011)
25. Qin, K.Y., Pei, Z.: On the Topological Properties of Fuzzy Rough Sets. Fuzzy Sets and Systems 151, 601–613 (2005)
26. Qin, K.Y., Yang, J., Pei, Z.: Generalized Rough Sets Based on Reflexive and Transitive Relations. Information Sciences 178, 4138–4141 (2008)
27. Radzikowska, A.M., Kerre, E.E.: A Comparative Study of Fuzzy Rough Sets. Fuzzy Sets and Systems 126, 137–155 (2002)
28. Thiele, H.: On Axiomatic Characterisation of Fuzzy Approximation Operators II, the Rough Fuzzy Set Based Case. In: Proceedings of the 31st IEEE International Symposium on Multiple-Valued Logic, pp. 330–335 (2001)

29. Wiweger, R.: On Topological Rough Sets. Bulletin of Polish Academy of Sciences: Mathematics 37, 89–93 (1989)
30. Wu, Q.E., Wang, T., Huang, Y.X., Li, J.S.: Topology Theory on Rough Sets. IEEE Transactions on Systems, Man and Cybernetics Part B—Cybernetics 38, 68–77 (2008)
31. Wu, W.-Z.: A Study on Relationship between Fuzzy Rough Approximation Operators and Fuzzy Topological Spaces. In: Wang, L., Jin, Y. (eds.) FSKD 2005. LNCS (LNAI), vol. 3613, pp. 167–174. Springer, Heidelberg (2005)
32. Wu, W.-Z.: On Some Mathematical Structures of T-fuzzy Rough Set Algebras in Infinite Universes of Discourse. Fundamenta Informaticae 108, 337–369 (2011)
33. Wu, W.-Z., Leung, Y., Mi, J.-S.: On Characterizations of $(\mathcal{I}, \mathcal{T})$-Fuzzy Rough Approximation Operators. Fuzzy Sets and Systems 154, 76–102 (2005)
34. Wu, W.-Z., Leung, Y., Zhang, W.-X.: On Generalized Rough Fuzzy Approximation Operators. In: Peters, J.F., Skowron, A. (eds.) Transactions on Rough Sets V. LNCS, vol. 4100, pp. 263–284. Springer, Heidelberg (2006)
35. Wu, W.-Z., Zhang, W.-X.: Constructive and Axiomatic Approaches of Fuzzy Approximation Operators. Information Sciences 159, 233–254 (2004)
36. Wu, W.-Z., Zhou, L.: On Intuitionistic Fuzzy Topologies Based on Intuitionistic Fuzzy Reflexive and Transitive Relations. Soft Computing 15, 1183–1194 (2011)
37. Yang, L.Y., Xu, L.S.: Topological Properties of Generalized Approximation Spaces. Information Sciences 181, 3570–3580 (2011)
38. Yao, Y.Y.: Constructive and Algebraic Methods of the Theory of Rough Sets. Journal of Information Sciences 109, 21–47 (1998)
39. Zhang, H.-P., Yao, O.Y., Wang, Z.D.: Note on "Generlaized Rough Sets Based on Reflexive and Transitive Relations". Information Sciences 179, 471–473 (2009)
40. Zhou, L., Wu, W.-Z., Zhang, W.-X.: On Intuitionistic Fuzzy Rough Sets and Their Topological Structures. International Journal of General Systems 38, 589–616 (2009)
41. Zhu, W.: Topological Approaches to Covering Rough Sets. Information Sciences 177, 1499–1508 (2007)

Approximation of Sets
Based on Partial Covering

Zoltán Ernő Csajbók

Department of Health Informatics, Faculty of Health, University of Debrecen,
Sóstói út 2-4, H-4400 Nyíregyháza, Hungary
csajbok.zoltan@foh.unideb.hu

Abstract. In classic Pawlakian rough set theory the sets used to approximations are equivalence classes which are pairwise disjoint and cover the
universe. In this article we give up not only the pairwise disjoint property
but also the covering of the universe.

After a historical and philosophical background, we define a general
set theoretic approximation framework. First, we reconstruct the rough
set theory and partly restate its some well–known facts in the language
of this framework.

Next, we present a special approximation scheme. It is based on the
partial covering of the universe which is called the base system and de-
noted by $\mathfrak{B}$. $\mathfrak{B}$-definable sets and lower and upper $\mathfrak{B}$-approximations are
straightforward point–free generalizations of Pawlakian ones. We study
such notions as single–layered base systems, $\mathfrak{B}$-representations of $\mathfrak{B}$-
definable sets, and the exactness of sets. It is a well-known fact that
the Pawlakian upper and lower approximations form a Galois connection. We clarify which conditions have to be satisfied by the upper and
lower $\mathfrak{B}$-approximations so that they form a (regular) Galois connection.
Excluding the cases when the empty set is the upper $\mathfrak{B}$-approximation
of certain nonempty sets gives rise to a partial upper $\mathfrak{B}$-approximation
map. We also clear up that a partial upper $\mathfrak{B}$-approximation map and a
total lower $\mathfrak{B}$-approximation map form a partial Galois connection.

In order to demonstrate the effectiveness of our approach we present
three real–life examples in the last section.

Keywords: Rough set theory, approximation of sets, partial covering,
Galois connections.

Szüleim emlékének, Feleségemnek, Ilonka néninek

1 Introduction

This article is based on my Ph.D. thesis [1]. It incorporates a number of aspects
of an uncommon generalization of rough set theory which relies on the partial
covering of the universe.

J.F. Peters et al. (Eds.): Transactions on Rough Sets XVI, LNCS 7736, pp. 144–220, 2013.
© Springer-Verlag Berlin Heidelberg 2013

1.1   A Historical Outline

The rough set theory (RST), among others, is a mathematical tool to manage inexact, uncertain, incomplete and imperfect data. It was invented by the Polish mathematician, Zdzisław Pawlak in the early 1980s [2,3].

The starting point is a nonempty *finite* set U of distinguishable objects, called the *universe of discourse*, and an *equivalence relation* ε on U [4]. The partition of U generated by ε is denoted by U/ε, and its elements are called ε-*elementary* sets (Fig. 1). An ε-elementary set can be viewed as a set of indiscernible objects characterized by the same available information about them [5,6]. In addition, any union of ε-elementary sets is referred to as *definable* set (Fig. 2).

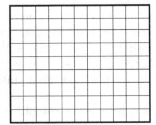   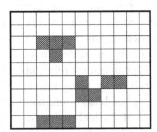

Fig. 1. ε-elementary sets   **Fig. 2.** Definable sets

Any subset $X \subseteq U$ can be naturally approximated by two sets called the lower and upper ε-approximations of X. The *lower ε-approximation* of X is the union of all the ε-elementary sets which are the subsets of X (Fig. 3), whereas the *upper ε-approximation* of X is the union of all the ε-elementary sets that have a nonempty intersection with X (Fig. 4).

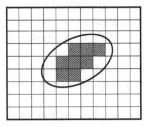   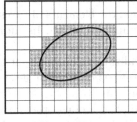   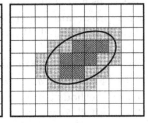

Fig. 3. Lower   **Fig. 4.** Upper   **Fig. 5.** Lower-upper
approximation   approximation   approximation

The difference between upper and lower ε-approximations is called the ε-*boundary* of X (Fig. 5). The subset X is ε-*crisp* (exact), if its ε-boundary is the empty set, ε-*rough* (inexact) otherwise.

Let $\sigma(U/\varepsilon)$ denote the extension of U/ε with all the unions of some ε-elementary sets and the empty set. It is easy to see that $\sigma(U/\varepsilon) \subseteq 2^U$ is a σ-algebra generated by U/ε, i.e., it is nonempty, closed under complementations and countable unions. In other words, $(U, \sigma(U/\varepsilon))$ is an Alexandrov topological space with the basis U/ε. $\sigma(U/\varepsilon)$ is the family of all open and closed sets [7].

In Pawlak's theory, the lower and upper ε-approximations can be defined by three equivalent forms. These three forms are based on elements, ε-elementary sets and the σ-algebra $\sigma(U/\varepsilon)$ [8,9,10]. In any case, both lower and upper ε-approximations of any subset $X \subseteq U$ belong to the σ-algebra $\sigma(U/\varepsilon)$.

The three equivalent definitions offer different interpretations of Pawlak's approximations. According to the element based formulation, the lower and upper approximation operators can be interpreted as the necessity and possibility operators of *modal logic* [11,12,13,14,15,16,17,18,19]. The σ-algebra based formulation relates them to interior and closure operators in *topological spaces* [20]. The formulation based on ε-elementary sets has been served as the "pattern" of *granular computing* developments [21,22,23,24,25]. Nowadays, granular computing is a fast developing branch of information technology.

The generalization of Pawlak's approximations can go along one of the three equivalent definitions mentioned above. A natural generalization of Pawlak's idea via the element based definition is that the equivalence relation is replaced by any other type of binary relation on U [26,27,28,29,30,31]. Another generalization can be obtained by using any covering of the universe and the imitation of the ε-elementary set based definition [20,32].

The case of σ-algebra based definition is a little more complicated. In the language of Alexandrov topological spaces, the σ-algebra $\sigma(U/\varepsilon)$ is the family of clopen sets, *i.e.*, the family of open sets coincides with the family of closed sets. The family of open sets is related to the lower approximation or interior operator, whereas the family of closed sets is related to the upper approximation or closure operator. As a possible generalization, one may use two different subsystems of the powerset of U [33]. A subsystem for the lower approximation which must be closed under unions and contains the empty set, and another subsystem for the upper approximation which, in turn, must be closed under intersections and contains U. Moreover, in order to keep the duality of lower and upper approximation operators, the elements of two subsystems must be related to each other through the complementation. In addition, this latter restriction can also be removed [8].

For the beginning of generalizations of rough set theory, see [5,34,35,36].

A list of some *research directions* on the rough set foundations and the rough set based methods can be found in [37].

Rough set theory can be applied among others in the areas of artificial intelligence, cognitive sciences, medicine and economics. It provides a powerful foundation to reveal and discover important structures and patterns in data and to classify complex objects. One of the main advantages of rough set theory is that it does not need any preliminary or additional information about data [38,39]. This attractive property of rough set theory is of especial importance for instance to data mining, machine learning, decision analysis, knowledge management, expert systems, patter recognition, medicine, engineering, banking, financial and market analysis [38,39,40].

For some additional recent general work, tutorials and historical review on rough set theory, see [41,42,43,44,45,46,47,48].

1.2   Basic Philosophical Background

There is a philosophical interpretation of the rough set theory too. It may also be seen as a relatively new possible mathematical approach to *vagueness* [49,50,37,51]. According to the entry for "vagueness" in the *Stanford Encyclopedia of Philosophy*:

> There is wide agreement that a term is vague to the extent that it has borderline cases. This makes the notion of a borderline case crucial in accounts of vagueness. ([52], the two introductory sentences.)
>
> Vagueness is standardly defined as the possession of borderline cases. For example, "tall" is vague because a man who is 1.8 meters in height is neither clearly tall nor clearly non-tall. No amount of *conceptual analysis* or *empirical investigation* can settle whether a 1.8 meter man is tall. ([52], Chapter 1. *The italics are mine.*)
>
> Borderline cases are inquiry resistant. Indeed, the inquiry resistance typically recurses. For in addition to the unclarity of the borderline case, there is normally unclarity as to where the unclarity begins. In other words "borderline case" has borderline cases. This higher order vagueness shows that "vague" is vague. ([52], Chapter 1.)

That is vague terms lack well-defined extensions—there is no sharp boundary between tall people and the rest [49]. In other words a set of objects is *vague* if objects exist that cannot be classified as belonging to either the set or its complement [5]. It should immediately be noted that in this context the notion "set" is used in a pre-theoretic sense.

It is an important fact that in rough set theory the extension of a set in question is known. Nevertheless, in practice, if we want to determine its extension we use some tools. In this way, however, the extension of an observed set in general cannot be determined exactly. Due to the limited nature of the measurement, the exact attributes which characterize the set in question, *i.e.*, its intension is not known. Rough set theory uses two sets whose intension, consequently their extension is exactly known in order to approximate observed sets.

The "vagueness" is a more than two thousand-year-old problem. Its origins go back to the so–called *Sorites paradox* [49,53,54,55] attributed to Aristotle's contemporary Eubulides of Miletus (4th c. BC), the Megarian logician. The word "sorites" in Greek means "heap". (To be more precise, the paradox derives its name from the Greek word *soros.*) Note that the far known Liar paradox in its purest form is also attributed to Eubulides.

One of the forms of the Sorites paradox is the following. Of course, one stone does not make a heap. Adding only one stone to what is not yet a heap surely cannot make a heap. Repeating this step adding stones one by one we arrive at the conclusion that heaps do not exist not even if they consist of more than, say, 100,000 stones. Then, where do we draw the line between what is a heap of stones and what is not?

I agree with Priest [54]: the Sorites is a very hard paradox, possibly harder than the Liar. For the Liar can be isolated, whereas the Sorites is everywhere and can take us anywhere. And I agree that the paradox is so hard because it systematically imposes upon us the existence of unbelievable or otherwise unacceptable *cut-off points*. No solution can avoid explaining why this happens. ([55], p. 24. *The italics are mine.*)

The counter-intuitiveness of Sorites phenomena lies in the fact that there must be a cut-off, *regardless* of where exactly it is located in the soritical sequence. ([55], p. 34. *The italics are the author's.*)

The Sorites paradox is not mere a curiosity such as R. Keefe and P. Smith remarked in [56]. To confirm this statement, let us look at the following example coming from medicine.

Example 1. Let us consider the fasting blood glucose test [57] which is used to screen for and diagnose diabetes. It is measured on a fast basis, *i.e.*, collected after an 8 and 10 hours fast. The test measures the amount of glucose in the blood right at the time of sample collection. On the clinical practice recommendations of the American Diabetes Association, the fasting glucose level is normal if the test result is between 3.9 mmol/L and 5.5 mmol/L, and indicates diabetes over 11.1 mmol/L on more than one testing occasion.

Now, for instance, the fasting glucose level 4.5 mmol/L is normal. Plausibly, increasing the normal fasting glucose level by 0.001 mmol/L (or 0.00001 mmol/L, if necessary) cannot make a difference. So, if the fasting glucose level 4.0 mmol/L is normal then 4.0 mmol/L plus 0.001 mmol/L is also normal. Now, since the fasting glucose level 4.001 mmol/L is normal, 4.001 mmol/L plus 0.001 mmol/L is also normal; and so on. Consequently, any fasting glucose level is normal, even if it is greater than, say, 25.0 mmol/L.

The Sorites paradox was not an attractive problem until the late 19th century. Next, numerous logicians and philosophers dealt with it. The anthology [56] collects for the first time the most important classical papers in the field.

The vagueness associated with the boundary region approach was first formulated in 1893 by G. Frege [58], next Peirce in 1902 [59].

Pawlak's fundamental view of vagueness can be characterized as "unable to classify" [60,37,61,5]. As Pawlak and Skowron wrote in [5]:

In contrast to odd numbers, the notion of a beautiful painting is vague, because we are *unable to classify* uniquely all paintings into two classes: beautiful and not beautiful. Some paintings cannot be decided whether they are beautiful or not and thus they remain in the doubtful area. Thus, beauty is not a precise but a vague concept. ([5], p. 5. *The italics are mine.*)

However, in spite of the fact that vagueness is very interesting phenomenon in philosophy, it is not allowed within standard mathematics.

Pawlak's information-based solution concerning vagueness is the following:[1]

> [...] in the proposed approach, we assume that any vague concept is replaced
> by *a pair of precise concepts*—called the lower and the upper approximation
> of the vague concept. The lower approximation consists of all objects which
> surely belong to the concept and the upper approximation contains all objects
> which possibly belong to the concept. The difference between the upper and
> the lower approximation constitutes the boundary region of the vague concept.
> Approximations are two basic operations in rough set theory.
> Hence, rough set theory expresses vagueness *not by means of membership, but
> by employing a boundary region of a set*. If the boundary region of a set is
> empty it means that the set is crisp, otherwise the set is rough (inexact). A
> non-empty boundary region of a set means that our knowledge about the set
> is not sufficient to define the set precisely. ([5], p. 6. *The italics are mine.*)

In sum, Pawlak's approach can be viewed as a specific implementation of Frege's
idea of vagueness [58], *i.e.*, imprecision is expressed by a boundary region of a
set [5].

1.3   Our Approach

There are many possibilities to generalize the rough set theory. To sum up, our
approach has three main foundation-stones:

1. *"unable to classify"* as the base of vagueness,
2. its presentation in a *point-free* manner, and
3. *partiality* of our knowledge about the universe.

Ad 1. Rough set theory has served as a "pattern" for granular computing (GrC).
However, there are fundamental differences between them. Granular computing
and also rough set theory have three semantic views, in particular, uncertainty
theory, knowledge engineering and how-to-solve/compute-it [64,65]. The most
important difference between the two theories is best illustrated in connection
with the uncertainty theory. Pawlak uses "unable to classify" as the *base of uncertainty*, while the granular computing regards a granule as a *unit of uncertainty*
[65].

Ad 2. The philosophy of rough set theory relies on the assumption that some
information (data, knowledge) are associated with every object of the universe
of discourse. Objects characterized by the same information are *indiscernible* or
similar in view of the available information about them. A set of all indiscernible
or similar objects form a unit of the basic knowledge. Such a unit can be seen in

---

[1] There is another contemporary information-based solution proposal concerning
vagueness, namely, Zadeh's fuzzy set theory [62]. "Zadeh's introduction of fuzzy
sets was not meant to be a contribution to the philosophy of vagueness. It was motivated by the need for a computational representation for linguistic terms appearing
in statements, which are often intended to provide synthetic information about complex situations." ([63], p. 893). Fuzzy set theory is complementary to rough set theory.
In this article, this aspect is only mentioned here.

a *point-wise* manner, *i.e.*, the content of the unit is visible, and in a *point-free* manner, *i.e.*, the content of the unit is hidden. We abstract each unit into a point. Such a collection of points is called the quotient structure. We will work on quotient structures, in other words we manage units in the point-free manner.

For more details concerning points 1 and 2, see [64,65,21].

Ad 3. In real life, information being at our disposal is generally insufficient. Consequently, it is natural to assume that there may be objects which we are unable to characterize at all. Moreover, there are features with which we can form a set of objects effectively, but we cannot form its complement effectively at the same time. For instance, the complements of a recursively enumerable set is not necessarily a recursively enumerable set as well [66].

In rough set theory, the sets used to approximation are the equivalence classes which are pairwise disjoint and cover the universe of discourse. If we give up the requirement of the pairwise disjointedness, we get a kind of generalization of the theory (Fig. 6). Its detailed elaboration can be found in the literature (see, *e.g.*, [67] and references therein).

The main question of the article is what would happen if we gave up not only the pairwise disjoint property but also the covering of the universe (Fig 7). The resulting system is called the *approximation of sets based on partial covering*, or *partial approximation of sets* for short.

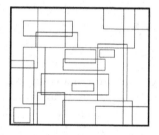   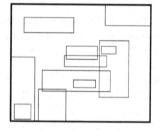

Fig. 6. Giving up the pairwise disjoint property   **Fig. 7.** Giving up the covering: partial base system

In the article we will examine the properties of the approximation of sets under these unusual conditions. At the most general abstraction level, we make the only essential condition that *the lower approximation of any set must be included in its upper approximation* [68,69].

Let our starting point be an arbitrary nonempty family $\mathfrak{B}$ of subsets of a nonempty universe of discourse U [70,71,72,73]. Its elements are called $\mathfrak{B}$-*sets*. On the analogy of the definition of the σ-algebra $\sigma(U/\varepsilon)$, let $\mathfrak{D}_{\mathfrak{B}}$ denote the extension of $\mathfrak{B}$ with the empty set and all the unions of some $\mathfrak{B}$-sets. In other words, $\mathfrak{D}_{\mathfrak{B}}$ is closed under arbitrary unions and contains every set in $\mathfrak{B}$ and the empty set. However, $\mathfrak{D}_{\mathfrak{B}}$ *neither covers the universe* (*i.e.*, it does not contain U) *nor forms σ-algebra* in general. Similarly to the rough set theory, any union of $\mathfrak{B}$-sets is referred to as $\mathfrak{B}$-*definable* set.

Our notion of lower and upper approximations are straightforward point-free generalizations of Pawlak's same approximation operators imitating the ε-elementary set based formulae and both of them belong to $\mathfrak{D}_{\mathfrak{B}}$. So, our lower and upper approximation operators are of the form $2^U \to \mathfrak{D}_{\mathfrak{B}}$. This approach corresponds to the *a priori* attitude in the sense of [74]. Here, $\mathfrak{D}_{\mathfrak{B}}$ is the family of the fundamental sets of our framework which can be seen as the *tools* which we use in order to approximate any subset of U. However, we have to emphasize that $\mathfrak{D}_{\mathfrak{B}}$ is just the set of definable sets, not the set of exact sets (in the sense of Section 5).

Our discussion will be within an overall approximation framework whose scope ranges from the weak approximation pair of maps on U [68] to the notion of Galois connection on 2^U [75,76,31]. Along this framework, the common features of both rough set theory and our approach can be treated uniformly. In addition, most notions of Pawlak's rough set theory constitute compound ones and they are split into two or more parts in our approach. This framework helps us to understand the state of their compound nature and to specify their constituents in a more general context.

Last but not least, it has been proved that the partial approximation of sets can be applied to solving practical problems [71,77,78,79].

1.4 An Overview of the Article

The article can be divided into three main parts. (1) Section 1–2 are two introductory sections; (2) Section 3–6 contain our theoretical results; (3) Section 7 presents different rea–life applications of our approach.

Section 1 is an introduction. It contains a historical outline, a philosophical background, and a brief summary of our approach.

Section 2 summarizes the basic concepts and notations used throughout the article.

Section 3 defines two general approximation frameworks, a large–scaled initial one, called the *initial approximation framework*, and a finer–scaled one, called the *general set theoretic approximation framework*. They allow us to treat the common features of classic rough set theory and its generalizations uniformly.

Section 4 is devoted to the basic concepts and properties of classic rough set theory relying on the general set theoretic approximation framework. We partly restate some well–known facts in the language of our approximation framework and provide new point–free proofs for a few of them.

Section 5 presents a special approximation framework based on the partial covering of the universe. It is fully integrated into the general set theoretic approximation framework. After some introductory remarks, Subsections 5.1 and 5.2 define the most fundamental concepts of our approach, the base system $\mathfrak{B}$ and the family of $\mathfrak{B}$-definable subsets.

Subsection 5.3 introduces a constrained version of $\mathfrak{B}$, called the single-layered base system. This allows us to prove some properties of our approximation framework which, in a sense, are similar to the properties of classic rough set theory.

Subsection 5.4 defines the so–called lower and upper 𝔅-approximations based on partial covering of the universe. First, we prove that they fit into the general set theoretic approximation framework. Lower 𝔅-approximation is always contractive, but upper 𝔅-approximation is extensive if and only if the base system 𝔅 covers the universe. We also show that the 𝔅-definable property is generally not equivalent to the equality of lower and upper 𝔅-approximations unlike Pawlakian rough set theory. The universe, the family of 𝔅-definable sets, the lower and upper 𝔅-approximations form a 𝔅-approximation space together.

Subsection 5.5 discusses the 𝔅-representations of 𝔅-definable sets. A subset D is 𝔅-representable, if there exists exactly one family of 𝔅-sets in such a way that its union equals to D. We prove that all 𝔅-definable subset of the universe are 𝔅-representable if and only if the base system 𝔅 is single-layered. We also give the explicit 𝔅-representations of 𝔅-definable subsets, among others, the lower and upper 𝔅-approximations, when the base system 𝔅 is single-layered.

Subsection 5.6 is about a special important notion of approximation spaces, the exactness. In Pawlakian approximation spaces the notions of "crisp" (i.e., the exactness) and "definable" are synonymous to each other. However, a 𝔅-definable subset is not necessarily 𝔅-crisp. Consequently, the notions of "definable" and "crisp" are not synonymous to each other in 𝔅-approximation spaces.

Last, Subsection 5.7 gives a possible interpretation of our approach.

In Section 6, we investigate what conditions have to be satisfied by the upper and lower 𝔅-approximations so that they form a Galois connection on $(2^U, \subseteq)$.

In Subsection 6.1, we prove that the upper and lower 𝔅-approximations form a Galois connection on $(2^U, \subseteq)$ if and only if the base system 𝔅 is a partition of the universe U.

Subsection 6.2 deals with partial lower and upper 𝔅-approximations. The empty set may be the lower 𝔅-approximation of certain nonempty subsets provided that all singletons are not 𝔅-definable. Excluding to allow that the empty set to be the lower 𝔅-approximation of a nonempty subset, we obtain a partial variant of the lower 𝔅-approximation. We show that under well–defined conditions there exists a unique total extension of the partial lower 𝔅-approximation which is exactly the lower 𝔅-approximation.

The empty set may be the upper 𝔅-approximation of certain nonempty subsets provided that the base system does not cover the universe. Excluding these uncommon cases we obtain a partial variant of the upper 𝔅-approximation. We prove that the partial upper 𝔅-approximation and the lower 𝔅-approximation form a partial Galois connection in the sense of Miné if and only if the 𝔅-sets are pairwise disjoint.

Section 7, to demonstrate the effectiveness of our approach, presents three real–life applications.

Subsection 7.1. The first application shows the relationship of our approach with natural computing via a biological example. In particular, we show how our approach helps us to understand some behavioral features of the natural vegetation heritage of Hungary.

Subsection 7.2. The second example models Intrusion Detection Systems (IDS) in computer security. Two separated approximation spaces are defined for anomalies and misuses at the same time. Thus, anomalies and misuses can be detected simultaneously.

Subsection 7.3. It generalizes the method presented in Subsection 7.2. In practice, two relevant groups of observed objects can be separated. A group whose elements really possess some features in question and another group whose elements do not substantially possess the same features. To model this situation, two separated approximation spaces are defined over the universe. Then, any collections of the observed objects can *simultaneously* be approximated in the two approximation spaces.

2   Basic Concepts

2.1   Basic Notations

Let U be any nonempty set. Let $\mathfrak{A} \subseteq 2^U$ be a family of sets whose elements are subsets of U.

The union and intersection of $\mathfrak{A}$ are $\bigcup \mathfrak{A} = \{x \mid \exists A \in \mathfrak{A}(x \in A)\}$ and $\bigcap \mathfrak{A} = \{x \mid \forall A \in \mathfrak{A}(x \in A)\}$, respectively.

If $\mathfrak{A}$ is empty we define $\bigcup \emptyset = \emptyset$ and $\bigcap \emptyset = U$.[2]

If $\epsilon \subseteq U \times U$ is an arbitrary binary relation on U, let $[x]_\epsilon$ denote the ϵ-related elements to x, i.e., $[x]_\epsilon = \{y \in U \mid (x, y) \in \epsilon\}$. They are called ϵ-elementary sets, and the family of $[x]_\epsilon$ is denoted by U/ϵ.

Binary relations are not necessarily symmetric, so it makes sense to consider its inverses. If ϵ^{-1} denote the inverse of a binary relation ϵ, we can also define ϵ^{-1}-elementary sets $[x]_{\epsilon^{-1}} = \{y \in U \mid (y, x) \in \epsilon\}$ and U/ϵ^{-1} as before.

Let X and Y be nonempty sets and $f : X \to Y$ be a map. If $\mathrm{dom} f = X$, f is *total*, if $\mathrm{dom} f \subsetneq X$, f is *partial*. If f is a partial map, then $\mathrm{dom} f = \emptyset$ is allowed. For the purpose of simplicity we will talk about partial maps without direct references to their partiality. However, statements with respect to partial maps always concern their restrictions to their domains.

A nonempty set P together with a partial order $\leq$ on P is called a *partial ordered set* or a *poset*, in symbol $(P, \leq)$ [80,75,81,82,31]. Any subset of a poset is in itself a poset which is partially ordered by the same (relative or induced) partial ordering relation.

The elements $x, y \in P$ are *comparable* if $x \leq y$ or $y \leq x$. Otherwise x and y are *incomparable*.

Let $S \subseteq P$. An element $m \in S$ is a *minimal* element of S, if

$$\forall x \in S \, (x \leq m \Rightarrow m = x), \text{ in other words } \nexists x \in S \, (x < m).$$

---

[2] "The equality $\bigcap \emptyset = U$ can be interpreted so that every element of U belongs to all sets in $\emptyset$ because the empty family $\emptyset$ contains no sets. The equality $\bigcup \emptyset = \emptyset$ is more obvious since $\emptyset$ has no elements." ([31], p. 402)

An element $M \in S$ is a *maximal* element of S, if

$$\forall x \in S \, (M \leq x \Rightarrow M = x), \text{ in other words } \nexists x \in S \, (M < x).$$

Note that minimal and maximal elements, provided they exist, are not necessarily comparable with all the elements of S in general.

A self-map $f : P \to P$ on $(P, \leq)$ is

- *extensive* if $x \leq f(x)$;
- *contractive* if $f(x) \leq x$;
- *idempotent* if $f(f(x)) = f(x)$;
- *normalized* if $f(m) = m$, when the minimal element $m \in P$ exists;
- *co-normalized* if $f(M) = M$, when the maximal element $M \in P$ exists.

If $(P, \leq_P)$ and $(Q, \leq_Q)$ are two posets, a map $f : P \to Q$ is *monotone* or *order–preserving* when $x \leq_P y \Rightarrow f(x) \leq_Q f(y)$, and *antitone* or *order–reversing* when $x \leq_P y \Rightarrow f(y) \leq_Q f(x)$.

A map $f : P \to Q$ is the *order isomorphism* between $(P, \leq_P)$ and $(Q, \leq_Q)$ if f is a bijection and both f and f^{-1} are monotone. In this case, it is said that P and Q are *order–isomorphic*, or *isomorphic* for short.

2.2 Galois Connections

Let $(P, \leq_P)$ and $(Q, \leq_Q)$ be two posets. Let the quadruple (P, f, g, Q) denote the pair of maps $f : P \to Q$ and $g : Q \to P$.

Definition 1. *The pair of maps (P, f, g, Q) is a (regular) Galois connection between P and Q, in notation $\mathbb{G}(P, f, g, Q)$, if*

$$\forall p \in P \, \forall q \in Q \, (f(p) \leq_Q q \Leftrightarrow p \leq_P g(q)).$$

The map f is called the lower adjoint *and g is called the* upper adjoint *of the Galois connection.*

If $P = Q$, $\mathbb{G}(P, f, g, P)$ is said a Galois connection on P.

The following theorem gives a useful characterization of Galois connections (see, e.g., [31], Lemma 79).

Proposition 1. *The pair of maps (P, f, g, Q) is a Galois connection between P and Q if and only if*

1. $p \leq_P g(f(p))$ for all $p \in P$ and $f(g(q)) \leq_Q q$ for all $q \in Q$;
2. the maps f and g are monotone.

Remark 1. Here we adopted the definition of Galois connection in which the maps are monotone. It is also called a monotone or covariant form. For more details on Galois connections, see ,e.g., [75,76,81].

Remark 2. Since Galois connections are not necessarily symmetric, the order of the maps in $\mathbb{G}(P, f, g, P)$ is important.

Finally, we will need the following notion.

Definition 2 ([83], Definition 2.2.2). *A pair of maps* (P, f, g, Q) *is the partial Galois connection between* P *and* Q, *denoted by* $\partial\mathbb{G}(P, f, g, Q)$, *if*

1. $f : P \to Q$ *is a monotone partial map,*
2. $g : Q \to P$ *is a monotone total map,*
3. $f(g(q))$ *exists for all* $q \in Q$, *and*
4. $\forall p \in P$ *and* $\forall q \in Q$ *such that* $f(p)$ *is defined,* $f(p) \leq_Q q \Leftrightarrow p \leq_P g(q)$.

Remark 3. In [83], A. Miné actually introduced the concept of $\mathcal{F}$-partial Galois connection $\partial\mathbb{G}(P, f, g, Q)$ between the concrete domain P and the abstract domain Q, where $\mathcal{F}$ is a set of concrete operators. We will apply this notion in the simplest form: $P = Q = 2^U$ and $\mathcal{F} = \emptyset$ which is allowed by Miné's definition.

3   General Approximation Frameworks

In order to be able to discuss the common features of both rough set theory and its possible generalizations uniformly, we define two general approximation frameworks, a large–scaled and a finer–scaled set theoretic one.

3.1   An Initial Approximation Framework

Let U be a nonempty set and $\langle l, u \rangle$ be an ordered pair of maps

$$l, u : 2^U \to 2^U$$

on $(2^U, \subseteq)$. Of course, the maps l and u are intended to be the *lower* and *upper approximations* of any subset $X \subseteq U$, respectively. Hence, the ordered pair $\langle l, u \rangle$ is called the *approximation pair*.

The most essential features of an approximation pair $\langle l, u \rangle$ can be summarized as follows.

0. *(Definability)* The subsets of a set are approximated by the beforehand given family of subsets of the set itself. The members of the beforehand given family of subsets are called *well defined*. More concretely, the maps l and u are of the form

$$l, u : 2^U \to \mathfrak{D}(\subseteq 2^U),$$

where $\mathfrak{D}$ is a family of well defined subsets of U.

Hereupon, the nature of an approximation pair $\langle l, u \rangle$ depends on how the lower and upper approximations are related to each other and the subset itself to be approximated.

1. *(Monotonicity)* The maps l and u are monotone with respect to the inclusion relation $\subseteq$ on 2^U.

2. (*Weak approximation property*) The approximation pair $\langle l, u \rangle$ is a *weak approximation pair* on U if

$$\forall X \in 2^U \; (l(X) \subseteq u(X)).$$

3. (*Strong approximation property*) The approximation pair $\langle l, u \rangle$ is a *strong approximation pair* on U, if each subset $X \in 2^U$ is bounded by $l(X)$ and $u(X)$:

$$\forall X \in 2^U \; (l(X) \subseteq X \subseteq u(X)).$$

4. (*Approximation hypothesis*) The pair of maps $(2^U, u, l, 2^U)$ forms a *Galois connection* on $(2^U, \subseteq)$, in notation $\mathbb{G}(2^U, u, l, 2^U)$, if

$$\forall X \in 2^U \; \forall Y \in 2^U \; (u(X) \subseteq Y \Leftrightarrow X \subseteq l(Y)).$$

Remark 4. Ad (0). It gives the most fundamental characterization of the approximation pair $\langle l, u \rangle$.

Ad (1). This property is a common and reasonable assumption.

Ad (2). The constraint $l(X) \subseteq u(X)$ seems to be the weakest condition for a sensible concept of set approximations [74,68].

Ad (3). This property is meaningful because the domain and codomain of l, u are the same [74].

Ad (4). In [84], a new hypothesis about approximation has been drawn up recently. According to this assumption, "[...] the notion of an 'approximation' may be mathematically modelled by the notion of a Galois connection" ([84], p. vii).

A finer–scaled characterization of the nature of set approximations can be obtained with *further specifications concerning the family of well defined subsets*. These additional specifications will be performed in the next subsection.

3.2   A General Set Theoretic Approximation Framework

Let U be an arbitrary nonempty set called the *universe of discourse*.

The first definition gives us the family of fundamental sets of the framework which can be considered as *primary tools*.

Definition 3. *Let $\mathfrak{B} = \{B_i \mid i \in I\} \subseteq 2^U$ be a nonempty family of nonempty subsets of U, where I denotes an index set.*

$\mathfrak{B}$ is called the base system, *its members are the* $\mathfrak{B}$-sets.

Some extensions of the base system $\mathfrak{B}$ can be defined.

Definition 4. *Let $\mathfrak{D}_\mathfrak{B} \subseteq 2^U$ be an extension of $\mathfrak{B}$ in such a way that*

1. $\mathfrak{B} \subseteq \mathfrak{D}_\mathfrak{B}$;
2. $\emptyset \in \mathfrak{D}_\mathfrak{B}$.

The members of $\mathfrak{D}_\mathfrak{B}$ are called definable, *while the members of $2^U \setminus \mathfrak{D}_\mathfrak{B}$ are* undefinable.

Any extension $\mathfrak{D}_\mathfrak{B}$ of $\mathfrak{B}$ can be seen as *derived tools*.

Example 2. The simplest extension of $\mathfrak{B}$ is $\mathfrak{D}_\mathfrak{B} = \mathfrak{B} \cup \{\emptyset\}$.

Example 3. Let $\mathfrak{D}_\mathfrak{B} \subseteq 2^U$ be an extension of $\mathfrak{B}$ in such a way that

1. $\emptyset \in \mathfrak{D}_\mathfrak{B}$;
2. for any index set $I' \subseteq I$, if $\mathfrak{B}' = \{B_i \mid i \in I'\} \subseteq \mathfrak{B}$, then $\bigcup \mathfrak{B}' \in \mathfrak{D}_\mathfrak{B}$.

Notice that $\mathfrak{B} \subseteq \mathfrak{D}_\mathfrak{B}$, and $\mathfrak{D}_\mathfrak{B}$ is closed under arbitrary unions.

If the universe U is finite, and $\mathfrak{B} = U/\varepsilon$, where U/ε is a partition of U generated by an equivalence relation ε on U, then $\mathfrak{D}_\mathfrak{B} = \sigma(U/\varepsilon)$. In this case, this extension procedure is just the scheme which is in Pawlakian rough set theory.

Example 4. Let $\mathfrak{D}_\mathfrak{B} \subseteq 2^U$ be an extension of $\mathfrak{B}$ in such a way that

1. $\emptyset \in \mathfrak{D}_\mathfrak{B}$;
2. if $B_1, B_2 \in \mathfrak{B}$ then a) $B_1 \cup B_2 \in \mathfrak{D}_\mathfrak{B}$; b) $B_1 \cap B_2 \in \mathfrak{D}_\mathfrak{B}$.

Notice that $\mathfrak{B} \subseteq \mathfrak{D}_\mathfrak{B}$, $\mathfrak{D}_\mathfrak{B}$ is closed under finite unions and intersections, and $\bigcup \mathfrak{B}, \bigcap \mathfrak{B} \in \mathfrak{D}_\mathfrak{B}$ do not hold necessarily when the cardinality of I is not finite.

Example 5. If $\sigma(\mathfrak{B})$ is the σ-algebra generated by $\mathfrak{B}$ then $\sigma(\mathfrak{B})$ is an extension of $\mathfrak{B}$ since $\emptyset \in \mathfrak{B}$ and $\mathfrak{B} \subseteq \sigma(\mathfrak{B})$.

We want to approximate any subset $S \in 2^U$ from "lower side" and "upper side"— no matter what they mean at this time. We have the only requirement at the highest level of abstraction that is to let the lower and upper approximations of subsets S be *definable*. We look at definable sets as *tools* to approximate subsets of the universe U.

If we look at the sets belonging to $\mathfrak{B}$ as primary tools, it is a highly reasonable requirement that they should *exactly* be approximated by themselves from "lower side". This property is called the *(lower) granularity* of $\mathfrak{B}$. If we gave it up, the roles of the primary tools would be depreciated. In Pawlakian rough set theory, however, not merely the granularity of U/ε but also the granularity of $\sigma(U/\varepsilon)$ fulfills. It can be proved (*cf.*, Proposition 7, Corollary 2) that if $D \in \sigma(U/\varepsilon)$, then $\underline{\varepsilon}(D) = D$ due to the particular construction of $\mathfrak{D}_{U/\varepsilon}$ and definition of $\underline{\varepsilon}$.

A lower approximation is called *standard* if not only the primary tools in $\mathfrak{B}$, but also the derived tools in $\mathfrak{D}_\mathfrak{B}$ are its fixpoints. In this article, we solely deal with standard lower approximations.

Remark 5. There is an asymmetry between lower and upper approximations, especially when they are Pawlakian type (*cf.*, Def. 9). In this case, the lower approximation of any primary tool is equal to itself, independently of whether the primary tools are pairwise disjoint or not. However, upper approximations behave in different ways. If the primary tools are pairwise disjoint, the upper approximations of primary tools are also equal to themselves, otherwise it is not reasonable to assume that.

The following definition, at the next level of abstraction, is about the *minimum requirements* of lower and upper approximations.

Definition 5. *Let $\langle l, u \rangle$ be an approximation pair $l, u : 2^U \to 2^U$ on $(2^U, \subseteq)$.*

It is said that an approximation pair $\langle l, u \rangle$ is the weak (generalized) *approximation pair on U if*

(C0) $l(2^U), u(2^U) \subseteq \mathfrak{D}_\mathfrak{B}$ (definability of l and u);[3]

(C1) l and u are monotone (monotonicity of l and u);

(C2) $u(\emptyset) = \emptyset$ (normality of u);

(C3) if $D \in \mathfrak{D}_\mathfrak{B}$, then $l(D) = D$ (granularity of $\mathfrak{D}_\mathfrak{B}$, i.e., l is standard);

(C4) if $S \in 2^U$, then $l(S) \subseteq u(S)$ (approximation property).

Informally, the intended meaning of the maps l and u, of course, is to express the lower and upper approximations of any subset of the universe U with the help of the beforehand given definable sets as *tools*.

Clearly, if $\langle l, u \rangle$ is a weak approximation pair on U, the maps l, u are total and many-to-one in general.

Proposition 2. *Let $\langle l, u \rangle$ be a weak approximation pair on U.*

1. *$l(\emptyset) = \emptyset$ (normality of l);*
2. *$\forall X \in 2^U \, (l(l(X)) = l(X))$ (idempotency of l).*
3. *$S \in \mathfrak{D}_\mathfrak{B}$ if and only if $l(S) = S$.*
4. *$u(2^U) \subseteq l(2^U) = \mathfrak{D}_\mathfrak{B}$.*

Proof.

1. By definition, $\emptyset \in \mathfrak{D}_\mathfrak{B}$ and so $l(\emptyset) = \emptyset$ by condition *(C3)*.
2. $l(X) \in \mathfrak{D}_\mathfrak{B}$ and so $l(l(X)) = l(X)$ by condition *(C3)*.
3. ($\Rightarrow$) It is just the same as the condition *(C3)*.
   ($\Leftarrow$) Since $l(S) \in \mathfrak{D}_\mathfrak{B}$ by condition *(C0)*, and so $l(S) = S \in \mathfrak{D}_\mathfrak{B}$.
4. $l(2^U) \subseteq \mathfrak{D}_\mathfrak{B}$ by condition *(C0)* and $\mathfrak{D}_\mathfrak{B} \subseteq l(2^U)$ by condition *(C3)*, thus $l(2^U) = \mathfrak{D}_\mathfrak{B}$.

   Let $S \in u(2^U) \subseteq \mathfrak{D}_\mathfrak{B}$. By the condition *(C3)*, $S = l(S) \in \mathfrak{D}_\mathfrak{B} = l(2^U)$, i.e., $u(2^U) \subseteq l(2^U)$.
   To show that the inclusion $u(2^U) \subseteq l(2^U)$ may be proper, let
   - $U = \{a, b\}$,
   - $\mathfrak{B} = \{\{a\}\}$,
   - $\mathfrak{D}_\mathfrak{B} = \{\emptyset, \{a\}, \{a, b\}\}$,
   - and $l, u : 2^U \to \mathfrak{D}_\mathfrak{B}$ be as follows:

$$X \mapsto l(X) = \begin{cases} \emptyset, & \text{if } X = \emptyset; \\ \{a\}, & \text{if } X = \{a\}; \\ \{a, b\}, & \text{otherwise.} \end{cases}$$

$$X \mapsto u(X) = \begin{cases} \emptyset, & \text{if } X = \emptyset; \\ \{a, b\}, & \text{otherwise.} \end{cases}$$

---

[3] As usual, $l(2^U), u(2^U)$ denote the ranges of the maps l and u.

Conditions *(C0)–(C4)* can easily be checked, however,

$$u(2^U) = \{\emptyset, \{a, b\}\} \subsetneq \{\emptyset, \{a\}, \{a, b\}\} = l(2^U) = \mathfrak{D}_\mathfrak{B}.$$

∎

The following example shows that for a weak approximation pair $\langle l, u \rangle$ on U each condition *(C0)–(C4)* is independent of the other four.

Example 6. Let U be a nonempty set. Let us assume that there exist $B_1, B_2(\neq \emptyset) \in 2^U$ in such a way that neither $B_1 \subseteq B_2$ nor $B_2 \subseteq B_1$ holds, and there exists a proper superset S of B_1 (i.e., $\emptyset \neq B_1 \subsetneq S \neq U$).

0. Let $\mathfrak{B} = \{B_1\}$, $\mathfrak{D}_\mathfrak{B} = \{\emptyset, B_1\}$ and l, u be the identity maps, i.e., $l, u : 2^U \to 2^U, X \mapsto X$. These l and u trivially satisfy all the five conditions except *(C0)*.

1. Let $\mathfrak{B} = \{B_1, B_2\}$, $\mathfrak{D}_\mathfrak{B} = \{\emptyset, B_1, B_2, B_1 \cup B_2\}$ and $l, u : 2^U \to \mathfrak{D}_\mathfrak{B}$ be as follows:

$$X \mapsto l(X) = \begin{cases} B_1, & \text{if } X = B_1; \\ B_2, & \text{if } X = B_2; \\ B_1 \cup B_2, & \text{if } X = B_1 \cup B_2, U; \\ \emptyset, & \text{otherwise.} \end{cases}$$

$$X \mapsto u(X) = \begin{cases} \emptyset, & \text{if } X = \emptyset; \\ B_1, & \text{if } X = B_1; \\ B_1 \cup B_2, & \text{if } X = B_1 \cup B_2, U; \\ B_2, & \text{otherwise.} \end{cases}$$

Conditions *(C0)*, *(C2)*, *(C3)* trivially hold. Let us check the condition (C4):

$$l(\emptyset) = \emptyset \subseteq \emptyset = u(\emptyset)$$
$$l(B_1) = B_1 \subseteq B_1 = u(B_1)$$
$$l(B_2) = B_2 \subseteq B_2 = u(B_2)$$
$$l(B_1 \cup B_2) = B_1 \cup B_2 \subseteq B_1 \cup B_2 = u(B_1 \cup B_2)$$
$$l(U) = B_1 \cup B_2 \subseteq B_1 \cup B_2 = u(U)$$
$$l(S) = \emptyset \subseteq B_2 = u(S)$$

and if $S'(\neq \emptyset, B_1, B_2, B_1 \cup B_2, S, U) \in 2^U$, then
$$l(S') = \emptyset \subseteq B_2 = u(S').$$

That is the condition *(C4)* also holds. However, in the case $B_1 \subsetneq S$

$$l(B_1) = B_1 \not\subseteq \emptyset = l(S)$$
$$u(B_1) = B_1 \not\subseteq B_2 = u(S).$$

Therefore, these l and u satisfy all the five conditions except *(C1)*.

2. Let $\mathfrak{B} = \{B_1\}$, $\mathfrak{D}_\mathfrak{B} = \{\emptyset, B_1\}$ and $l, u : 2^U \to \mathfrak{D}_\mathfrak{B}$ be as follows:
$$X \mapsto l(X) = \begin{cases} \emptyset, & \text{if } X = \emptyset; \\ B_1, & \text{otherwise.} \end{cases}$$

$X \mapsto u(X) = B_1$.

Conditions *(C0)*, *(C1)*, *(C3)* and *(C4)* hold, but $u(\emptyset) = B_1 \neq \emptyset$.

Therefore, these l and u satisfy all the five conditions except *(C2)*.

3. Let $\mathfrak{B} = \{B_1\}$, $\mathfrak{D}_\mathfrak{B} = \{\emptyset, B_1, B_1 \cup B_2\}$ and $l, u : 2^U \to \mathfrak{D}_\mathfrak{B}$ be as follows:

$$X \mapsto l(X) = \begin{cases} \emptyset, & \text{if } X = \emptyset; \\ B_1, & \text{if } X(\neq \emptyset) \subseteq B_2; \\ B_1 \cup B_2, & \text{otherwise.} \end{cases}$$

$$X \mapsto u(X) = \begin{cases} \emptyset, & \text{if } X = \emptyset; \\ B_1 \cup B_2, & \text{otherwise.} \end{cases}$$

Conditions *(C0)*, *(C1)*, *(C2)* trivially hold. Let us check the condition *(C4)*:

$$l(\emptyset) = \emptyset \subseteq \emptyset = u(\emptyset)$$

if $S \in 2^U$ in such a way that $S(\neq \emptyset) \subseteq B_2$, then

$$l(S) = B_1 \subseteq B_1 \cup B_1 = u(S)$$

if $S \in 2^U$ in such a way that $S(\neq \emptyset) \not\subseteq B_2$, then

$$l(S) = B_1 \cup B_2 \subseteq B_1 \cup B_2 = u(S).$$

That is the condition *(C4)* also holds. However, $l(B_1) = B_1 \cup B_2 \neq B_1$.

Therefore, these l and u satisfy all the five conditions except *(C3)*.

4. Let $\mathfrak{B} = \{B_1\}$, $\mathfrak{D}_\mathfrak{B} = \{\emptyset, B_1\}$ and $l, u : 2^U \to \mathfrak{D}_\mathfrak{B}$ be as follows:

$$X \mapsto l(X) = \begin{cases} \emptyset, & \text{if } X = \emptyset; \\ B_1, & \text{otherwise.} \end{cases}$$

$X \mapsto u(X) = \emptyset$.

These l and u trivially satisfy all the five conditions except *(C4)*.

The next definition classifies the approximation pairs as how the lower and upper approximations of a subset are related to the subset itself to be approximated.

Definition 6. *Let $\langle l, u \rangle$ be an approximation pair $l, u : 2^U \to \mathfrak{D}_\mathfrak{B}$.*

It is said that the approximation pair $\langle l, u \rangle$ is

(C5) *a lower semi–strong approximation pair on U if it is weak and if $S \in 2^U$, then $l(S) \subseteq S$ (l is contractive);*

(C6) *an upper semi–strong approximation pair on U if it is weak and if $S \in 2^U$, then $S \subseteq u(S)$ (u is extensive);*

(C7) *a strong approximation pair on U if it is lower semi–strong and upper semi–strong at the same time, i.e., each subset $S \in 2^U$ is bounded by $l(S)$ and $u(S)$: $\forall S \in 2^U$ ($l(S) \subseteq S \subseteq u(S)$).*

If U is a nonempty set, and $\mathfrak{D}_\mathfrak{B} = 2^U$, it is straightforward that the approximation pair $l, u : 2^U \to 2^U$, $X \mapsto X$ is strong.

The next example shows that there are weak approximation pairs which are neither lower semi–strong nor upper semi–strong, not lower semi–strong but upper semi–strong, lower semi–strong but not upper semi–strong.

Example 7. Let $U = \{a, b\}$ and $\mathfrak{B} = \{\{a\}\}$ be the base system.

1. Let $\mathfrak{D}_{\mathfrak{B}} = \{\emptyset, \{a\}\}$, and the maps $l, u : 2^U \to \mathfrak{D}_{\mathfrak{B}}$ be as follows:
$$X \mapsto l(X), u(X) = \begin{cases} \emptyset, & \text{if } X = \emptyset; \\ \{a\}, & \text{otherwise.} \end{cases}$$
Conditions *(C0)–(C4)* can easily be checked:
(C0) $l(2^U), u(2^U) = \{\emptyset, \{a\}\} = \mathfrak{D}_{\mathfrak{B}}$.
(C1) l is monotone:
$\quad \emptyset \subset \{a\}, \{b\}, \{a, b\} \Rightarrow l(\emptyset) = \emptyset \subset \{a\} = l(\{a\}), l(\{b\}), l(\{a, b\})$.
$\quad \{a\}, \{b\} \subset \{a, b\} \Rightarrow l(\{a\}), l(\{b\}) = \{a\} \subseteq \{a\} = l(\{a, b\})$.
$\quad$ The monotonicity of u can be proved in the same way.
(C2) $u(\emptyset) = \emptyset$.
(C3) $l(\emptyset) = \emptyset$, $l(\{a\}) = \{a\}$.
(C4) l and u are the same maps.
However, for $\{b\} \in 2^U$

$$l(\{b\}) = \{a\} \not\subseteq \{b\}; \{b\} \not\subseteq u(\{b\}) = \{a\}. \tag{1}$$

Therefore, the approximation pair $\langle l, u \rangle$ is neither lower semi–strong nor upper semi–strong.

2. Let $\mathfrak{D}_{\mathfrak{B}} = \{\emptyset, \{a\}, \{a, b\}\}$, and the maps $l, u : 2^U \to \mathfrak{D}_{\mathfrak{B}}$ be as follows:
$$X \mapsto l(X), u(X) = \begin{cases} \emptyset, & \text{if } X = \emptyset; \\ \{a\}, & \text{if } X = \{a\}; \\ \{a, b\}, & \text{otherwise.} \end{cases}$$
Conditions *(C0)–(C4)* can easily be checked:
(C0) $l(2^U), u(2^U) = \{\emptyset, \{a\}, \{a, b\}\} = \mathfrak{D}_{\mathfrak{B}}$.
(C1) l is monotone:
$\quad \emptyset \subset \{a\} \Rightarrow l(\emptyset) = \emptyset \subset \{a\} = l(\{a\})$.
$\quad \emptyset \subset \{b\}, \{a, b\} \Rightarrow l(\emptyset) = \emptyset \subset \{a, b\} = l(\{b\}), l(\{a, b\})$.
$\quad \{a\} \subset \{a, b\} \Rightarrow l(\{a\}) = \{a\} \subset \{a, b\} = l(\{a, b\})$.
$\quad \{b\} \subset \{a, b\} \Rightarrow l(\{b\}) = \{a, b\} \subseteq \{a, b\} = l(\{a, b\})$.
$\quad$ The monotonicity of u can be proved in the same way.
(C2) $u(\emptyset) = \emptyset$.
(C3) $l(\emptyset) = \emptyset$, $l(\{a\}) = \{a\}$, $l(\{a, b\}) = \{a, b\}$.
(C4) l and u are the same maps.
Let us check that u is extensive:
$\quad - \emptyset \subseteq \emptyset = u(\emptyset)$;
$\quad - \{a\} \subseteq \{a\} = u(\{a\})$;
$\quad - \{b\} \subseteq \{a, b\} = u(\{b\})$;
$\quad - \{a, b\} \subseteq \{a, b\} = u(\{a, b\})$.
However, in the case $\{b\} \in 2^U$,

$$l(\{b\}) = \{a, b\} \not\subseteq \{b\}. \tag{2}$$

Therefore, the approximation pair $\langle l, u \rangle$ is not lower semi–strong, but upper semi–strong.

3. Let $\mathfrak{D}_{\mathfrak{B}} = \{\emptyset, \{a\}, \{a, b\}\}$, and the maps $l, u : 2^U \to \mathfrak{D}_{\mathfrak{B}}$ be as follows:
$$X \mapsto l(X), u(X) = \begin{cases} \emptyset, & \text{if } X = \emptyset, \{b\}; \\ \{a\}, & \text{if } X = \{a\}; \\ \{a, b\}, & \text{otherwise.} \end{cases}$$
Conditions *(C0)–(C4)* can easily be checked:
(C0) $l(2^U), u(2^U) = \{\emptyset, \{a\}, \{a, b\}\} = \mathfrak{D}_{\mathfrak{B}}$.
(C1) l is monotone:

$\emptyset \subset \{a\} \Rightarrow l(\emptyset) = \emptyset \subset \{a\} = l(\{a\})$,
$\emptyset \subset \{b\} \Rightarrow l(\emptyset) = \emptyset \subseteq \emptyset = l(\{b\})$,
$\emptyset \subset \{a, b\} \Rightarrow l(\emptyset) = \emptyset \subset \{a, b\} = l(\{a, b\})$,
$\{a\} \subset \{a, b\} \Rightarrow l(\{a\}) = \{a\} \subset \{a, b\} = l(\{a, b\})$,
$\{b\} \subset \{a, b\} \Rightarrow l(\{b\}) = \emptyset \subset \{a, b\} = l(\{a, b\})$.

The monotonicity of u can be proved in the same way.
(C2) $u(\emptyset) = \emptyset$.
(C3) $l(\emptyset) = \emptyset$, $l(\{a\}) = \{a\}$, $l(\{a, b\}) = \{a, b\}$.
(C4) l and u are the same maps.
Let us check that l is contractive:

- $l(\emptyset) = \emptyset \subseteq \emptyset$;
- $l(\{a\}) = \{a\} \subseteq \{a\}$;
- $l(\{b\}) = \emptyset \subset \{b\}$;
- $l(\{a, b\}) = \{a, b\} \subseteq \{a, b\}$.

However, in the case $\{b\} \in 2^U$,

$$\{b\} \not\subseteq \emptyset = u(\{b\}). \tag{3}$$

Therefore, the approximation pair $\langle l, u \rangle$ is lower semi–strong, but not upper semi–strong.

Using the preliminary notations, the notion of the generalized approximation space can be defined.

Definition 7. *The ordered quadruple* $\langle U, \mathfrak{D}_{\mathfrak{B}}, l, u \rangle$ *is a weak/lower semi–strong /upper semi–strong/strong (generalized) approximation space, if the approximation pair* $\langle l, u \rangle$ *is weak/lower semi–strong/upper semi–strong/strong, respectively.*

Proposition 3. *Let* $\langle U, \mathfrak{D}_{\mathfrak{B}}, l, u \rangle$ *be a generalized approximation space.*

1. *If* $\langle U, \mathfrak{D}_{\mathfrak{B}}, l, u \rangle$ *is weak, then*
   (a) $l(U) \subseteq \bigcup \mathfrak{D}_{\mathfrak{B}}$;
   (b) $l(U) = \bigcup \mathfrak{D}_{\mathfrak{B}}$ *if and only if* $\bigcup \mathfrak{D}_{\mathfrak{B}} \in \mathfrak{D}_{\mathfrak{B}}$.
   (c) $u(U) \subseteq \bigcup \mathfrak{D}_{\mathfrak{B}}$.
2. *If* $\langle U, \mathfrak{D}_{\mathfrak{B}}, l, u \rangle$ *is upper semi–strong, then* $u(U) = \bigcup \mathfrak{D}_{\mathfrak{B}} = U$.

Proof.

1. (a) By definability of l, $l(U) \in \mathfrak{D}_{\mathfrak{B}}$ and so $l(U) \subseteq \bigcup \mathfrak{D}_{\mathfrak{B}}$.

(b) ($\Rightarrow$) By definability of l, $l(U) = \bigcup \mathfrak{D}_\mathfrak{B} \in \mathfrak{D}_\mathfrak{B}$.

($\Leftarrow$) Let us assume that $\bigcup \mathfrak{D}_\mathfrak{B} \in \mathfrak{D}_\mathfrak{B}$. Since $\bigcup \mathfrak{D}_\mathfrak{B} \subseteq U$, then by condition *(C3)* and monotonicity of l, $l(\bigcup \mathfrak{D}_\mathfrak{B}) = \bigcup \mathfrak{D}_\mathfrak{B} \subseteq l(U)$. Comparing it with 1/(a), we obtain $l(U) = \bigcup \mathfrak{D}_\mathfrak{B}$.

(c) By definability of u, $u(U) \in \mathfrak{D}_\mathfrak{B}$ and so $u(U) \subseteq \bigcup \mathfrak{D}_\mathfrak{B}$.

2. By point 1/(c), $u(U) \subseteq \bigcup \mathfrak{D}_\mathfrak{B}$. On the other hand, since u is extensive and monotone, $\bigcup \mathfrak{D}_\mathfrak{B} \subseteq U$ implies $\bigcup \mathfrak{D}_\mathfrak{B} \subseteq u(\bigcup \mathfrak{D}_\mathfrak{B}) \subseteq u(U)$. Consequently, $u(U) = \bigcup \mathfrak{D}_\mathfrak{B}$.

Clearly, $u(U) \subseteq U$. Since u is extensive, thus $U \subseteq u(U)$ also holds. Therefore, $u(U) = U$. ∎

In generalized approximation spaces the notion of *well approximated sets* can be introduced. These sets are called crisp.

Definition 8. *Let $\langle U, \mathfrak{D}_\mathfrak{B}, l, u \rangle$ be a weak generalized approximation space and $S \in 2^U$.*
The subset S is crisp, if $l(S) = u(S)$.

Proposition 4. *Let $\langle U, \mathfrak{D}_\mathfrak{B}, l, u \rangle$ be a strong generalized approximation space. If $S \in 2^U$ is crisp, then S is definable.*

Proof. $\langle U, \mathfrak{D}_\mathfrak{B}, l, u \rangle$ is strong, thus $l(S) \subseteq S \subseteq u(S)$. Since S is crisp, therefore $l(S) = S = u(S)$, and so $S \in \mathfrak{D}_\mathfrak{B}$ by Proposition 2, point 3. ∎

In general, the crisp property of a set does not imply its definability in not strong generalized approximation spaces. One can check that in all three cases of Example 7, the set $\{b\}$ is crisp (because of l and u are the same maps, and so $l(\{b\}) = u(\{b\})$ trivially holds), but $\{b\}$ is not definable in any cases (*i.e.*, $\{b\} \notin \mathfrak{D}_\mathfrak{B}$). Of course, its lower and upper approximations are definable (*i.e.*, $l(\{b\}), u(\{b\}) \in \mathfrak{D}_\mathfrak{B}$).

4   Fundamentals of Rough Set Theory

The basic concepts and properties of rough set theory can be found, *e.g.*, in [85,3,4]. Here we will cite only notions and statements which are required in our subsequent work. Moreover, we partly restate these well–known facts in the language of the set theoretic approximation framework. On the other hand, we provide new point-free proofs for a few of them (see, especially, Section 4.2).

4.1   Basic Notions

Let U be a nonempty set and ε be an equivalence relation on U. In Pawlakian rough set theory the base system is the partition U/ε. Its extension $\mathfrak{D}_{U/\varepsilon}$ contains U/ε, the empty set and closed under arbitrary unions. The members of $\mathfrak{D}_{U/\varepsilon}$ are called ε-definable, while the members of $2^U \setminus \mathfrak{D}_{U/\varepsilon}$ are called ε-undefinable.

Remark 6. By special structure of U/ε, $\mathfrak{D}_{U/\varepsilon}$ is nonempty, closed under arbitrary unions, intersections and complementations. In other words, $(U, \mathfrak{D}_{U/\varepsilon})$ is an Alexandrov topological space with the basis U/ε.

Having given the definable sets, the Pawlakian approximation pair $\langle \underline{\varepsilon}, \overline{\varepsilon} \rangle$ can be defined in three equivalent forms [8,9,10] as follows.

Definition 9. *Let $\langle \underline{\varepsilon}, \overline{\varepsilon} \rangle$ be an approximation pair $\underline{\varepsilon}, \overline{\varepsilon} : 2^U \to 2^U$ on $(2^U, \subseteq)$.*
$\langle \underline{\varepsilon}, \overline{\varepsilon} \rangle$ *is a* Pawlakian approximation pair on U*, if*
the lower ε-approximation *of a subset $X \in 2^U$ is*

$$\underline{\varepsilon}(X) = \{x \in U \mid [x]_\varepsilon \subseteq X\} \tag{4a}$$

$$= \bigcup\{Y \mid Y \in U/\varepsilon, Y \subseteq X\} \tag{4b}$$

$$= \bigcup\{D \mid D \in \mathfrak{D}_{U/\varepsilon}, D \subseteq X\}, \tag{4c}$$

and the upper ε-approximation *of a subset $X \in 2^U$ is*

$$\overline{\varepsilon}(X) = \{x \in U \mid [x]_\varepsilon \cap X \neq \emptyset\} \tag{5a}$$

$$= \bigcup\{Y \mid Y \in U/\varepsilon Y \cap X \neq \emptyset\} \tag{5b}$$

$$= \bigcap\{D \mid D \in \mathfrak{D}_{U/\varepsilon}, X \subseteq D\}. \tag{5c}$$

Remark 7. The above equations respectively emphasize the *local* (Eqs. (4a) and (5a)), *global* (Eqs. (4b) and (5b)) and *topological* (Eqs. (4c) and (5c)) nature of Pawlakian approximations. From another point of view, the local approach is *point-wise* and the two latter ones are *point-free* in nature.

Our approach relies on the generalization of formulae Eqs. (4b), (5b) when the base sets are not pairwise disjoint and they do not necessarily cover U.

Proposition 5. *Let $\langle \underline{\varepsilon}, \overline{\varepsilon} \rangle$ be a Pawlakian approximation pair on U. Then*

1. the formulae 4b and 4c are equivalent, i.e.,

$$\bigcup\{Y \mid Y \in U/\varepsilon, Y \subseteq X\} = \bigcup\{D \mid D \in \mathfrak{D}_{U/\varepsilon}, D \subseteq X\};$$

2. the formulae 5b and 5c are equivalent, i.e.,

$$\bigcup\{Y \mid Y \in U/\varepsilon, Y \cap X \neq \emptyset\} = \bigcap\{D \mid D \in \mathfrak{D}_{U/\varepsilon}, X \subseteq D\}.$$

Proof. 1. It follows from the fact that every $D(\subseteq X) \in \mathfrak{D}_{U/\varepsilon}$ is of the form

$$D = \bigcup\{Y \mid Y \in U/\varepsilon, Y \subseteq X\}.$$

2. Since $Y \cap D = Y$ or $\emptyset$ for any $Y \in U/\varepsilon$ and $D \in \mathfrak{D}_{U/\varepsilon}$, thus

$$\bigcup\{Y \mid Y \in U/\varepsilon, Y \cap X \neq \emptyset\} \subseteq \bigcup\{Y \mid Y \in U/\varepsilon, Y \cap D \neq \emptyset\} = D$$

for any $D \in \mathfrak{D}_{U/\varepsilon}$ where $X \subseteq D$.

In other words,

$$\bigcup\{Y \mid Y \in U/\varepsilon, Y \cap X \neq \emptyset\} \subseteq D$$

for all $D \in \{D \mid D \in \mathfrak{D}_{U/\varepsilon}, X \subseteq D\}$.
In addition, by definition, $X \subseteq \bigcup\{Y \mid Y \in U/\varepsilon, Y \cap X \neq \emptyset\} \in \mathfrak{D}_{U/\varepsilon}$, i.e.,

$$\bigcup\{Y \mid Y \in U/\varepsilon, Y \cap X \neq \emptyset\} \in \{D \mid D \in \mathfrak{D}_{U/\varepsilon}, X \subseteq D\}.$$

∎

Remark 8. In Eq. (5c), contrary to expectations, the formula

$$\bigcap\{D \mid D \in \mathfrak{D}_{U/\varepsilon}, X \subseteq D\}$$

cannot be replaced with the formula $\bigcup\{D \mid D \in \mathfrak{D}_{U/\varepsilon}, X \cap D \neq \emptyset\}$ as the next example shows.

Let $U = \{x_1, x_2\}, U/\varepsilon = \{\{x_1\}, \{x_2\}\}, \mathfrak{D}_{U/\varepsilon} = \{\emptyset, \{x_1\}, \{x_2\}, \{x_1, x_2\}\}$. Then
$\bar{\varepsilon}(\{x_1\}) = \bigcup\{\{x_1\}\} = \{x_1\}$ \qquad (according to Eq. (5b)),
$\bar{\varepsilon}(\{x_1\}) = \bigcap\{\{x_1\}, \{x_1, x_2\}\} = \{x_1\}$ (according to Eq. (5c)),
however, according to the formula $\bigcup\{D \mid D \in \mathfrak{D}_{U/\varepsilon}, X \cap D \neq \emptyset\}$ we do not obtain the correct result: $\bar{\varepsilon}(\{x_1\}) \neq \bigcup\{\{x_1\}, \{x_1, x_2\}\} = \{x_1, x_2\}$.

Proposition 6. *Let $\langle \underline{\varepsilon}, \bar{\varepsilon} \rangle$ be a Pawlakian approximation pair on U. Then*

1. $\underline{\varepsilon}(2^U), \bar{\varepsilon}(2^U) \subseteq \mathfrak{D}_{U/\varepsilon}$ *(definability of $\underline{\varepsilon}$ and $\bar{\varepsilon}$), and $\underline{\varepsilon}, \bar{\varepsilon}$ are total and generally many-to-one.*
2. *If $X \subseteq Y$, then $\underline{\varepsilon}(X) \subseteq \underline{\varepsilon}(Y)$ and $\bar{\varepsilon}(X) \subseteq \bar{\varepsilon}(Y)$ ($\underline{\varepsilon}, \bar{\varepsilon}$ are monotone).*
3. $\underline{\varepsilon}(\emptyset) = \bar{\varepsilon}(\emptyset) = \emptyset$ *($\underline{\varepsilon}, \bar{\varepsilon}$ are normalized).*
4. $\forall X \in 2^U$ *($\underline{\varepsilon}(X) \subseteq X \subseteq \bar{\varepsilon}(X)$)   ($\underline{\varepsilon}$ is contractive, $\bar{\varepsilon}$ is extensive).*

Proof. Statement 1 is straightforward by Def. 9. Statements 2, 3, 4 are in [4], Proposition 2.2 (5–6), Proposition 2.2 (2), Proposition 2.2 (1), respectively. ∎

According to Proposition 6 (1–4), rough set theory fulfills the conditions *(C0)*, *(C1), (C2), (C4)* and *(C7)* concerning an approximation pair within the general set theoretic approximation framework.

Proposition 7. *([4], Proposition 2.1 (a)) Let $\langle \underline{\varepsilon}, \bar{\varepsilon} \rangle$ be a Pawlakian approximation pair on U. Then $X \in \mathfrak{D}_{U/\varepsilon}$ if and only if $\underline{\varepsilon}(X) = \bar{\varepsilon}(X)$.*

Corollary 1. *If $D \in \mathfrak{D}_{U/\varepsilon}$, then $\underline{\varepsilon}(D) = D$.*

Proof. It immediately follows from Proposition 6 (4) and Proposition 7.

∎

According to Corollary 1, rough set theory fulfills the condition *(C3)*, as well.
Summing up the above results, in the language of the general set theoretic approximation framework, a Pawlakian approximation pair $\langle \underline{\varepsilon}, \bar{\varepsilon} \rangle$ is a strong one. Consequently, the quadruple $\langle U, \mathfrak{D}_{U/\varepsilon}, \underline{\varepsilon}, \bar{\varepsilon} \rangle$ forms a strong approximation space. It is also called *Pawlakian approximation space*.

Remark 9. Note that the idea of approximation *space* is a bit older than Pawlak's initial works. For the evolutionary survey of approximation spaces, see [6].

The next statement is a characteristic feature of rough set theory.

Corollary 2. $\underline{\varepsilon}(X) = X$ *if and only if* $X = \overline{\varepsilon}(X)$.

Proof. Since $\underline{\varepsilon}(X) \in \mathfrak{D}_{U/\varepsilon}$ ($\overline{\varepsilon}(X) \in \mathfrak{D}_{U/\varepsilon}$), then $X = \underline{\varepsilon}(X) \in \mathfrak{D}_{U/\varepsilon}$ ($X = \overline{\varepsilon}(X) \in \mathfrak{D}_{U/\varepsilon}$) by Proposition 6 (4) and Proposition 7, and so $X = \underline{\varepsilon}(X) = \overline{\varepsilon}(X)$ ($X = \overline{\varepsilon}(X) = \underline{\varepsilon}(X)$) by Proposition 7. ∎

The next properties of $\underline{\varepsilon}$ and $\overline{\varepsilon}$ partly follows from Proposition 2. Of course, they can easily be proved by Def. 9 directly.

Proposition 8. *Let* $\langle U, \mathfrak{D}_{U/\varepsilon}, \underline{\varepsilon}, \overline{\varepsilon} \rangle$ *be a Pawlakian approximation space.*

1. $\underline{\varepsilon}(U) = \overline{\varepsilon}(U) = U$ ($\underline{\varepsilon}$, $\overline{\varepsilon}$ *are co-normalized*).
2. $\forall X \in 2^U$ ($\underline{\varepsilon}(\underline{\varepsilon}(X)) = \underline{\varepsilon}(X) \wedge \overline{\varepsilon}(\overline{\varepsilon}(X)) = \overline{\varepsilon}(X)$) ($\underline{\varepsilon}$, $\overline{\varepsilon}$ *are idempotent*).
3. $\underline{\varepsilon}(2^U) = \overline{\varepsilon}(2^U) = \mathfrak{D}_{U/\varepsilon}$.

Proof. Statements 1 and 2 are in [4], Proposition 2.2 (2) and Proposition 2.2 (11–12), respectively. Statement 3 is an immediate consequence of Proposition 7. ∎

Definition 10. *Let* $\langle U, \mathfrak{D}_{U/\varepsilon}, \underline{\varepsilon}, \overline{\varepsilon} \rangle$ *be a Pawlakian approximation space and* $X \subseteq U$. *The* ε-*boundary of* X *is*

$$B_\varepsilon(X) = \overline{\varepsilon}(X) \setminus \underline{\varepsilon}(X).$$

X *is* ε-crisp, *if* $B_\varepsilon(X) = \emptyset$, *otherwise* X *is* ε-rough.

Proposition 9 ([86], Proposition 4.14). *Let* $\langle U, \mathfrak{D}_{U/\varepsilon}, \underline{\varepsilon}, \overline{\varepsilon} \rangle$ *be a Pawlakian approximation space and* $X \subseteq U$.

1. X *is* ε-crisp *if and only if* X *is* ε-definable.
2. X *is* ε-rough *if and only if* X *is* ε-undefinable.

Proof.

1. X is ε-crisp $\Leftrightarrow B_\varepsilon(X) = \overline{\varepsilon}(X) \setminus \underline{\varepsilon}(X) = \emptyset \Leftrightarrow \overline{\varepsilon}(X) \subseteq \underline{\varepsilon}(X)$. However, $\underline{\varepsilon}(X) \subseteq \overline{\varepsilon}(X)$ holds for all $X \in 2^U$ by Proposition 6 (4), and so X is ε-crisp $\Leftrightarrow \underline{\varepsilon}(X) = \overline{\varepsilon}(X)$. According to Proposition 7, $\underline{\varepsilon}(X) = \overline{\varepsilon}(X) \Leftrightarrow X \in \mathfrak{D}_{U/\varepsilon}$.
2. It is the contrapositive version of (1). ∎

As a consequence of Proposition 9, in Pawlakian approximation spaces the notions "ε-crisp" and "ε-definable" are synonymous to each other, and so are "ε-rough" and "ε-undefinable". However, the notions "ε-crisp" and "ε-definable" are two *different* notions, they are inherently one and the same only in Pawlakian approximation spaces. As we will see, in partial approximation of sets this compound notion splits into two parts.

4.2 Granularity Aspects of Rough Set Theory

The following statement is elementary, however, in the context of Pawlakian rough set theory it is an important fact. For the sake of simple reference, it is formulated in a lemma. It follows from just the fact that the partition U/ε consists of nonempty pairwise disjoint subsets of U.

Lemma 1. $\forall \mathfrak{X} \in 2^{U/\varepsilon} \ \forall X \in U/\varepsilon \ (X \subseteq \bigcup \mathfrak{X} \Leftrightarrow X \in \mathfrak{X}).$[4]

Proposition 10 ([70], Theorem 8). *Let $\langle U, \mathfrak{D}_{U/\varepsilon}, \underline{\varepsilon}, \overline{\varepsilon} \rangle$ be a Pawlakian approximation space.*

Then the posets $(2^{U/\varepsilon}, \subseteq)$ and $(\mathfrak{D}_{U/\varepsilon}, \subseteq)$ are order isomorphic via the map $i_\varepsilon : 2^{U/\varepsilon} \to \mathfrak{D}_{U/\varepsilon}, \mathfrak{X} \mapsto \bigcup \mathfrak{X}.$

Proof. We will show that the map i_ε is a bijection and both i_ε and i_ε^{-1} are monotone.

Let $\mathfrak{X}_1, \mathfrak{X}_2 \in 2^{U/\varepsilon}$ be in such a way that $\bigcup \mathfrak{X}_1 = \bigcup \mathfrak{X}_2 \in \mathfrak{D}_{U/\varepsilon}$. By Lemma 1,

$$\forall X \in U/\varepsilon \ (X \in \mathfrak{X}_1 \Leftrightarrow X \subseteq \bigcup \mathfrak{X}_1 = \bigcup \mathfrak{X}_2 \Leftrightarrow X \in \mathfrak{X}_2),$$

i.e., $\mathfrak{X}_1 = \mathfrak{X}_2$, thus i_ε is injective. By definition of $\mathfrak{D}_{U/\varepsilon}$, i_ε is surjective. Consequently, i_ε is a bijection.

Clearly, the map i_ε is monotone, since $\mathfrak{X}_1, \mathfrak{X}_2 \in 2^{U/\varepsilon}$, $\mathfrak{X}_1 \subseteq \mathfrak{X}_2$ immediately implies $\bigcup \mathfrak{X}_1 \subseteq \bigcup \mathfrak{X}_2$.

Now, let $D_1, D_2 \in \mathfrak{D}_{U/\varepsilon}$ be in such a way that $D_1 \subseteq D_2$. Since i_ε is a bijection, there exist unique $i_\varepsilon^{-1}(D_1) = \mathfrak{X}_1 \in 2^{U/\varepsilon}$ and $i_\varepsilon^{-1}(D_2) = \mathfrak{X}_2 \in 2^{U/\varepsilon}$ such that $D_1 = \bigcup \mathfrak{X}_1, D_2 = \bigcup \mathfrak{X}_2$. By Lemma 1, if $X \in \mathfrak{X}_1$, then

$$X \subseteq \bigcup \mathfrak{X}_1 = D_1 \subseteq D_2 = \bigcup \mathfrak{X}_2 \Leftrightarrow X \in \mathfrak{X}_2,$$

i.e., $\mathfrak{X}_1 \subseteq \mathfrak{X}_2$, and so i_ε^{-1} is also monotone. ∎

Corollary 3 ([86], Corollary 3.5). *Any ε-definable subset D of U can be written uniquely in the following form:*

$$D = \bigcup \mathfrak{X}, \text{ where } \mathfrak{X} = \{X \mid X \in U/\varepsilon, X \subseteq D\} \in 2^{U/\varepsilon},$$

that is, there is no other $\mathfrak{X}' \in 2^{U/\varepsilon}$ satisfying $D = \bigcup \mathfrak{X}'$.

Proof. Since $D \in \mathfrak{D}_{U/\varepsilon}$, thus $D = \bigcup \{X \mid X \in U/\varepsilon, X \subseteq D\}$ immediately holds. However, i_ε is a bijection, and so $i_\varepsilon^{-1}(D)$ always exists and

$$i_\varepsilon^{-1}(D) = \{X \mid X \in U/\varepsilon, X \subseteq D\} \in 2^{U/\varepsilon}$$

is unique. ∎

---

[4] $2^{U/\varepsilon}$ denotes the power set of U/ε.

Proposition 11 ([86], Proposition 3.7). *Let $\langle U, \mathfrak{D}_{U/\varepsilon}, \underline{\varepsilon}, \overline{\varepsilon} \rangle$ be a Pawlakian approximation space and X be a subset of U.*

Then the sets $\underline{\varepsilon}(X)$, $\overline{\varepsilon}(X)$ can be written uniquely in the following forms:

$$\underline{\varepsilon}(X) = \bigcup \underline{\mathfrak{X}}, \text{ where } \underline{\mathfrak{X}} = \{Y \mid Y \in U/\varepsilon, Y \subseteq X\} \in 2^{U/\varepsilon},$$

$$\overline{\varepsilon}(X) = \bigcup \overline{\mathfrak{X}}, \text{ where } \overline{\mathfrak{X}} = \{Y \mid Y \in U/\varepsilon, Y \cap X \neq \emptyset\} \in 2^{U/\varepsilon},$$

that is, there are no other $\underline{\mathfrak{X}}'$, $\overline{\mathfrak{X}}' \in 2^{U/\varepsilon}$ satisfying $\underline{\varepsilon}(X) = \bigcup \underline{\mathfrak{X}}'$ and $\overline{\varepsilon}(X) = \bigcup \overline{\mathfrak{X}}'$.

Proof. According to Eqs. (4b) and (5b) in Def. 9, we only have to prove the uniqueness.

$\underline{\varepsilon}(X)$, $\overline{\varepsilon}(X) \in \mathfrak{D}_{U/\varepsilon}$, and so, by Proposition 10, $i_\varepsilon^{-1}(\underline{\varepsilon}(X))$ and $i_\varepsilon^{-1}(\overline{\varepsilon}(X))$ are unique and, by Lemma 1, we get

$$\begin{aligned}
i_\varepsilon^{-1}(\underline{\varepsilon}(X)) &= \{Y \mid Y \in U/\varepsilon, Y \subseteq \underline{\varepsilon}(X)\} \\
&= \{Y \mid Y \in U/\varepsilon, Y \subseteq \bigcup \{Y' \mid Y' \in U/\varepsilon, Y' \subseteq X\}\} \\
&= \{Y \mid Y \in U/\varepsilon, Y \in \{Y' \mid Y' \in U/\varepsilon, Y' \subseteq X\}\} \\
&= \{Y \mid Y \in U/\varepsilon, Y \subseteq X\} = \underline{\mathfrak{X}}. \\
i_\varepsilon^{-1}(\overline{\varepsilon}(X)) &= \{Y \mid Y \in U/\varepsilon, Y \subseteq \overline{\varepsilon}(X)\} \\
&= \{Y \mid Y \in U/\varepsilon, Y \subseteq \bigcup \{Y' \mid Y' \in U/\varepsilon, Y' \cap X \neq \emptyset\}\} \\
&= \{Y \mid Y \in U/\varepsilon, Y \in \bigcup \{Y' \mid Y' \in U/\varepsilon, Y' \cap X \neq \emptyset\}\} \\
&= \{Y \mid Y \in U/\varepsilon, Y \cap X \neq \emptyset\} = \overline{\mathfrak{X}}.
\end{aligned}$$

∎

4.3 Galois Connection of Upper and Lower Approximations

It is a well known fact that $\overline{\varepsilon}$ and $\underline{\varepsilon}$ form a $\mathbb{G}(2^U, \overline{\varepsilon}, \underline{\varepsilon}, 2^U)$ Galois connection. Now, let us investigate this connection in a wider context.

Lower and upper ε-approximations can be generalized via their element based definitions (4a) and (5a) relying on *arbitrary* binary relations ϵ on U [31].

Definition 11. *Let ϵ be an arbitrary binary relation on U and $X \in 2^U$.*
The lower ϵ-approximation *of X is*

$$\underline{\epsilon}(X) = \{x \in U \mid [x]_\epsilon \subseteq X\},$$

and the upper ϵ-approximation *of X is*
$$\overline{\epsilon}(X) = \{x \in U \mid [x]_\epsilon \cap X \neq \emptyset\}.$$

If ϵ^{-1} denotes the inverse relation of ϵ, in the same manner one can also define the lower and upper ϵ^{-1}-approximations of X.

Proposition 12. ([85], Proposition 134) *Let ϵ be an arbitrary binary relation on U. Then $\mathbb{G}(2^U, \overline{\epsilon}, \underline{\epsilon}^{-1}, 2^U)$ and $\mathbb{G}(2^U, \overline{\epsilon^{-1}}, \underline{\epsilon}, 2^U)$ are Galois connections on $(2^U, \subseteq)$.*

The next corollary is an immediate consequence of Proposition 12.

Corollary 4. *Let ϵ be an arbitrary binary relation on U.*
The pair $(\bar{\epsilon}, \underline{\epsilon})$ is a Galois connection on $(2^U, \subseteq)$ if and only if ϵ is symmetric.
In particular, if ε is an equivalence relation on U, $\mathbb{G}(2^U, \bar{\varepsilon}, \underline{\varepsilon}, 2^U)$ is a Galois connection on 2^U.

The next examples show that even if the relation ϵ is symmetric, it is not sufficient that the upper and lower ϵ-approximations relying on point-free definitions form Galois connection.

Example 8 ([77], Example 3.10). Let $U = \{x_1, x_2, x_3\}$ and

$$\epsilon = \{(x_1, x_1), (x_1, x_2), (x_2, x_1), (x_2, x_3), (x_3, x_2)\} \subset U \times U$$

be a symmetric binary relation on U.
    We define the straightforward generalizations of U/ε and $\mathfrak{D}_{U/\varepsilon}$ as follows.

$$[x_1]_\epsilon = \{u \in U \mid (x_1, u) \in \epsilon\} = \{x_1, x_2\},$$
$$[x_2]_\epsilon = \{u \in U \mid (x_2, u) \in \epsilon\} = \{x_1, x_3\},$$
$$[x_3]_\epsilon = \{u \in U \mid (x_3, u) \in \epsilon\} = \{x_2\},$$
$$U/\epsilon = \{[x_1]_\epsilon, [x_2]_\epsilon, [x_3]_\epsilon\} = \{\{x_1, x_2\}, \{x_1, x_3\}, \{x_2\}\},$$
$$\mathfrak{D}_{U/\epsilon} = \{\emptyset, [x_1]_\epsilon, [x_2]_\epsilon, [x_3]_\epsilon, [x_1]_\epsilon \cup [x_2]_\epsilon, [x_1]_\epsilon \cup [x_3]_\epsilon, [x_2]_\epsilon$$
$$\cup[x_3]_\epsilon, [x_1]_\epsilon \cup [x_2]_\epsilon \cup [x_3]_\epsilon\}$$
$$= \{\emptyset, \{x_1, x_2\}, \{x_1, x_3\}, \{x_2\}, \underbrace{\{x_1, x_2\} \cup \{x_1, x_3\}}_{\{x_1, x_2, x_3\}}, \underbrace{\{x_1, x_2\} \cup \{x_2\}}_{\{x_1, x_2\}},$$
$$\underbrace{\{x_1, x_3\} \cup \{x_2\}}_{\{x_1, x_2, x_3\}}, \underbrace{\{x_1, x_2\} \cup \{x_1, x_3\} \cup \{x_2\}}_{\{x_1, x_2, x_3\}}\}$$
$$= \{\emptyset, \{x_1, x_2\}, \{x_1, x_3\}, \{x_2\}, \{x_1, x_2, x_3\}\}.$$

Note that $\mathfrak{D}_{U/\epsilon} \subseteq 2^U$ is a subsystem of 2^U which contains the empty set and closed under unions but is not a σ-algebra [10]. For instance, $\{x_1, x_2\} \cap \{x_1, x_3\} = \{x_1\} \notin \mathfrak{D}_{U/\epsilon}$.

Case 1. Elementary set based definitions relying on U/ε.
    Let us define the lower and upper ϵ-approximations taking the pattern by Eqs. (4b) and (5b), respectively:

$$\underline{\epsilon}_e : 2^U \to \mathfrak{D}_{U/\epsilon}, \quad X \mapsto \bigcup \{Y \mid Y \in U/\epsilon, Y \subseteq X\}.$$
$$\bar{\epsilon}_e : 2^U \to \mathfrak{D}_{U/\epsilon}, \quad X \mapsto \bigcup \{Y \mid Y \in U/\epsilon, Y \cap X \neq \emptyset\}.$$

Clearly, the maps $\underline{\epsilon}_e$ and $\bar{\epsilon}_e$ are monotone.

For instance, for the set $\{x_2\} \in 2^U$:

$$\underline{\epsilon}_e(\{x_2\}) = \bigcup\{Y \mid Y \in U/\epsilon, Y \subseteq \{x_2\}\} = \bigcup\{\{x_2\}\} = \{x_2\}.$$
$$\overline{\epsilon}_e(\{x_2\}) = \bigcup\{Y \mid Y \in U/\epsilon, Y \cap \{x_2\} \neq \emptyset\}$$
$$= \bigcup\{\{x_1, x_2\}, \{x_2\}\} = \{x_1, x_2\}.$$

Do the relations $\{x_2\} \subseteq \underline{\epsilon}_e(\overline{\epsilon}_e(\{x_2\}))$ and/or $\overline{\epsilon}_e(\underline{\epsilon}_e(\{x_2\})) \subseteq \{x_2\}$ hold?

$$\underline{\epsilon}_e(\overline{\epsilon}_e(\{x_2\})) = \underline{\epsilon}_e(\{x_1, x_2\}) = \bigcup\{Y \mid Y \in U/\epsilon, Y \subseteq \{x_1, x_2\}\}$$
$$= \bigcup\{\{x_1, x_2\}, \{x_2\}\} = \{x_1, x_2\} \supseteq \{x_2\}.$$
$$\overline{\epsilon}_e(\underline{\epsilon}_e(\{x_2\})) = \overline{\epsilon}_e(\{x_2\}) = \{x_1, x_2\} \not\subseteq \{x_2\}.$$

That is, by Proposition 1, $(2^U, \overline{\epsilon}_e, \underline{\epsilon}_e, 2^U)$ does not form Galois connection.

Case 2. *Subsystem based definitions relying on* $\mathfrak{D}_{U/\epsilon}$.

Let us define the lower and upper ϵ-approximations after the pattern of the Eqs. (4c) and (5c), respectively (note that, $\mathfrak{D}_{U/\epsilon}$ is closed under unions, but not closed under intersections):

$$\underline{\epsilon}_s : 2^U \to \mathfrak{D}_{U/\epsilon}, \quad X \mapsto \bigcup\{Y \mid Y \in \mathfrak{D}_{U/\epsilon}, Y \subseteq X\},$$
$$\overline{\epsilon}_s : 2^U \to 2^U, \quad X \mapsto \bigcap\{Y \mid Y \in \mathfrak{D}_{U/\epsilon}, X \subseteq Y\}.$$

Clearly, the map $\underline{\epsilon}_s$ is monotone.

The map $\overline{\epsilon}_s$ is also monotone. Namely, let $X_1 \subseteq X_2$ be subsets of U.

- If $\{Y \mid Y \in \mathfrak{D}_{U/\epsilon}, X_1 \subseteq Y\} = \emptyset$, then $\{Y \mid Y \in \mathfrak{D}_{U/\epsilon}, X_2 \subseteq Y\} = \emptyset$ also holds, and so $\overline{\epsilon}_s(X_1) = \overline{\epsilon}_s(X_2) = \bigcap \emptyset = U$.
- If $\{Y \mid Y \in \mathfrak{D}_{U/\epsilon}, X_1 \subseteq Y\} \neq \emptyset$ and $\{Y \mid Y \in \mathfrak{D}_{U/\epsilon}, X_2 \subseteq Y\} = \emptyset$, then $\overline{\epsilon}_s(X_1) \subseteq \overline{\epsilon}_s(X_2) = \bigcap \emptyset = U$.
- If $\{Y \mid Y \in \mathfrak{D}_{U/\epsilon}, X_1 \subseteq Y\}, \{Y \mid Y \in \mathfrak{D}_{U/\epsilon}, X_2 \subseteq Y\} \neq \emptyset$, then

$$\{Y \mid Y \in \mathfrak{D}_{U/\epsilon}, X_2 \subseteq Y\} \subseteq \{Y \mid Y \in \mathfrak{D}_{U/\epsilon}, X_1 \subseteq Y\},$$

and so

$$\overline{\epsilon}_s(X_1) = \bigcap\{Y \mid Y \in \mathfrak{D}_{U/\epsilon}, X_1 \subseteq Y\}$$
$$\subseteq \bigcap\{Y \mid Y \in \mathfrak{D}_{U/\epsilon}, X_2 \subseteq Y\} = \overline{\epsilon}_s(X_2).$$

For instance, for the set $\{x_1\}$:

$$\underline{\epsilon}_s(\{x_1\}) = \bigcup\{Y \mid Y \in \mathfrak{D}_{U/\epsilon}, Y \subseteq \{x_1\}\} = \bigcup\{\emptyset\} = \emptyset,$$
$$\overline{\epsilon}_s(\{x_1\}) = \bigcap\{Y \mid Y \in \mathfrak{D}_{U/\epsilon}, \{x_1\} \subseteq Y\}$$
$$= \bigcap\{\{x_1, x_2\}, \{x_1, x_3\}, \{x_1, x_2, x_3\}\} = \{x_1\}.$$

Do the relations $\{x_1\} \subseteq \underline{\epsilon}_s(\overline{\epsilon}_s(\{x_1\}))$ and/or $\overline{\epsilon}_s(\underline{\epsilon}_s(\{x_1\})) \subseteq \{x_1\}$ hold?

$$\underline{\epsilon}_s(\overline{\epsilon}_s(\{x_1\})) = \underline{\epsilon}_s(\{x_1\}) = \emptyset \not\supseteq \{x_1\},$$

$$\overline{\epsilon}_s(\underline{\epsilon}_s(\{x_1\})) = \overline{\epsilon}_s(\emptyset) = \bigcap\{Y \mid Y \in \mathfrak{D}_{U/\epsilon}, \emptyset \subseteq Y\} = \bigcap \mathfrak{D}_{U/\epsilon} = \emptyset \subsetneqq \{x_1\}.$$

That is, by Proposition 1, $(2^U, \overline{\epsilon}_s, \underline{\epsilon}_s, 2^U)$ does not form a Galois connection.

Case 3. *Point-wise definitions.* Now let us check that the sets $\{x_1\}$ and $\{x_2\}$ fulfill the conditions of Proposition 1 in the case of point-wise definitions of the approximations.

Let us define the lower and upper ϵ-approximations of $\{x_1\}$ and $\{x_2\}$ in the point-wise manner due to Eqs. (4a) and (5a):

$$\underline{\epsilon}_p(\{x_1\}) = \{x \in U \mid [x]_\epsilon \subseteq \{x_1\}\} = \emptyset,$$
$$\overline{\epsilon}_p(\{x_1\}) = \{x \in U \mid [x]_\epsilon \cap \{x_1\} \neq \emptyset\} = \{x_1, x_2\},$$
$$\underline{\epsilon}_p(\{x_2\}) = \{x \in U \mid [x]_\epsilon \subseteq \{x_2\}\} = \{x_3\},$$
$$\overline{\epsilon}_p(\{x_2\}) = \{x \in U \mid [x]_\epsilon \cap \{x_2\} \neq \emptyset\} = \{x_1, x_3\}.$$

Of course, by Corollary 4, the maps $\underline{\epsilon}_p$ and $\overline{\epsilon}_p$ are monotone. Moreover, the formulae

$$\{x_1\} \subseteq \underline{\epsilon}_p(\overline{\epsilon}_p(\{x_1\})) \text{ and } \overline{\epsilon}_p(\underline{\epsilon}_p(\{x_1\})) \subseteq \{x_1\},$$
$$\{x_2\} \subseteq \underline{\epsilon}_p(\overline{\epsilon}_p(\{x_2\})) \text{ and } \overline{\epsilon}_p(\underline{\epsilon}_p(\{x_3\})) \subseteq \{x_3\}$$

must hold. Indeed,

$$\{x_1\} \subseteq \underline{\epsilon}_p(\overline{\epsilon}_p(\{x_1\})) = \underline{\epsilon}_p(\{x_1, x_2\}) = \{x \in U \mid [x]_\epsilon \subseteq \{x_1, x_2\}\} = \{x_1, x_3\},$$
$$\overline{\epsilon}_p(\underline{\epsilon}_p(\{x_1\})) = \overline{\epsilon}_p(\emptyset) = \{x \in U \mid [x]_\epsilon \cap \emptyset \neq \emptyset\} = \emptyset \subseteq \{x_1\},$$

and

$$\{x_2\} \subseteq \underline{\epsilon}_p(\overline{\epsilon}_p(\{x_2\})) = \underline{\epsilon}_p(\{x_1, x_3\}) = \{x \in U \mid [x]_\epsilon \subseteq \{x_1, x_3\}\} = \{x_2\},$$
$$\overline{\epsilon}_p(\underline{\epsilon}_p(\{x_2\})) = \overline{\epsilon}_p(\{x_3\}) = \{x \in U \mid [x]_\epsilon \cap \{x_3\} \neq \emptyset\} = \{x_2\} \subseteq \{x_2\}.$$

5 Approximation of Sets Based on Partial Covering

5.1 Introduction

In practice, there are objects which cannot be characterized by certain features directly.

Some illustrative examples:

- Bald men cannot be characterized with the property "color of hair".
- An infinite set is investigated via a finite family of its finite subsets. For instance, a number theorist studies the regularities of natural numbers using computers.

– Security policies are partial–natured in corporate information security. Typically some policies may only apply to specific hardware appliances, software applications or type of information.

Moreover, there are features with which *a set and its complement* cannot be treated simultaneously. For instance, complements of recursively enumerable sets are not necessarily recursively enumerable. The membership of recursively enumerable sets can effectively be determined by a *finite amount* of information, while the determination of their non-membership requires an *infinite amount* of information [66]. That is, the complement of a recursively enumerable set cannot necessarily be determined effectively. In other words, the recursively enumerable sets can be managed by computers (e.g., via a special rewriting system, the Markov algorithm [87]), while its complement not necessarily. Thus, this is an important *practical* partial approximation problem: how can we approximate an arbitrary set with recursively enumerable sets?

Another question is the point-freeness. Let us suppose that we study a collection of groups of individuals. In some cases it is important to distinguish individuals in these groups, whereas in other cases the differentiation is irrelevant. For instance, in genotype-phenotype investigations for understanding evolution it is reasonable to distinguish individuals (for a generalized point-set topological theory, see, *e.g.*, [88]). On the other hand, during the investigation of spreading of different types of floral zones in a given geographical area, the distinction of the individuals has no relevance.

Moreover, these floral zones overlap each other. In addition they generally do not cover the entire area, e.g. on lands of desert, or when we investigate the spreading of woodlands excluding the underwood. As another example, in the game of go there are two groups of stones, black and white. Black stones are inherently undistinguishable, so are the white ones. In addition, the black and white zones overlap each other, and even together they never cover the entire game table.

Throughout this section let U be a nonempty set called the *universe of discourse*.

5.2   Base Systems

According to the general set theoretic approximation framework, let $\mathfrak{B} \subseteq 2^U$ be a base system, *i.e.*, a nonempty family of nonempty subsets of U. Its members, the $\mathfrak{B}$-sets, are considered as our *primary tools* because we want the subsets of U to be approximated with their help.

Now, let us define our *derived tools*, *i.e.*, an extension of $\mathfrak{B}$ as follows.

Definition 12 ([77], Definition 4.1). *A nonempty subset $X \in 2^U$ is $\mathfrak{B}$-definable if there exists a family of sets $\mathfrak{D} \subseteq \mathfrak{B}$ in such a way that $X = \bigcup \mathfrak{D}$, otherwise X is $\mathfrak{B}$-undefinable.*

The empty set is considered to be a $\mathfrak{B}$-definable set.

Let $\mathfrak{D}_{\mathfrak{B}}$ denote the family of $\mathfrak{B}$-definable sets of U.

5.3 Single–Layered Base Systems

Some properties of rough set theory can partly be preserved with the help of the next constrained version of the base system.

Definition 13 ([77], Definition 4.2). *The base system* $\mathfrak{B} \subseteq 2^U$ *is single–layered, if*

$$\forall B \in \mathfrak{B} \;\; \forall \mathfrak{B}' \subseteq \mathfrak{B} \setminus \{B\} \, (B \cap \bigcup \mathfrak{B}' \neq B),$$

and one–layered, *if*

$$\forall B \in \mathfrak{B} \;\; \forall \mathfrak{B}' \subseteq \mathfrak{B} \setminus \{B\} \, (B \cap \bigcup \mathfrak{B}' = \emptyset).$$

Informally, a base system $\mathfrak{B}$ is single–layered if every nonempty $\mathfrak{B}$-definable subset has at least one element which can be characterized by exactly one primary tool, whereas $\mathfrak{B}$ is one–layered if every element of the universe can be characterized by at most one primary tool.

Remark 10. According to [89], it can be said that a $\mathfrak{B}$-set B is *semi–reducible* with respect to the base system $\mathfrak{B}$, if there exists a $\mathfrak{B}' \subseteq \mathfrak{B}$ in such a way that $B \subseteq \bigcup \mathfrak{B}'$, otherwise, B is *semi–irreducible*. Note that, in [89] Def. 6, $\bigcup \mathfrak{B} = U$, *i.e.*, the base system is a covering of the universe, and $B = \bigcup \mathfrak{B}'$. Clearly, a $B \in \mathfrak{B}$ is semi–irreducible, if

$$\forall \mathfrak{B}' \subseteq \mathfrak{B} \setminus \{B\} \, (B \cap \bigcup \mathfrak{B}' \neq B).$$

In other words, a base system $\mathfrak{B}$ is single–layered if every $\mathfrak{B}$-set is semi–irreducible. See also [90].

An important question is how can we form a single–layered base system from an arbitrary one. In general, this problem is reduced by the practice to finite base systems ($|\mathfrak{B}| < \infty$).

The simplest way to construct a single/one–layered base system from an arbitrary one is to form its *intersection structure* as a starting point. Formally, a nonempty family $\mathfrak{S}$ of subsets of the universe U is an intersection structure if $\forall \mathfrak{S}'(\neq \emptyset) \subseteq \mathfrak{S} \, (\bigcap \mathfrak{S}' \in \mathfrak{S})$, *i.e.*, it is closed under intersections [75]. Note that the intersection structure $\mathfrak{S}$ is a closure system, if $U \in \mathfrak{S}$.

Let us take an arbitrary base system $\mathfrak{B}$ and create its intersection structure $C(\mathfrak{B})$ as the smallest set which satisfies the following two properties:

1. $\mathfrak{B} \subseteq C(\mathfrak{B})$;
2. if $B, B' \in C(\mathfrak{B})$, then $B \cap B' \in C(\mathfrak{B})$.

Note that any intersections of primary tools are also considered primary tools, *i.e.*, new "combined" primary tools appear in $C(\mathfrak{B})$. In other words, the intersection structure $C(\mathfrak{B})$ is a collection of all original and all possible "combined" primary tools.

Having given the intersection structure $C(\mathfrak{B})$, first, we can create a single–layered base system $SC(\mathfrak{B})$ as the smallest set which satisfies the following two properties:

1. $SC(\mathfrak{B}) = \bigcap \mathfrak{B}$ is a single–layered base system;
2. if $B, B' \in C(\mathfrak{B})$ in such a way that $B \subset B'$, then let $B, B' \setminus B \in SC(\mathfrak{B})$.

Next, having given a single–layered base system $SC(\mathfrak{B})$, we can create a one–layered base system $OSC(\mathfrak{B})$ as the smallest set which satisfies the following two properties:

1. $OSC(\mathfrak{B}) = \bigcap \mathfrak{B}$ is a one–layered base system;
2. if $B, B' \in SC(\mathfrak{B})$ in such a way that $B \cap B' \neq \emptyset$, then let the differences $B \setminus B', B' \setminus B \in OSC(\mathfrak{B})$.

Proposition 13 ([77], Proposition 4.3). *Let $\mathfrak{B} \subseteq 2^U$ be a base system. Then the map $i_{\mathfrak{B}} : 2^{\mathfrak{B}} \to \mathfrak{D}_{\mathfrak{B}}, \mathfrak{D} \mapsto \bigcup \mathfrak{D}$ is a bijection if and only if $\mathfrak{B}$ is single–layered.*[5]

Proof. If $|\mathfrak{B}| = 1$, the base system $\mathfrak{B} = \{B\}$ is single–layered and

$$i_{\mathfrak{B}} : \{\emptyset, \{B\}\} \to \{\emptyset, B\}, \quad \emptyset \mapsto \bigcup \emptyset = \emptyset, \ \{B\} \mapsto \bigcup \{B\} = B$$

is a bijection evidently. Now, let us suppose that $|\mathfrak{B}| > 1$.

($\Rightarrow$) Let us assume, by contradiction, that the base system $\mathfrak{B}$ is not single–layered. If so,

$$\exists B \in \mathfrak{B} \ \exists \mathfrak{B}' \subseteq \mathfrak{B} \setminus \{B\} \, (B \subseteq \bigcup \mathfrak{B}').$$

Hence, $\mathfrak{B}', \mathfrak{B}' \cup \{B\} \in 2^{\mathfrak{B}}$ and $\bigcup \mathfrak{B}' = \bigcup(\mathfrak{B}' \cup \{B\}) \in \mathfrak{D}_{\mathfrak{B}}$, but $\mathfrak{B}' \neq \mathfrak{B}' \cup \{B\}$ because of $\mathfrak{B}' \subseteq \mathfrak{B} \setminus \{B\}$. This, however, contradicts the assumption that the map $i_{\mathfrak{B}}$ is injective.

($\Leftarrow$) Clearly, by Def. 12, the map $i_{\mathfrak{B}}$ is onto.

By contradiction, let us suppose that the map $i_{\mathfrak{B}}$ is not injective. In this case,

$$\exists \mathfrak{B}_1, \mathfrak{B}_2 \subseteq \mathfrak{B} \, (\mathfrak{B}_1 \neq \mathfrak{B}_2 \wedge \bigcup \mathfrak{B}_1 = \bigcup \mathfrak{B}_2).$$

Since $\mathfrak{B}_1 \neq \mathfrak{B}_2$, there exists $B \in \mathfrak{B}$ in such a way that B is an element of either one or the other. Without any loss of generality we can assume that $B \in \mathfrak{B}_1$ and $B \notin \mathfrak{B}_2$. Clearly, $B \subseteq \bigcup \mathfrak{B}_1 = \bigcup \mathfrak{B}_2$. Hence, $B \in \mathfrak{B}$, $\mathfrak{B}_2 \subseteq \mathfrak{B} \setminus \{B\}$ but $B \cap \bigcup \mathfrak{B}_2 = B$, which, however, contradicts the assumption that the base system $\mathfrak{B}$ is single–layered. ∎

The following two statements, provided that the base system is single–layered, present certain properties that Pawlakian rough set theory has.

Lemma 2 ([77], Lemma 4.3). *For a base system $\mathfrak{B} \subseteq 2^U$*

$$\forall B \in \mathfrak{B} \ \forall \mathfrak{B}' \subseteq \mathfrak{B} \, (B \subseteq \bigcup \mathfrak{B}' \Leftrightarrow B \in \mathfrak{B}')$$

if and only if the base system $\mathfrak{B}$ is single–layered.

---

[5] $2^{\mathfrak{B}}$ denotes the power set of $\mathfrak{B}$.

Proof. ($\Rightarrow$) Let us suppose, by contradiction, that the base system $\mathfrak{B}$ is not single–layered, that is $\exists B \in \mathfrak{B} \land \exists \mathfrak{B}' \subseteq \mathfrak{B} \setminus \{B\}(B \subseteq \bigcup \mathfrak{B}')$.

Hence, $B \subseteq \bigcup \mathfrak{B}'$ but $B \notin \mathfrak{B}'$. This contradicts the assumption that

$$\forall B \in \mathfrak{B} \; \forall \mathfrak{B}' \subseteq \mathfrak{B} \; (B \subseteq \bigcup \mathfrak{B}' \Rightarrow B \in \mathfrak{B}').$$

($\Leftarrow$) Of course, the statement $B \in \mathfrak{B}' \Rightarrow B \subseteq \bigcup \mathfrak{B}'$ is trivial. Thus we only have to prove that $\forall B \in \mathfrak{B} \; \forall \mathfrak{B}' \subseteq \mathfrak{B} \; (B \subseteq \bigcup \mathfrak{B}' \Rightarrow B \in \mathfrak{B}')$. Contrary to this statement, let us assume that $\exists B \in \mathfrak{B} \; \exists \mathfrak{B}' \subseteq \mathfrak{B}(B \subseteq \bigcup \mathfrak{B}' \land B \notin \mathfrak{B}')$.

Hence, $\mathfrak{B}' \subseteq \mathfrak{B} \setminus \{B\}$ and $B \subseteq \bigcup \mathfrak{B}'$ which, however, contradicts the assumption that the base system $\mathfrak{B}$ is single-layered. ∎

Proposition 14 ([77], Proposition 4.5). *Let $\mathfrak{B} \subseteq 2^U$ be a base system. Then the posets $(2^{\mathfrak{B}}, \subseteq)$ and $(\mathfrak{D}_{\mathfrak{B}}, \subseteq)$ are order isomorphic via the map*

$$i_{\mathfrak{B}} : 2^{\mathfrak{B}} \to \mathfrak{D}_{\mathfrak{B}}, \; \mathfrak{X} \mapsto \bigcup \mathfrak{X}$$

if and only if the base system $\mathfrak{B}$ is single–layered.

Proof. By Proposition 13, the map $i_{\mathfrak{B}}$ is a bijection if and only if the base system $\mathfrak{B}$ is single–layered.

The monotonicity of $i_{\mathfrak{B}}$ is trivial. The monotonicity of $i_{\mathfrak{B}}^{-1}$ can similarly be proved to Proposition 10 changing the reference to Lemma 1 for the reference to Lemma 2. ∎

5.4   Lower and Upper $\mathfrak{B}$-Approximations

Let us define the lower and upper approximations based on partial covering. Recall that $\mathfrak{B}$ does not cover the universe necessarily.

Definition 14 ([77], Definition 4.6). *Let $\mathfrak{B} \subseteq 2^U$ be a base system and X be any subset of U.*

The lower *$\mathfrak{B}$-approximation of X (Fig. 8.) is*

$$\mathfrak{C}_{\mathfrak{B}}^{\flat}(X) = \bigcup \{Y \mid Y \in \mathfrak{B}, Y \subseteq X\},$$

the upper *$\mathfrak{B}$-approximation of X (Fig. 9.) is*

$$\mathfrak{C}_{\mathfrak{B}}^{\sharp}(X) = \bigcup \{Y \mid Y \in \mathfrak{B}, Y \cap X \neq \emptyset\}.$$

The ordered pair $\langle \mathfrak{C}_{\mathfrak{B}}^{\flat}, \mathfrak{C}_{\mathfrak{B}}^{\sharp} \rangle$ is called a $\mathfrak{B}$-approximation pair on U.

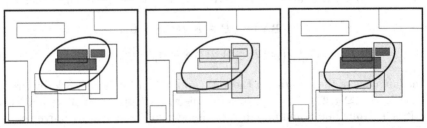

Fig. 8. Lower approximation   **Fig. 9.** Upper approximation   **Fig. 10.** Lower and upper approximations

Notice that $\mathfrak{C}_{\mathfrak{B}}^{\flat}$ and $\mathfrak{C}_{\mathfrak{B}}^{\sharp}$ are the straightforward point-free generalizations of lower and upper ε-approximations relying on ε-elementary sets.

Clearly, $\mathfrak{C}_{\mathfrak{B}}^{\flat}(X), \mathfrak{C}_{\mathfrak{B}}^{\sharp}(X) \in \mathfrak{D}_{\mathfrak{B}}$, and the maps $\mathfrak{C}_{\mathfrak{B}}^{\flat}, \mathfrak{C}_{\mathfrak{B}}^{\sharp} : 2^U \to \mathfrak{D}_{\mathfrak{B}}$ are total and generally many-to-one.

Proposition 15. *Let $\langle \mathfrak{C}_{\mathfrak{B}}^{\flat}, \mathfrak{C}_{\mathfrak{B}}^{\sharp} \rangle$ be a $\mathfrak{B}$-approximation pair on U. Then*

1. *$\langle \mathfrak{C}_{\mathfrak{B}}^{\flat}, \mathfrak{C}_{\mathfrak{B}}^{\sharp} \rangle$ is a $\mathfrak{C}_{\mathfrak{B}}^{\flat}$-semi-strong approximation pair on U;*
2. *$\langle \mathfrak{C}_{\mathfrak{B}}^{\flat}, \mathfrak{C}_{\mathfrak{B}}^{\sharp} \rangle$ is a strong approximation pair on U if and only if the base system $\mathfrak{B}$ covers the universe U.*

In other words, the maps $\mathfrak{C}_{\mathfrak{B}}^{\flat}$, $\mathfrak{C}_{\mathfrak{B}}^{\sharp}$ fulfill the following conditions:

(C0) $\mathfrak{C}_{\mathfrak{B}}^{\flat}(2^U), \mathfrak{C}_{\mathfrak{B}}^{\sharp}(2^U) \subseteq \mathfrak{D}_{\mathfrak{B}}$ *(definability of $\mathfrak{C}_{\mathfrak{B}}^{\flat}$ and $\mathfrak{C}_{\mathfrak{B}}^{\sharp}$).*
(C1) $\mathfrak{C}_{\mathfrak{B}}^{\flat}$ *and* $\mathfrak{C}_{\mathfrak{B}}^{\sharp}$ *are monotone (monotonicity of $\mathfrak{C}_{\mathfrak{B}}^{\flat}$ and $\mathfrak{C}_{\mathfrak{B}}^{\sharp}$).*
(C2) $\mathfrak{C}_{\mathfrak{B}}^{\sharp}(\emptyset) = \emptyset$ *(normality of $\mathfrak{C}_{\mathfrak{B}}^{\sharp}$).*
(C3) If $D \in \mathfrak{D}_{\mathfrak{B}}$, then $\mathfrak{C}_{\mathfrak{B}}^{\flat}(D) = D$ ($\mathfrak{C}_{\mathfrak{B}}^{\flat}$ is standard).
(C4) If $S \in 2^U$, then $\mathfrak{C}_{\mathfrak{B}}^{\flat}(S) \subseteq \mathfrak{C}_{\mathfrak{B}}^{\sharp}(S)$ (approximation property).
(C5) $\mathfrak{C}_{\mathfrak{B}}^{\flat}$ *is contractive.*
(C6) $\mathfrak{C}_{\mathfrak{B}}^{\sharp}$ *is extensive if and only if $\mathfrak{B}$ covers the universe U.*

Proof. The conditions *(C0)*, *(C1)*, *(C2)* and *(C4)*, *(C5)* are straightforward by Def. 14.

(C3) Clearly, if $\emptyset \in \mathfrak{D}_{\mathfrak{B}}$, then $\mathfrak{C}_{\mathfrak{B}}^{\flat}(\emptyset) = \emptyset$.

If $D(\neq \emptyset) \in \mathfrak{D}_{\mathfrak{B}}$, there exists at least one nonempty family of sets $\mathfrak{B}' \subseteq \mathfrak{B}$ in such a way that

$$D = \bigcup \mathfrak{B}' = \bigcup\{B \mid B \in \mathfrak{B}', B \subseteq D\} \subseteq \bigcup\{B \mid B \in \mathfrak{B}, B \subseteq D\} = \mathfrak{C}_{\mathfrak{B}}^{\flat}(D).$$

On the other hand, we have $\mathfrak{C}_{\mathfrak{B}}^{\flat}(D) \subseteq D$ by condition *(C5)*. Thus $\mathfrak{C}_{\mathfrak{B}}^{\flat}(D) = D$.

(C6) ($\Rightarrow$) If $\mathfrak{C}_{\mathfrak{B}}^{\sharp}$ is extensive, then

$$U \subseteq \mathfrak{C}_{\mathfrak{B}}^{\sharp}(U) = \bigcup\{B \mid B \in \mathfrak{B}, B \subseteq U\} = \bigcup \mathfrak{B}.$$

Of course, $\bigcup \mathfrak{B} \subseteq U$, and so $\bigcup \mathfrak{B} = U$.

($\Leftarrow$) If $\mathfrak{B}$ covers the universe, then $\forall S \in 2^U (S \subseteq U = \bigcup \mathfrak{B})$. Thus we get

$$S \subseteq \bigcup(\mathfrak{B} \setminus \{B \mid B \in \mathfrak{B}, B \cap S = \emptyset\})$$
$$= \bigcup\{B \mid B \in \mathfrak{B}, B \cap S \neq \emptyset\} = \mathfrak{C}_{\mathfrak{B}}^{\sharp}(S). \qquad \blacksquare$$

In the language of the general set theoretic approximation framework, by Proposition 15, $\langle 2^U, \mathfrak{D}_{\mathfrak{B}}, \mathfrak{C}_{\mathfrak{B}}^{\flat}, \mathfrak{C}_{\mathfrak{B}}^{\sharp} \rangle$ is a lower semi–strong approximation framework, and it is a strong one if and only if the base system $\mathfrak{B}$ covers the universe.

The next properties of $\mathfrak{C}_{\mathfrak{B}}^{\flat}$ and $\mathfrak{C}_{\mathfrak{B}}^{\sharp}$ immediately follow from Proposition 2. Of course, they can easily be proved by Def. 14 directly.

Proposition 16. *Let* $\langle \mathfrak{C}_{\mathfrak{B}}^{\flat}, \mathfrak{C}_{\mathfrak{B}}^{\sharp} \rangle$ *be a* $\mathfrak{B}$-*approximation pair on* U. *Then*

1. $\mathfrak{C}_{\mathfrak{B}}^{\flat}(\emptyset) = \emptyset$ *(normality of* $\mathfrak{C}_{\mathfrak{B}}^{\flat}$ *).*
2. $\forall S \in 2^U \ (\mathfrak{C}_{\mathfrak{B}}^{\flat}(\mathfrak{C}_{\mathfrak{B}}^{\flat}(S)) = \mathfrak{C}_{\mathfrak{B}}^{\flat}(S))$ *(idempotency of* $\mathfrak{C}_{\mathfrak{B}}^{\flat}$ *).*
3. $\mathfrak{C}_{\mathfrak{B}}^{\flat}(2^U) = \mathfrak{D}_{\mathfrak{B}}$ *(* $\mathfrak{C}_{\mathfrak{B}}^{\flat}$ *is surjective).*
4. $\mathfrak{C}_{\mathfrak{B}}^{\sharp}(2^U) \subseteq \mathfrak{C}_{\mathfrak{B}}^{\flat}(2^U) = \mathfrak{D}_{\mathfrak{B}}$.

Proof. We only have to show that in (4), the inclusion $\mathfrak{C}_{\mathfrak{B}}^{\sharp}(2^U) \subseteq \mathfrak{C}_{\mathfrak{B}}^{\flat}(2^U)$ may be proper because of the particular constructions of lower and upper approximation maps.

To do this, let $U = \{a, b\}$, $\mathfrak{B} = \{\{a\}, \{a, b\}\}$ and $\mathfrak{D}_{\mathfrak{B}} = \{\emptyset, \{a\}, \{a, b\}\}$. Then

$$\mathfrak{C}_{\mathfrak{B}}^{\flat} : 2^U \to \mathfrak{D}_{\mathfrak{B}}, \ X \mapsto \mathfrak{C}_{\mathfrak{B}}^{\flat}(X) = \begin{cases} \emptyset, & \text{if } X = \emptyset, \{b\}; \\ \{a\}, & \text{if } X = \{a\}; \\ \{a, b\}, & \text{if } X = \{a, b\}, \end{cases}$$

and

$$\mathfrak{C}_{\mathfrak{B}}^{\sharp} : 2^U \to \mathfrak{D}_{\mathfrak{B}}, \ X \mapsto \mathfrak{C}_{\mathfrak{B}}^{\sharp}(X) = \begin{cases} \emptyset, & \text{if } X = \emptyset; \\ \{a, b\}, & \text{if } X = \{a\}, \{b\}, \{a, b\}. \end{cases}$$

Conditions *(C1)–(C5)* can easily be checked. However, $\mathfrak{C}_{\mathfrak{B}}^{\sharp}$ in not surjective:

$$\mathfrak{C}_{\mathfrak{B}}^{\sharp}(2^U) = \{\emptyset, \{a, b\}\} \subsetneq \{\emptyset, \{a\}, \{a, b\}\} = \mathfrak{C}_{\mathfrak{B}}^{\flat}(2^U) = \mathfrak{D}_{\mathfrak{B}}. \qquad \blacksquare$$

Unlike Pawlakian approximation spaces (*cf.*, Proposition 7), the $\mathfrak{B}$-definable property is generally not equivalent to the condition $\mathfrak{C}_{\mathfrak{B}}^{\flat}(X) = \mathfrak{C}_{\mathfrak{B}}^{\sharp}(X)$.

Proposition 17 ([77], Proposition 4.7). *Let* $\mathfrak{B} \subseteq 2^U$ *be a base system. Then*

1. $X \in 2^U$ *is* $\mathfrak{B}$-*definable if and only if* $\mathfrak{C}_{\mathfrak{B}}^{\flat}(X) = X$.
2. $X \in 2^U$ *is* $\mathfrak{B}$-*undefinable if and only if* $\mathfrak{C}_{\mathfrak{B}}^{\flat}(X) \neq X$.

Proof.

1. It is straightforward, when $X = \emptyset$. Let $X \neq \emptyset$.
   ($\Rightarrow$) If $X \in \mathfrak{D}_{\mathfrak{B}}$, there exists at least one nonempty family of sets $\mathfrak{B}' \subseteq \mathfrak{B}$ in such a way that

$$X = \bigcup \mathfrak{B}' = \bigcup \{Y \mid Y \in \mathfrak{B}', Y \subseteq X\} \subseteq \bigcup \{Y \mid Y \in \mathfrak{B}, Y \subseteq X\} = \mathfrak{C}_{\mathfrak{B}}^{\flat}(X).$$

   On the other hand, $\mathfrak{C}_{\mathfrak{B}}^{\flat}(X) \subseteq X$, thus $X = \mathfrak{C}_{\mathfrak{B}}^{\flat}(X)$.
   ($\Leftarrow$) $X = \mathfrak{C}_{\mathfrak{B}}^{\flat}(X) \in \mathfrak{D}_{\mathfrak{B}}$.
2. It is the contrapositive version of (1). $\qquad \blacksquare$

5.5   Representation of Sets

Clearly, for a $\mathfrak{B}$-definable subset $D \in \mathfrak{D}_{\mathfrak{B}}$, there may exist two or more families of $\mathfrak{B}$-sets such that their unions are equal to D. For instance, let $\mathfrak{B} = \{B_1, B_2\}$ $(B_1, B_2 \in 2^U)$ be a base system in such a way that $B_1 \subsetneq B_2$. If $\mathfrak{F}_1 = \{B_1, B_2\}$,

$\mathfrak{F}_2 = \{B_2\}$, then $\mathfrak{F}_1 \neq \mathfrak{F}_2$ but $\bigcup \mathfrak{F}_1 = \bigcup \mathfrak{F}_2 = B_2$. Of course, the same is true for lower and upper $\mathfrak{B}$-approximations in general.

If $D \in \mathfrak{D}_\mathfrak{B}$ is a $\mathfrak{B}$-definable set, then let $\mathfrak{F}_\mathfrak{B}(D) \subseteq \mathfrak{B}$ denote a possible family of $\mathfrak{B}$-sets so that $\bigcup \mathfrak{F}_\mathfrak{B}(D) = D$. $\mathfrak{F}_\mathfrak{B}(D)$ is called a (possible) $\mathfrak{B}$-*composition* of D. Unlike Pawlakian approximation spaces, the $\mathfrak{B}$-compositions of $\mathfrak{B}$-definable sets are generally not unique.

Definition 15. *Let $\mathfrak{B} \subseteq 2^U$ be a base system.*

The $\mathfrak{B}$-definable set $D \in \mathfrak{D}_\mathfrak{B}$ is $\mathfrak{B}$-representable, if there exists exactly one $\mathfrak{B}$-composition $\mathfrak{F}_\mathfrak{B}(D)(\subseteq \mathfrak{B})$ of D in such a way that $D = \bigcup \mathfrak{F}_\mathfrak{B}(D)$.

In this case, it is said that $\mathfrak{F}_\mathfrak{B}(D)$ is the $\mathfrak{B}$-representation of D.

Proposition 18. *Let $\mathfrak{B} \subseteq 2^U$ be a base system.*

All $\mathfrak{B}$-definable subsets $D \in \mathfrak{D}_\mathfrak{B}$ of U are $\mathfrak{B}$-representable if and only if the base system $\mathfrak{B}$ is single-layered.

Proof. All $\mathfrak{B}$-definable subsets of U are $\mathfrak{B}$-representable if and only if the map $\mathfrak{F}_\mathfrak{B} : \mathfrak{D}_\mathfrak{B} \to 2^\mathfrak{B}$, $\bigcup \mathfrak{D} \mapsto \mathfrak{D}$ is the inverse of $i_\mathfrak{B} : 2^\mathfrak{B} \to \mathfrak{D}_\mathfrak{B}, \mathfrak{D} \mapsto \bigcup \mathfrak{D}$. A map has an inverse map if and only if it is a bijection. Consequently, all $\mathfrak{B}$-definable subsets of U are $\mathfrak{B}$-representable if and only if the map $i_\mathfrak{B} : 2^\mathfrak{B} \to \mathfrak{D}_\mathfrak{B}, \mathfrak{D} \mapsto \bigcup \mathfrak{D}$ is a bijection. And so, this proposition is just a restatement of Proposition 13. ∎

Corollary 5. *Let $\mathfrak{B} \subseteq 2^U$ be a base system.*

All $\mathfrak{B}$-definable subsets $D \in \mathfrak{D}_\mathfrak{B}$ of U are $\mathfrak{B}$-representable in the following form

$$D = \bigcup \mathfrak{F}_\mathfrak{B}(D), \text{ where } \mathfrak{F}_\mathfrak{B}(D) = \{Y \mid Y \in \mathfrak{B}, Y \subseteq D\},$$

if and only if the base system $\mathfrak{B}$ is single-layered.

Proof. According to Proposition 18, $\mathfrak{B}$ is single-layered if and only if all $\mathfrak{B}$-definable subsets are $\mathfrak{B}$-representable. And so, we only have to show that the $\mathfrak{B}$-representations of all $\mathfrak{B}$-definable subsets are of the form

$$\mathfrak{F}_\mathfrak{B}(D) = \{Y \mid Y \in \mathfrak{B}, Y \subseteq D\}.$$

Since, by definition, $D = \bigcup \{Y \mid Y \in \mathfrak{B}, Y \subseteq D\}$ satisfies for all $\mathfrak{B}$-definable subsets $D \in \mathfrak{D}_\mathfrak{B}$, the claim immediately follows from the uniqueness of $\mathfrak{B}$-representation. ∎

Proposition 19. *Let $\langle 2^U, \mathfrak{D}_\mathfrak{B}, \mathfrak{C}^\flat_\mathfrak{B}, \mathfrak{C}^\sharp_\mathfrak{B} \rangle$ be a lower semi–strong approximation space and X be a subset of U.*

Then the sets $\mathfrak{C}^\flat_\mathfrak{B}(X)$ and $\mathfrak{C}^\sharp_\mathfrak{B}(X)$ are $\mathfrak{B}$-representable in the forms

$$\mathfrak{C}^\flat_\mathfrak{B}(X) = \bigcup \mathfrak{F}^\flat_\mathfrak{B}(X), \text{ where } \mathfrak{F}^\flat_\mathfrak{B}(X) = \{Y \mid Y \in \mathfrak{B}, Y \subseteq X\},$$

$$\mathfrak{C}^\sharp_\mathfrak{B}(X) = \bigcup \mathfrak{F}^\sharp_\mathfrak{B}(X), \text{ where } \mathfrak{F}^\sharp_\mathfrak{B}(X) = \{Y \mid Y \in \mathfrak{B}, Y \cap X \neq \emptyset\},$$

if and only if the base system $\mathfrak{B}$ is single–layered.

Proof. Since $\mathfrak{C}_{\mathfrak{B}}^{\flat}(X), \mathfrak{C}_{\mathfrak{B}}^{\sharp}(X) \in \mathfrak{D}_{\mathfrak{B}}$, by Corollary 5, they are $\mathfrak{B}$-representable if and only if the base system $\mathfrak{B}$ is single layered. And so, we only have to show that $\mathfrak{F}_{\mathfrak{B}}(\mathfrak{C}_{\mathfrak{B}}^{\flat}(X))$ and $\mathfrak{F}_{\mathfrak{B}}(\mathfrak{C}_{\mathfrak{B}}^{\sharp}(X))$ are of the forms

$$\mathfrak{F}_{\mathfrak{B}}(\mathfrak{C}_{\mathfrak{B}}^{\flat}(X)) = \{Y \mid Y \in \mathfrak{B}, Y \subseteq X\},$$
$$\mathfrak{F}_{\mathfrak{B}}(\mathfrak{C}_{\mathfrak{B}}^{\sharp}(X)) = \{Y \mid Y \in \mathfrak{B}, Y \cap X \neq \emptyset\}.$$

By Corollary 5 and Lemma 2, we have

$$
\begin{aligned}
\mathfrak{F}_{\mathfrak{B}}(\mathfrak{C}_{\mathfrak{B}}^{\flat}(X)) &= \{Y \mid Y \in \mathfrak{B}, Y \subseteq \mathfrak{C}_{\mathfrak{B}}^{\flat}(X)\} \\
&= \{Y \mid Y \in \mathfrak{B}, Y \subseteq \bigcup\{Y' \mid Y' \in \mathfrak{B}, Y' \subseteq X\}\} \\
&= \{Y \mid Y \in \mathfrak{B}, Y \in \{Y' \mid Y' \in \mathfrak{B}, Y' \subseteq X\}\} \\
&= \{Y \mid Y \in \mathfrak{B}, Y \subseteq X\}, \\
\mathfrak{F}_{\mathfrak{B}}(\mathfrak{C}_{\mathfrak{B}}^{\sharp}(X)) &= \{Y \mid Y \in \mathfrak{B}, Y \subseteq \mathfrak{C}_{\mathfrak{B}}^{\sharp}(X)\} \\
&= \{Y \mid Y \in \mathfrak{B}, Y \subseteq \bigcup\{Y' \mid Y' \in \mathfrak{B}, Y' \cap X \neq \emptyset\}\} \\
&= \{Y \mid Y \in \mathfrak{B}, Y \in \{Y' \mid Y' \in \mathfrak{B}, Y' \cap X \neq \emptyset\}\} \\
&= \{Y \mid Y \in \mathfrak{B}, Y \cap X \neq \emptyset\}.
\end{aligned}
$$

∎

Remark 11. Of course, the equations

$$\mathfrak{C}_{\mathfrak{B}}^{\flat}(X) = \bigcup \mathfrak{F}_{\mathfrak{B}}^{\flat}(X) \text{ and } \mathfrak{C}_{\mathfrak{B}}^{\sharp}(X) = \bigcup \mathfrak{F}_{\mathfrak{B}}^{\sharp}(X)$$

trivially satisfy, they are just the definition of lower and upper $\mathfrak{B}$-approximations. Proposition 19, therefore, claims nothing else that there are no other set families $\mathfrak{X}_1, \mathfrak{X}_2 \subseteq \mathfrak{B}$ satisfying the equations $\mathfrak{C}_{\mathfrak{B}}^{\flat}(X) = \bigcup \mathfrak{X}_1$ and $\mathfrak{C}_{\mathfrak{B}}^{\sharp}(X) = \bigcup \mathfrak{X}_2$ if and only if the base system $\mathfrak{B}$ is single-layered.

5.6   Exactness in $\mathfrak{B}$-Approximation Spaces

In Pawlakian approximation spaces, the notions of "ε-crisp" and "ε-definable" are inherently one and the same, they are are synonymous to each other.

> The R-definable sets are those subsets of the universe which can be exactly defined in the knowledge base K, whereas the R-undefinable sets cannot be defined in this knowledge base.
> The R-definable sets will be also called R-exact sets, and R-undefinable sets will be also said to be R-inexact or R-rough. ([4], p. 9. *The italics are the author's.* Here, R is an equivalence relation on a finite universe U, pp. 3–4.)

The equivalence of "ε-crisp" and "ε-definable" formally is drawn up by Proposition 9 (1). Moreover, a subset $X \subseteq U$ is ε-definable, and consequently ε-crisp as well, if and only if its lower ε-approximation is equal to its upper ε-approximation according to Proposition 9.

In our approach, however, the compound notion of "crisp" and "definable" splits into two parts.

Definition 16. *Let* $\langle 2^U, \mathfrak{D}_{\mathfrak{B}}, \mathfrak{C}^{\flat}_{\mathfrak{B}}, \mathfrak{C}^{\sharp}_{\mathfrak{B}} \rangle$ *be a lower semi–strong* $\mathfrak{B}$*-approximation space and* $X \subseteq U$.
   The subset X *is* $\mathfrak{B}$*-crisp, if* $\mathfrak{C}^{\flat}_{\mathfrak{B}}(X) = \mathfrak{C}^{\sharp}_{\mathfrak{B}}(X)$, *otherwise* X *is* $\mathfrak{B}$*-rough.*

Definition 17. *Let* $\langle 2^U, \mathfrak{D}_{\mathfrak{B}}, \mathfrak{C}^{\flat}_{\mathfrak{B}}, \mathfrak{C}^{\sharp}_{\mathfrak{B}} \rangle$ *be a lower semi–strong* $\mathfrak{B}$*-approximation space and* $X \subseteq U$.
   The set $\mathfrak{N}_{\mathfrak{B}}(X) = \mathfrak{C}^{\sharp}_{\mathfrak{B}}(X) \setminus \mathfrak{C}^{\flat}_{\mathfrak{B}}(X)$ *is called the* $\mathfrak{B}$*-boundary of* X.

Unlike Pawlakian rough set theory, the $\mathfrak{B}$-boundary $\mathfrak{N}_{\mathfrak{B}}(X)$ is not necessarily $\mathfrak{B}$-definable.
   The next elementary facts are formulated in propositions for the sake of simple reference.

Proposition 20. *Let* $\langle 2^U, \mathfrak{D}_{\mathfrak{B}}, \mathfrak{C}^{\flat}_{\mathfrak{B}}, \mathfrak{C}^{\sharp}_{\mathfrak{B}} \rangle$ *be a lower semi–strong* $\mathfrak{B}$*-approximation space and* $X \subseteq U$.
   The subset X *is* $\mathfrak{B}$*-crisp if and only if the* $\mathfrak{B}$*-boundary* $\mathfrak{N}_{\mathfrak{B}}(X) = \emptyset$.

Proof. $\mathfrak{N}_{\mathfrak{B}}(X) = \mathfrak{C}^{\sharp}_{\mathfrak{B}}(X) \setminus \mathfrak{C}^{\flat}_{\mathfrak{B}}(X) = \emptyset \Leftrightarrow \mathfrak{C}^{\sharp}_{\mathfrak{B}}(X) \subseteq \mathfrak{C}^{\flat}_{\mathfrak{B}}(X)$. However, $\mathfrak{C}^{\flat}_{\mathfrak{B}}(X) \subseteq \mathfrak{C}^{\sharp}_{\mathfrak{B}}(X)$ holds for all $X \in 2^U$ by approximation property *(C4)* (*cf.*, Proposition 15), and so $\mathfrak{N}_{\mathfrak{B}}(X) = \emptyset \Leftrightarrow \mathfrak{C}^{\flat}_{\mathfrak{B}}(X) = \mathfrak{C}^{\sharp}_{\mathfrak{B}}(X) \Leftrightarrow X$ is $\mathfrak{B}$-crisp. $\blacksquare$

Proposition 21. *Let* $\langle 2^U, \mathfrak{D}_{\mathfrak{B}}, \mathfrak{C}^{\flat}_{\mathfrak{B}}, \mathfrak{C}^{\sharp}_{\mathfrak{B}} \rangle$ *be a strong* $\mathfrak{B}$*-approximation space and* $X \subseteq U$.
   The subset X *is* $\mathfrak{B}$*-crisp if and only if* $\mathfrak{C}^{\flat}_{\mathfrak{B}}(X) = \mathfrak{C}^{\sharp}_{\mathfrak{B}}(X) = X$.

Proof. In strong $\mathfrak{B}$-approximation spaces every subset $X \subseteq U$ is bounded by its lower and upper $\mathfrak{B}$-approximations: $\mathfrak{C}^{\flat}_{\mathfrak{B}}(X) \subseteq X \subseteq \mathfrak{C}^{\sharp}_{\mathfrak{B}}(X)$. Hence, X is crisp $\Leftrightarrow \mathfrak{C}^{\flat}_{\mathfrak{B}}(X) = \mathfrak{C}^{\sharp}_{\mathfrak{B}}(X) \Leftrightarrow \mathfrak{C}^{\flat}_{\mathfrak{B}}(X) = \mathfrak{C}^{\sharp}_{\mathfrak{B}}(X) = X$. $\blacksquare$

Proposition 22. *Let* $\langle 2^U, \mathfrak{D}_{\mathfrak{B}}, \mathfrak{C}^{\flat}_{\mathfrak{B}}, \mathfrak{C}^{\sharp}_{\mathfrak{B}} \rangle$ *be a strong* $\mathfrak{B}$*-approximation space and* $X \subseteq U$. *If* X *is* $\mathfrak{B}$*-crisp, then* X *is* $\mathfrak{B}$*-definable.*

Proof. By Proposition 21, X is $\mathfrak{B}$-*crisp* if and only if $\mathfrak{C}^{\flat}_{\mathfrak{B}}(X) = \mathfrak{C}^{\sharp}_{\mathfrak{B}}(X) = X$, and X is $\mathfrak{B}$-definable if and only if $X = \mathfrak{C}^{\flat}_{\mathfrak{B}}(X)$ by Proposition 17 (1).

A subset $X \in 2^U$ is $\mathfrak{B}$-definable if and only if $X = \mathfrak{C}^{\flat}_{\mathfrak{B}}(X)$ by Proposition 17 (1). However, as the next simple example shows, $X = \mathfrak{C}^{\flat}_{\mathfrak{B}}(X)$ generally does not imply $X = \mathfrak{C}^{\sharp}_{\mathfrak{B}}(X)$ even though the $\mathfrak{B}$-approximation space is a strong one. Let $\mathfrak{B} = \{B_1, B_2\}$ be a base system, where $B_1 \subsetneqq B_2$ ($B_1, B_2 \in 2^U$). Then $\mathfrak{C}^{\flat}_{\mathfrak{B}}(B_1) = B_1 \subsetneqq B_2 = \mathfrak{C}^{\sharp}_{\mathfrak{B}}(B_1)$.
   In other words, a $\mathfrak{B}$-definable subset is not necessarily $\mathfrak{B}$-crisp not even in strong $\mathfrak{B}$-approximation spaces. The converse statement only holds in strong $\mathfrak{B}$-approximation spaces by Proposition 22. Consequently, in our approach, the notions of "definable" and "crisp" are not synonymous to each other in the sense of Pawlak.

5.7 A Possible Interpretation of Our Approach

Let us suppose that we observe a collection of objects which is modeled as an abstract set, called the universe of discourse.

In real life, when we observe objects we cannot decide directly whether an object possesses a certain feature or not. Therefore we need a *tool* to be at our disposal with which we are able to judge easily and unambiguously whether an object possesses a *property* ascertained by the tool or not. It is expected that all tools can be used simply and quickly. The objects which are classified by a tool are modeled as a crisp subset of the universe. With a slight abuse of terminology, these subsets are simply called tools as well.

In sum, we model an object of interest as the element of an abstract set, called the universe, and the fact that "it possesses a property" as "it is the element of a suitable crisp subset of the universe".

Different tools usually form different subsets, but they are not necessarily disjoint. Notice that the complement of a tool is not necessarily a tool at the same time because the complement may not be used simply and quickly. For instance, let us take the tools being recursively enumerable. However, the complement of a recursively enumerable set is not necessarily recursively enumerable [66]. This significant fact confirms the partial nature of our approach [91].

Properties in $\mathfrak{B}$ are our primary tools which serve as fundamental building blocks of knowledge about the universe. Properties in $\mathfrak{D}_\mathfrak{B}$ are our derived tools which are formed from primary tools. To characterize any subset of the universe we want to use $\mathfrak{D}_\mathfrak{B}$. It is said that a property $D \in \mathfrak{D}_\mathfrak{B}$ *characterizes* a subset X of the universe, if $D \subseteq X$, and X is *characterized in terms of* $\mathfrak{D}_\mathfrak{B}$, if X is $\mathfrak{B}$-definable.

However, apart from the derived tools themselves, any other subsets cannot be characterized in terms of $\mathfrak{D}_\mathfrak{B}$. Therefore, their description is replaced by a pair of derived tools, in particular, their lower and upper approximations.

The universe can be divided into the following parts by means of lower and upper approximations concerning a subset $X \subseteq U$ [4,5]:

- $\mathfrak{B}$-*positive region of* X: $\mathfrak{C}_\mathfrak{B}^\flat(X) = \bigcup\{Y \mid Y \in \mathfrak{B}, Y \subseteq X\}$, *i.e.*, the lower $\mathfrak{B}$-approximation of X.
  $\{Y \mid Y \in \mathfrak{B}, Y \subseteq X\}$ is the family of all properties which *certainty characterize* X with respect to the current derived tools $\mathfrak{D}_\mathfrak{B}$.
- *Upper* $\mathfrak{B}$-*approximation of* X: $\mathfrak{C}_\mathfrak{B}^\sharp(X) = \bigcup\{Y \mid Y \in \mathfrak{B}, Y \cap X \neq \emptyset\}$.
  $\{Y \mid Y \in \mathfrak{B}, Y \cap X \neq \emptyset\}$ is the family of all properties which *possibly characterize* X with respect to the current derived tools $\mathfrak{D}_\mathfrak{B}$.
- $\mathfrak{B}$-*negative region of* X: $\bigcup(\mathfrak{D}_\mathfrak{B} \setminus \{Y \mid Y \in \mathfrak{B}, Y \cap X \neq \emptyset\})$.
  $\mathfrak{D}_\mathfrak{B} \setminus \{Y \mid Y \in \mathfrak{B}, Y \cap X \neq \emptyset\}$ is the family of all properties which *certainty do not characterize* X with respect to the current derived tools $\mathfrak{D}_\mathfrak{B}$.
- $\mathfrak{B}$-*borderline region of* X:

$$\bigcup(\{Y \mid Y \in \mathfrak{B}, Y \cap X \neq \emptyset\} \setminus \{Y \mid Y \in \mathfrak{B}, Y \subseteq X\}).$$

$\{Y \mid Y \in \mathfrak{B}, Y \cap X \neq \emptyset\} \setminus \{Y \mid Y \in \mathfrak{B}, Y \subseteq X\}$ is the family of all properties which cannot be classified with certainty either as characterizing X or as not characterizing X with respect to the current derived tools $\mathfrak{D}_\mathfrak{B}$.

6   Galois Connections

Recall that for any arbitrary binary relation ϵ on U, pairs of maps $(2^U, \bar{\epsilon}, \underline{\epsilon^{-1}}, 2^U)$ and $(2^U, \overline{\epsilon^{-1}}, \underline{\epsilon}, 2^U)$ are Galois connections ($cf.$, Proposition 12). Especially, when ε is an equivalence relation on U, the upper and lower ε-approximations form a $\mathbb{G}(2^U, \bar{\varepsilon}, \underline{\varepsilon}, 2^U)$ Galois connection. Note that the left adjoint is the upper ε-approximation $\bar{\varepsilon}$ and the right adjoint is the lower ε-approximation $\underline{\varepsilon}$. Some further observations about upper and lower approximations as Galois connection see, $e.g.$, [92,31,93,84].

Let $\langle 2^U, \mathfrak{D}_\mathfrak{B}, \mathfrak{C}_\mathfrak{B}^\flat, \mathfrak{C}_\mathfrak{B}^\sharp \rangle$ be a lower semi–strong $\mathfrak{B}$-approximation space. In this Section we will investigate what conditions have to be satisfied by a lower semi–strong $\mathfrak{B}$-approximation space so that the pair of maps $(2^U, \mathfrak{C}_\mathfrak{B}^\sharp, \mathfrak{C}_\mathfrak{B}^\flat, 2^U)$ forms a Galois connection on $(2^U, \subseteq)$. To do this, we take up the assertions of Proposition 1 and examine its conditions under which they hold point by point.

6.1   Regular Galois Connection

Let $\langle 2^U, \mathfrak{D}_\mathfrak{B}, \mathfrak{C}_\mathfrak{B}^\flat, \mathfrak{C}_\mathfrak{B}^\sharp \rangle$ be a lower semi–strong $\mathfrak{B}$-approximation space.

The maps $\mathfrak{C}_\mathfrak{B}^\sharp$ and $\mathfrak{C}_\mathfrak{B}^\flat$ are trivially monotone, $i.e.$, Proposition 1 (2) immediately holds. Thus we have to examine only Proposition 1 (1) in detail.

The next proposition answers the *first half* of Proposition 1 (1).

Proposition 23 ([70], Theorem 20). *Let $\langle 2^U, \mathfrak{D}_\mathfrak{B}, \mathfrak{C}_\mathfrak{B}^\flat, \mathfrak{C}_\mathfrak{B}^\sharp \rangle$ be a lower semi–strong $\mathfrak{B}$-approximation space.*
Then $\forall X \in 2^U (X \subseteq \mathfrak{C}_\mathfrak{B}^\flat(\mathfrak{C}_\mathfrak{B}^\sharp(X)))$ if and only if $\bigcup \mathfrak{B} = U$.

Proof. ($\Rightarrow$) By contradiction, let us assume that $\bigcup \mathfrak{B} \neq U$. Accordingly, $\exists X'(\neq \emptyset) \subseteq U \setminus \bigcup \mathfrak{B}$. Hence, $\mathfrak{C}_\mathfrak{B}^\flat(\mathfrak{C}_\mathfrak{B}^\sharp(X')) = \emptyset$, which gives $\emptyset \neq X' \subseteq \mathfrak{C}_\mathfrak{B}^\flat(\mathfrak{C}_\mathfrak{B}^\sharp(X')) = \emptyset$, a contradiction.

($\Leftarrow$) $\mathfrak{C}_\mathfrak{B}^\sharp(X) \in \mathfrak{D}_\mathfrak{B}$, and so, by Proposition 16 (3), $\mathfrak{C}_\mathfrak{B}^\flat(\mathfrak{C}_\mathfrak{B}^\sharp(X)) = \mathfrak{C}_\mathfrak{B}^\sharp(X)$. Since $\bigcup \mathfrak{B} = U$, by condition *(C6)* ($cf.$, Proposition 15), $\mathfrak{C}_\mathfrak{B}^\sharp$ is extensive, thus $X \subseteq \mathfrak{C}_\mathfrak{B}^\sharp(X) = \mathfrak{C}_\mathfrak{B}^\flat(\mathfrak{C}_\mathfrak{B}^\sharp(X))$. ∎

Remark 12. Proposition 23 does not require that the base system $\mathfrak{B}$ should be single–layered.

Let us take up the question of the *second half* of Proposition 1 (1). In general, it also does not hold.

Proposition 24 ([70], Theorem 21). *Let* $\langle 2^U, \mathfrak{D}_{\mathfrak{B}}, \mathfrak{C}^{\flat}_{\mathfrak{B}}, \mathfrak{C}^{\sharp}_{\mathfrak{B}} \rangle$ *be a lower semi-strong* $\mathfrak{B}$-*approximation space, and let us assume that the base system* $\mathfrak{B}$ *is single–layered.*
Then

$$\forall X \in 2^U \ (\mathfrak{C}^{\sharp}_{\mathfrak{B}}(\mathfrak{C}^{\flat}_{\mathfrak{B}}(X)) \subseteq X)$$

if and only if the $\mathfrak{B}$-*sets are pairwise disjoint.*

Proof. ($\Rightarrow$) Let us suppose, by contradiction, that the $\mathfrak{B}$-sets are not pairwise disjoint. If so,

$$\exists B_1, B_2 \in \mathfrak{B} \ (B_1 \neq B_2 \wedge B_1 \cap B_2 \neq \emptyset),$$

where neither $B_1 \subseteq B_2$ nor $B_2 \subseteq B_1$ holds because of the base system $\mathfrak{B}$ is single-layered. Hence, e.g., for B_1, we get $\mathfrak{C}^{\sharp}_{\mathfrak{B}}(\mathfrak{C}^{\flat}_{\mathfrak{B}}(B_1)) = \mathfrak{C}^{\sharp}_{\mathfrak{B}}(B_1) \supseteq B_1 \cup B_2 \not\subseteq B_1$, a contradiction.

($\Leftarrow$) If $X = \emptyset$, then $\mathfrak{C}^{\sharp}_{\mathfrak{B}}(\mathfrak{C}^{\flat}_{\mathfrak{B}}(\emptyset)) = \mathfrak{C}^{\sharp}_{\mathfrak{B}}(\emptyset) = \emptyset \subseteq \emptyset$ trivially holds (independently of the $\mathfrak{B}$-sets are pairwise disjoint or not).

Let $\emptyset \neq X \in 2^U$.

If $\mathfrak{C}^{\flat}_{\mathfrak{B}}(X) = \emptyset$, then $\mathfrak{C}^{\sharp}_{\mathfrak{B}}(\emptyset) = \emptyset \subseteq X$.

Let $\emptyset \neq \mathfrak{C}^{\flat}_{\mathfrak{B}}(X) = \bigcup \mathfrak{B}' \subseteq X$ for a family of $\mathfrak{B}$-sets $\mathfrak{B}' \subseteq \mathfrak{B}$ (such a $\mathfrak{B}'$ exists because $\mathfrak{C}^{\flat}_{\mathfrak{B}}(X)$ is $\mathfrak{B}$-definable and $\mathfrak{C}^{\flat}_{\mathfrak{B}}$ is contractive). Since the $\mathfrak{B}$-sets are pairwise disjoint,

$$\{Y \mid Y \in \mathfrak{B}, Y \cap \mathfrak{C}^{\flat}_{\mathfrak{B}}(X) \neq \emptyset\} = \{Y \mid Y \in \mathfrak{B}, Y \subseteq \bigcup \mathfrak{C}^{\flat}_{\mathfrak{B}}(X)\}.$$

Hence, we get

$$\mathfrak{C}^{\sharp}_{\mathfrak{B}}(\mathfrak{C}^{\flat}_{\mathfrak{B}}(X)) = \bigcup \{Y \mid Y \in \mathfrak{B}, Y \cap \mathfrak{C}^{\flat}_{\mathfrak{B}}(X) \neq \emptyset\}$$
$$= \bigcup \{Y \mid Y \in \mathfrak{B}, Y \subseteq \mathfrak{C}^{\flat}_{\mathfrak{B}}(X)\}$$
$$= \mathfrak{C}^{\flat}_{\mathfrak{B}}(\mathfrak{C}^{\flat}_{\mathfrak{B}}(X)) = \mathfrak{C}^{\flat}_{\mathfrak{B}}(X) \subseteq X. \qquad \blacksquare$$

Proposition 25 ([70], Theorem 22). *Let* $\langle 2^U, \mathfrak{D}_{\mathfrak{B}}, \mathfrak{C}^{\flat}_{\mathfrak{B}}, \mathfrak{C}^{\sharp}_{\mathfrak{B}} \rangle$ *be a lower semi-strong* $\mathfrak{B}$-*approximation space, and let us assume that the base system* $\mathfrak{B}$ *is single–layered.*
The pair of maps $(2^U, \mathfrak{C}^{\sharp}_{\mathfrak{B}}, \mathfrak{C}^{\flat}_{\mathfrak{B}}, 2^U)$ *forms a Galois connection on* $(2^U, \subseteq)$ *if and only if the base system* $\mathfrak{B}$ *is a partition of* U.

Proof. The maps $\mathfrak{C}^{\sharp}_{\mathfrak{B}}$ and $\mathfrak{C}^{\flat}_{\mathfrak{B}}$ are monotone, and so by Proposition 23 and Proposition 24, the conditions Proposition 1 (1-2) are satisfied. $\qquad \blacksquare$

According to Proposition 25, Galois connection between the pair of maps $(2^U, \mathfrak{C}^{\sharp}_{\mathfrak{B}}, \mathfrak{C}^{\flat}_{\mathfrak{B}}, 2^U)$ was proved under the condition that the base system $\mathfrak{B}$ is single-layered. However, as we have seen in Proposition 23 (1), the fulfillment of the first half of Proposition 1 (1) does not require that the base system $\mathfrak{B}$ to be single-layered. Now we examine whether the condition that the base system $\mathfrak{B}$ is single–layered can be removed from Proposition 24.

First we need the following lemma.

Lemma 3 ([77], Lemma 4.11). *Let* $\langle 2^U, \mathfrak{D}_\mathfrak{B}, \mathfrak{C}_\mathfrak{B}^\flat, \mathfrak{C}_\mathfrak{B}^\sharp \rangle$ *be a lower semi–strong* $\mathfrak{B}$-*approximation space. If*

$$\forall X \in 2^U (\mathfrak{C}_\mathfrak{B}^\sharp (\mathfrak{C}_\mathfrak{B}^\flat (X)) \subseteq X),$$

the base system $\mathfrak{B}$ *is singled–layered.*

Proof. Let us suppose, by contradiction, that $\mathfrak{B}$ is not singled–layered. If so, $\exists B \in \mathfrak{B} \wedge \exists \mathfrak{B}' \subseteq \mathfrak{B} \setminus \{B\} (B \subseteq \bigcup \mathfrak{B}')$. Hence, $B \subseteq \bigcup \mathfrak{B}'$ but $B \notin \mathfrak{B}'$, and so there exists at least one $B \neq B' \in \mathfrak{B}'$ in such a way that $B' \cap B \neq \emptyset$.

We have to distinguish three cases:

Case (1) $B \subsetneq B'$: $\mathfrak{C}_\mathfrak{B}^\sharp (\mathfrak{C}_\mathfrak{B}^\flat (B)) = \mathfrak{C}_\mathfrak{B}^\sharp (B) \supseteq B' \supsetneq B$, a contradiction.

Case (2) $B' \subsetneq B$: $\mathfrak{C}_\mathfrak{B}^\sharp (\mathfrak{C}_\mathfrak{B}^\flat (B')) = \mathfrak{C}_\mathfrak{B}^\sharp (B') \supseteq B \supsetneq B'$, a contradiction.

Case (3) $B' \cap B \neq \emptyset$, but neither $B \subsetneq B'$ nor $B' \subsetneq B$ holds: $\mathfrak{C}_\mathfrak{B}^\sharp (\mathfrak{C}_\mathfrak{B}^\flat (B')) = \mathfrak{C}_\mathfrak{B}^\sharp (B') \supseteq B \cup B' \supsetneq B'$, a contradiction. ∎

Remark 13. The converse statement in Lemma 3 does not hold. Let $\mathfrak{B} = \{B_1, B_2\}$ be a base system in such a way that $B_1 \cap B_2 \neq \emptyset$ but $B_1 \not\subseteq B_2 \wedge B_2 \not\subseteq B_1$. Clearly, $\mathfrak{B}$ is single–layered, and, *e.g.*, $\mathfrak{C}_\mathfrak{B}^\sharp (\mathfrak{C}_\mathfrak{B}^\flat (B_1)) = \mathfrak{C}_\mathfrak{B}^\sharp (B_1) = B_1 \cup B_2 \not\subseteq B_1$.

Proposition 26 ([77], Proposition 4.13). *Let* $\langle 2^U, \mathfrak{D}_\mathfrak{B}, \mathfrak{C}_\mathfrak{B}^\flat, \mathfrak{C}_\mathfrak{B}^\sharp \rangle$ *be a lower semi–strong* $\mathfrak{B}$-*approximation space.*

Then

$$\forall X \in 2^U (\mathfrak{C}_\mathfrak{B}^\sharp (\mathfrak{C}_\mathfrak{B}^\flat (X)) \subseteq X)$$

if and only if the $\mathfrak{B}$-*sets are pairwise disjoint.*

Proof. ($\Rightarrow$) The base system $\mathfrak{B}$ is single–layered by Lemma 3. Hereafter the proof is the same as in Proposition 24.

($\Leftarrow$) The $\mathfrak{B}$-sets are pairwise disjoint which immediately implies that the base system $\mathfrak{B}$ is single-layered. Hereafter the proof is the same as in Proposition 24. ∎

Theorem 1 ([77], Theorem 4.14). *Let* $\langle 2^U, \mathfrak{D}_\mathfrak{B}, \mathfrak{C}_\mathfrak{B}^\flat, \mathfrak{C}_\mathfrak{B}^\sharp \rangle$ *be a lower semi–strong* $\mathfrak{B}$-*approximation space.*

The pair of maps $(2^U, \mathfrak{C}_\mathfrak{B}^\sharp, \mathfrak{C}_\mathfrak{B}^\flat, 2^U)$ *forms a Galois connection on* $(2^U, \subseteq)$ *if and only if the base system* $\mathfrak{B}$ *is a partition of* U.

Proof. The maps $\mathfrak{C}_\mathfrak{B}^\sharp$ and $\mathfrak{C}_\mathfrak{B}^\flat$ are monotone, and so by Proposition 23 and Proposition 26, the conditions Proposition 1 (1-2) satisfy. ∎

6.2   Partial Galois Connection

On Partial Lower $\mathfrak{B}$**-approximations.** If a nonempty $X \in 2^U$ does not include nonempty $\mathfrak{B}$-definable subsets, then $\mathfrak{C}_\mathfrak{B}^\flat (X) = \bigcup \emptyset = \emptyset \subseteq X$ holds. However, the inclusion $\emptyset \subseteq X (\neq \emptyset)$ *per se* does not provide new information about the relationship between X and $\mathfrak{B}$. This phenomenon appears in Pawlakian classic rough set theory, too.

Definition 18. *Let* $\langle 2^U, \mathfrak{D}_{\mathfrak{B}}, \mathfrak{C}_{\mathfrak{B}}^{\flat}, \mathfrak{C}_{\mathfrak{B}}^{\sharp} \rangle$ *be a lower semi–strong* $\mathfrak{B}$-*approximation space, and* X *be any subset of* U.

The partial lower $\mathfrak{B}$-approximation *of* X *is*

$$\partial \mathfrak{C}_{\mathfrak{B}}^{\flat}(X) = \begin{cases} \mathfrak{C}_{\mathfrak{B}}^{\flat}(X), & \text{if } X = \emptyset \vee (X \neq \emptyset \wedge \mathfrak{C}_{\mathfrak{B}}^{\flat}(X) \neq \emptyset); \\ \text{undefined,} & \text{otherwise.} \end{cases} \tag{6}$$

Informally, Eq. (6) excludes that the the empty set to be the lower $\mathfrak{B}$-approximation of a nonempty subset of U.

Remark 14. If $X \in 2^U$ is nonempty and its lower $\mathfrak{B}$-approximation $\mathfrak{C}_{\mathfrak{B}}^{\flat}(X)$ is empty at the same time, then its partial lower $\mathfrak{B}$-approximation $\partial\mathfrak{C}_{\mathfrak{B}}^{\flat}(X)$ is undefined by Def. 18. This implies that the map $\partial\mathfrak{C}_{\mathfrak{B}}^{\flat}$ is total only if the base system $\mathfrak{B}$ contains all singleton sets $\{x\}$ $(x \in U)$, in other words, if all singletons are $\mathfrak{B}$-definable. This is a rather special situation as well. That is to exclude that we allow the empty set to be the lower $\mathfrak{B}$-approximation of a nonempty subset of U might be problematic as well.

The formula $X = \emptyset \vee (X \neq \emptyset \wedge \mathfrak{C}_{\mathfrak{B}}^{\flat}(X) \neq \emptyset)$ is equal to the formula

$$X = \emptyset \vee \mathfrak{C}_{\mathfrak{B}}^{\flat}(X) \neq \emptyset \tag{7}$$

In the following the latter is used because it is far simpler.

There exists at least one nonempty $\mathfrak{B}$-set B by Def. 12. Then $\mathfrak{C}_{\mathfrak{B}}^{\flat}(B) = B \neq \emptyset$ according to Def. 14. Hence, $\partial\mathfrak{C}_{\mathfrak{B}}^{\flat}$ is defined on at least one nonempty subset of U.

A natural total extension of $\partial\mathfrak{C}_{\mathfrak{B}}^{\flat}$ is the lower $\mathfrak{B}$-approximation $\mathfrak{C}_{\mathfrak{B}}^{\flat}$. That is the map $\partial\mathfrak{C}_{\mathfrak{B}}^{\flat}$ can be made total if it is allowed that the empty set may be the lower $\mathfrak{B}$-approximation of a nonempty subset of U. Of course, any extension $\mathfrak{C}_{\mathfrak{B}}^{*}$ of $\partial\mathfrak{C}_{\mathfrak{B}}^{\flat}$ also has to be $\mathfrak{B}$-definable and contractive, *i.e.*, formally, the condition

$$\forall X \in 2^U (\mathfrak{C}_{\mathfrak{B}}^{*}(X) \in \mathfrak{D}_{\mathfrak{B}} \wedge \mathfrak{C}_{\mathfrak{B}}^{*}(X) \subseteq X)$$

has to be fulfilled by $\mathfrak{C}_{\mathfrak{B}}^{*}$. Under the previous assumptions, we will show that any extension of this type is unique.

Proposition 27. *Let* $\langle 2^U, \mathfrak{D}_{\mathfrak{B}}, \mathfrak{C}_{\mathfrak{B}}^{\flat}, \mathfrak{C}_{\mathfrak{B}}^{\sharp} \rangle$ *be a lower semi–strong* $\mathfrak{B}$-*approximation space, and* X *be any subset of* U.

The total extension $\mathfrak{C}_{\mathfrak{B}}^{\flat}$ *of* $\partial\mathfrak{C}_{\mathfrak{B}}^{\flat}$ *is unique under the conditions that*

1. *the empty set may be the lower* $\mathfrak{B}$-*approximation of nonempty subsets of* U;
2. $\forall X \in 2^U (\mathfrak{C}_{\mathfrak{B}}^{*}(X) \in \mathfrak{D}_{\mathfrak{B}} \wedge \mathfrak{C}_{\mathfrak{B}}^{*}(X) \subseteq X)$ *has to be fulfilled by any total extension* $\mathfrak{C}_{\mathfrak{B}}^{*}$ *of* $\mathfrak{C}_{\mathfrak{B}}^{\flat}$.

Proof. It is straightforward that $\mathfrak{C}_{\mathfrak{B}}^{\flat}$ is a total extension of $\partial\mathfrak{C}_{\mathfrak{B}}^{\flat}$ from dom $\partial\mathfrak{C}_{\mathfrak{B}}^{\flat}$ to 2^U, and (1–2) automatically satisfy.

In order to prove the uniqueness, let us suppose, by contradiction, that $\mathfrak{C}_{\mathfrak{B}}^{*}$ is an extension of $\partial\mathfrak{C}_{\mathfrak{B}}^{\flat}$ from dom $\partial\mathfrak{C}_{\mathfrak{B}}^{\flat}$ to 2^U which differs from $\mathfrak{C}_{\mathfrak{B}}^{\flat}$ and

$$\forall X \in 2^U (\mathfrak{C}_{\mathfrak{B}}^{*}(X) \in \mathfrak{D}_{\mathfrak{B}} \wedge \mathfrak{C}_{\mathfrak{B}}^{*}(X) \subseteq X)$$

holds.

Since $\mathfrak{C}_{\mathfrak{B}}^*$ is an extension of $\partial\mathfrak{C}_{\mathfrak{B}}^\flat$, thus $\mathfrak{C}_{\mathfrak{B}}^* = \partial\mathfrak{C}_{\mathfrak{B}}^\flat = \mathfrak{C}_{\mathfrak{B}}^\flat$ on dom $\partial\mathfrak{C}_{\mathfrak{B}}^\flat$, i.e., when $X = \emptyset \vee \mathfrak{C}_{\mathfrak{B}}^\flat(X) \neq \emptyset$ satisfies (see Eq. (7)). On the other hand, $\mathfrak{C}_{\mathfrak{B}}^*$ differs from $\mathfrak{C}_{\mathfrak{B}}^\flat$, thus there exists at least one nonempty $X' \in 2^U \setminus$ dom $\partial\mathfrak{C}_{\mathfrak{B}}^\flat$ in such a way that $\mathfrak{C}_{\mathfrak{B}}^*(X') \neq \mathfrak{C}_{\mathfrak{B}}^\flat(X')$.

From the formula

$$X \in 2^U \setminus \text{dom } \partial\mathfrak{C}_{\mathfrak{B}}^\flat \Leftrightarrow \neg(X = \emptyset \vee \mathfrak{C}_{\mathfrak{B}}^\flat(X) \neq \emptyset) \Leftrightarrow X \neq \emptyset \wedge \mathfrak{C}_{\mathfrak{B}}^\flat(X) = \emptyset,$$

we get that $\mathfrak{C}_{\mathfrak{B}}^\flat(X) = \emptyset$ for every nonempty subset $X \in 2^U \setminus \text{dom } \partial\mathfrak{C}_{\mathfrak{B}}^\flat$. In particular, $\mathfrak{C}_{\mathfrak{B}}^*(X') \neq \mathfrak{C}_{\mathfrak{B}}^\flat(X') = \emptyset$.

Since $\emptyset \neq \mathfrak{C}_{\mathfrak{B}}^*(X') \in \mathfrak{D}_{\mathfrak{B}}$, there exists a nonempty family of sets $\mathfrak{B}' \subseteq \mathfrak{B}$ in such a way that $\mathfrak{C}_{\mathfrak{B}}^*(X') = \bigcup \mathfrak{B}' \subseteq X'$. Hence,

$$\emptyset \neq \mathfrak{C}_{\mathfrak{B}}^*(X') = \bigcup \mathfrak{B}' = \bigcup\{Y \mid Y \in \mathfrak{B}', Y \subseteq X'\}$$
$$\subseteq \bigcup\{Y \mid Y \in \mathfrak{B}, Y \subseteq X'\}$$
$$= \mathfrak{C}_{\mathfrak{B}}^\flat(X') = \emptyset,$$

which is a contradiction.   ∎

Partial Upper $\mathfrak{B}$-approximations. According to Proposition 15, $\mathfrak{C}_{\mathfrak{B}}^\sharp$ is extensive if and only if the base system $\mathfrak{B}$ covers the universe. If $\bigcup \mathfrak{B} \neq U$, for all subsets $X \subseteq U \setminus \bigcup \mathfrak{B}$, $\mathfrak{C}_{\mathfrak{B}}^\sharp(X) = \bigcup \emptyset = \emptyset$ hold. In other words, the empty set may be the upper $\mathfrak{B}$-approximation of certain nonempty subsets of U. Indeed, if $\mathfrak{C}_{\mathfrak{B}}^\sharp(X) \neq \emptyset$, then $X \not\subseteq \mathfrak{C}_{\mathfrak{B}}^\sharp(X)$ is also possible.

Definition 19. X is $\mathfrak{B}$-approximable if $X \subseteq \mathfrak{C}_{\mathfrak{B}}^\sharp(X)$, otherwise it is said that X has a $\mathfrak{B}$-approximation gap.

The $\mathfrak{B}$-approximation gap may be interpreted so that our knowledge about the universe encoded in the base system is incomplete and not enough to approximate X. This phenomenon may be natural/necessary or not. In the latter case, in order to fulfill the inclusion $X \subseteq \mathfrak{C}_{\mathfrak{B}}^\sharp(X)$ as far as possible, the base system $\mathfrak{B}$ has to be augmented via taking into account additional features concerning the observed system. In both former and latter cases, another possible solution is that the upper $\mathfrak{B}$-approximation map is defined as a partial one excluding the $\mathfrak{B}$-approximation gaps.

Definition 20. Let $\langle 2^U, \mathfrak{D}_{\mathfrak{B}}, \mathfrak{C}_{\mathfrak{B}}^\flat, \mathfrak{C}_{\mathfrak{B}}^\sharp \rangle$ be a lower semi–strong $\mathfrak{B}$-approximation space, and X be any subset of U.

The partial upper $\mathfrak{B}$-approximation of X is

$$\partial\mathfrak{C}_{\mathfrak{B}}^\sharp(X) = \begin{cases} \mathfrak{C}_{\mathfrak{B}}^\sharp(X), & \text{if } X \text{ is } \mathfrak{B}\text{-approximable;} \\ \text{undefined}, & \text{otherwise.} \end{cases} \tag{8}$$

There exists at least one nonempty $\mathfrak{B}$-set B by Def. 12. Then $B \subseteq \mathfrak{C}_{\mathfrak{B}}^{\sharp}(B)$ according to Def. 14. Hence, $\partial\mathfrak{C}_{\mathfrak{B}}^{\sharp}$ is defined on at least one nonempty subset of U.

Notice that $\mathfrak{C}_{\mathfrak{B}}^{\flat}(X) \subseteq X \subseteq \partial\mathfrak{C}_{\mathfrak{B}}^{\sharp}(X)$ holds provided X is $\mathfrak{B}$-approximatable, i.e., on dom $\partial\mathfrak{C}_{\mathfrak{B}}^{\sharp}$.

As Theorem 1 shows, the pair of maps $(2^U, \mathfrak{C}_{\mathfrak{B}}^{\sharp}, \mathfrak{C}_{\mathfrak{B}}^{\flat}, 2^U)$ forms a Galois connection on $(2^U, \subseteq)$ if and only if the base system $\mathfrak{B}$ is a partition of U. The question naturally arises whether the Galois connection can be generalized so that the pair of maps $(2^U, \partial\mathfrak{C}_{\mathfrak{B}}^{\sharp}, \mathfrak{C}_{\mathfrak{B}}^{\flat}, 2^U)$ may form a Galois connection in some sense. Moreover, if the answer is yes, what conditions have to be fulfilled by a lower semi–strong $\mathfrak{B}$-approximation space $\langle 2^U, \mathfrak{D}_{\mathfrak{B}}, \mathfrak{C}_{\mathfrak{B}}^{\flat}, \mathfrak{C}_{\mathfrak{B}}^{\sharp}\rangle$ so that $(2^U, \partial\mathfrak{C}_{\mathfrak{B}}^{\sharp}, \mathfrak{C}_{\mathfrak{B}}^{\flat}, 2^U)$ forms a Galois connection of this special type. Recall that $\mathfrak{C}_{\mathfrak{B}}^{\flat}$ is a total map and $\partial\mathfrak{C}_{\mathfrak{B}}^{\sharp}$ is a partial map on 2^U, and so the notion of the *partial* Galois connection which is drawn up in Def. 2 may be suitable for our purpose. In the following, we take up the points (1–4) in Definition 2 and examine the conditions under which they hold point by point.

Clearly, the map $\partial\mathfrak{C}_{\mathfrak{B}}^{\sharp}$ is a monotone partial map and $\mathfrak{C}_{\mathfrak{B}}^{\flat}$ is a monotone total map. Hence, the conditions (1–2) in Def. 2 immediately hold. Thus we have to examine only the conditions (3–4) in Def. 2 in detail.

The next proposition answers the condition (3) in Def. 2.

Proposition 28 ([70], Theorem 25). *Let $\langle 2^U, \mathfrak{D}_{\mathfrak{B}}, \mathfrak{C}_{\mathfrak{B}}^{\flat}, \mathfrak{C}_{\mathfrak{B}}^{\sharp}\rangle$ be a lower semi–strong $\mathfrak{B}$-approximation space.*
    Then $\partial\mathfrak{C}_{\mathfrak{B}}^{\sharp}(\mathfrak{C}_{\mathfrak{B}}^{\flat}(X))$ is defined for all $X \in 2^U$.

Proof. Let $X \in 2^U$ be an arbitrary subset of U. By the idempotency property of $\mathfrak{C}_{\mathfrak{B}}^{\flat}(X)$ (*cf.*, Proposition 16 (2)), $\mathfrak{C}_{\mathfrak{B}}^{\flat}(\mathfrak{C}_{\mathfrak{B}}^{\flat}(X)) = \mathfrak{C}_{\mathfrak{B}}^{\flat}(X)$. Thus,

$$\mathfrak{C}_{\mathfrak{B}}^{\flat}(X) = \mathfrak{C}_{\mathfrak{B}}^{\flat}(\mathfrak{C}_{\mathfrak{B}}^{\flat}(X)) = \bigcup\{Y \mid Y \in \mathfrak{B}, Y \subseteq \mathfrak{C}_{\mathfrak{B}}^{\flat}(X)\}$$
$$\subseteq \bigcup\{Y \mid Y \in \mathfrak{B}, Y \cap \mathfrak{C}_{\mathfrak{B}}^{\flat}(X) \neq \emptyset\}$$
$$= \mathfrak{C}_{\mathfrak{B}}^{\sharp}(\mathfrak{C}_{\mathfrak{B}}^{\flat}(X)),$$

that is, by Def. 20, $\partial\mathfrak{C}_{\mathfrak{B}}^{\sharp}(\mathfrak{C}_{\mathfrak{B}}^{\flat}(X))$ is defined. ∎

The next two propositions deal with the condition (4) in Def. 2.

Proposition 29 ([70], Theorem 26). *Let $\langle 2^U, \mathfrak{D}_{\mathfrak{B}}, \mathfrak{C}_{\mathfrak{B}}^{\flat}, \mathfrak{C}_{\mathfrak{B}}^{\sharp}\rangle$ be a lower semi–strong $\mathfrak{B}$-approximation space.*
    Then for all $\mathfrak{B}$-approximable subsets $X \in 2^U$ and all subsets $Y \in 2^U$:

$$\partial\mathfrak{C}_{\mathfrak{B}}^{\sharp}(X) \subseteq Y \Rightarrow X \subseteq \mathfrak{C}_{\mathfrak{B}}^{\flat}(Y).$$

Proof. Let $X, Y \in 2^U$ be two subsets of U in such a way that X is $\mathfrak{B}$-approximatable, and $X \subseteq \mathfrak{C}_{\mathfrak{B}}^{\sharp}(X) = \partial\mathfrak{C}_{\mathfrak{B}}^{\sharp}(X) \subseteq Y$. Hence, by the standard and monotonicity properties of $\mathfrak{C}_{\mathfrak{B}}^{\flat}$, we get

$$X \subseteq \mathfrak{C}_{\mathfrak{B}}^{\sharp}(X) = \mathfrak{C}_{\mathfrak{B}}^{\flat}(\mathfrak{C}_{\mathfrak{B}}^{\sharp}(X)) \subseteq \mathfrak{C}_{\mathfrak{B}}^{\flat}(Y). \qquad \blacksquare$$

Lemma 4 ([77], Lemma 4.19). *Let* $\langle 2^U, \mathfrak{D}_{\mathfrak{B}}, \mathfrak{C}_{\mathfrak{B}}^{\flat}, \mathfrak{C}_{\mathfrak{B}}^{\sharp} \rangle$ *be a lower semi–strong* $\mathfrak{B}$*-approximation space. If*

$$X \subseteq \mathfrak{C}_{\mathfrak{B}}^{\flat}(Y) \Rightarrow \partial \mathfrak{C}_{\mathfrak{B}}^{\sharp}(X) \subseteq Y$$

holds for all $\mathfrak{B}$*-approximable subsets* $X \in 2^U$ *and all subsets* $Y \in 2^U$, *the base system* $\mathfrak{B}$ *is singled–layered.*

Proof. First, we note that, if for all $\mathfrak{B}$-approximable subsets $X \in 2^U$ and all subsets $Y \in 2^U$, the relationship $X \subseteq \mathfrak{C}_{\mathfrak{B}}^{\flat}(Y) \Rightarrow \partial \mathfrak{C}_{\mathfrak{B}}^{\sharp}(X) \subseteq Y$ is satisfied, then, of course, for all $\mathfrak{B}$-approximable subsets $X \in 2^U$,

$$X \subseteq \mathfrak{C}_{\mathfrak{B}}^{\flat}(X) \Rightarrow \partial \mathfrak{C}_{\mathfrak{B}}^{\sharp}(X) \subseteq X$$

also has to be satisfied.

Since $\mathfrak{C}_{\mathfrak{B}}^{\flat}$ is contractive, then $\mathfrak{C}_{\mathfrak{B}}^{\flat}(X) \subseteq X$ holds for all subsets $X \in 2^U$. Thus, if $X \subseteq \mathfrak{C}_{\mathfrak{B}}^{\flat}(X)$, then $\mathfrak{C}_{\mathfrak{B}}^{\flat}(X) = X$. Consequently, by Proposition 17(1), X is $\mathfrak{B}$-definable, *i.e.*, $X \in \mathfrak{D}_{\mathfrak{B}}$.

On the other hand, for all $\mathfrak{B}$-approximable subsets $X \in 2^U$, $X \subseteq \mathfrak{C}_{\mathfrak{B}}^{\sharp}(X)$. Thus if for all $\mathfrak{B}$-approximable subsets $X \in 2^U$, the inclusion $\partial \mathfrak{C}_{\mathfrak{B}}^{\sharp}(X) = \mathfrak{C}_{\mathfrak{B}}^{\sharp}(X) \subseteq X$ also holds, then $X = \mathfrak{C}_{\mathfrak{B}}^{\sharp}(X)$.

For all these reasons, we can restate the previous statement as follows. For all $\mathfrak{B}$-approximable subsets $X \in 2^U$, $X \in \mathfrak{D}_{\mathfrak{B}} \Rightarrow X = \mathfrak{C}_{\mathfrak{B}}^{\sharp}(X)$ has to be satisfied.

Now, let us suppose, by contradiction, that $\mathfrak{B}$ is not singled–layered, that is, $\exists B \in \mathfrak{B} \wedge \exists \mathfrak{B}' \subseteq \mathfrak{B} \setminus \{B\} \, (B \subseteq \bigcup \mathfrak{B}')$. Hence, $B \subseteq \bigcup \mathfrak{B}'$, but $B \notin \mathfrak{B}'$, and so there exists at least one $B \neq B' \in \mathfrak{B}'$ in such a way that $B' \cap B \neq \emptyset$. Of course, $B, B' \in \mathfrak{D}_{\mathfrak{B}}$, and $B \subseteq \mathfrak{C}_{\mathfrak{B}}^{\sharp}(B)$, $B' \subseteq \mathfrak{C}_{\mathfrak{B}}^{\sharp}(B')$, *i.e.*, B, B' are $\mathfrak{B}$-approximable.

We have to distinguish three cases:

Case (1) If $B \subsetneq B'$, then $\mathfrak{C}_{\mathfrak{B}}^{\sharp}(B) \supseteq B' \supsetneq B$, a contradiction.

Case (2) If $B' \subsetneq B$, then $\mathfrak{C}_{\mathfrak{B}}^{\sharp}(B') \supseteq B \supsetneq B'$, a contradiction.

Case (3) If $B' \cap B \neq \emptyset$, but neither $B \subsetneq B'$ nor $B' \subsetneq B$ holds, then $\mathfrak{C}_{\mathfrak{B}}^{\sharp}(B') \supseteq B \cup B' \supsetneq B'$, a contradiction. $\qquad \blacksquare$

Remark 15. The converse statement does not hold. Let $\mathfrak{B} = \{B_1, B_2\}$ be a base system such that $B_1 \cap B_2 \neq \emptyset$ but $B_1 \not\subseteq B_2 \wedge B_2 \not\subseteq B_1$. Clearly, $\mathfrak{B}$ is single–layered. Let $X \in 2^U$ in such a way that $X \subsetneq B_1$, $X \cap B_2 \neq \emptyset$ but $X \subsetneq B_2$. Then $X \subseteq B_1 \cup B_2 = \mathfrak{C}_{\mathfrak{B}}^{\sharp}(X)$, *i.e.*, X is $\mathfrak{B}$-approximable. Hence, $X \subseteq \mathfrak{C}_{\mathfrak{B}}^{\flat}(B_1) = B_1$, but $\partial \mathfrak{C}_{\mathfrak{B}}^{\sharp}(X) = \mathfrak{C}_{\mathfrak{B}}^{\sharp}(X) = B_1 \cup B_2 \not\subseteq B_1$.

Proposition 30 ([77], Proposition 4.21). *Let* $\langle 2^U, \mathfrak{D}_{\mathfrak{B}}, \mathfrak{C}_{\mathfrak{B}}^{\flat}, \mathfrak{C}_{\mathfrak{B}}^{\sharp} \rangle$ *be a lower semi–strong* $\mathfrak{B}$*-approximation space.*

Then for all $\mathfrak{B}$*-approximable subsets* $X \in 2^U$ *and all subsets* $Y \in 2^U$,

$$X \subseteq \mathfrak{C}_{\mathfrak{B}}^{\flat}(Y) \Rightarrow \partial \mathfrak{C}_{\mathfrak{B}}^{\sharp}(X) \subseteq Y,$$

if and only if the $\mathfrak{B}$*-sets are pairwise disjoint.*

Proof. ($\Rightarrow$) Let us suppose, by contradiction, that the $\mathfrak{B}$-sets are not pairwise disjoint, If so,

$$\exists B_1, B_2 \in \mathfrak{B} \ (B_1 \neq B_2 \wedge B_1 \cap B_2 \neq \emptyset).$$

By Lemma 4, the base system $\mathfrak{B}$ is single-layered, and so neither $B_1 \subseteq B_2$ nor $B_2 \subseteq B_1$ holds. Clearly, *e.g.*, $B_1 \subseteq \mathfrak{C}_{\mathfrak{B}}^{\sharp}(B_1)$, *i.e.*, B_1 is $\mathfrak{B}$-approximable. Hence, we get

$$B_1 \subseteq \mathfrak{C}_{\mathfrak{B}}^{\flat}(B_1), \text{ but } \partial\mathfrak{C}_{\mathfrak{B}}^{\sharp}(B_1) = \mathfrak{C}_{\mathfrak{B}}^{\sharp}(B_1) \supseteq B_1 \cup B_2 \nsubseteq B_1,$$

a contradiction.

($\Leftarrow$) Let $X, Y \in 2^U$ in such a way that X is $\mathfrak{B}$-approximable and $X \subseteq \mathfrak{C}_{\mathfrak{B}}^{\flat}(Y)$. Then, by the monotonicity of $\mathfrak{C}_{\mathfrak{B}}^{\sharp}$ and Proposition 24,

$$\partial\mathfrak{C}_{\mathfrak{B}}^{\sharp}(X) = \mathfrak{C}_{\mathfrak{B}}^{\sharp}(X) \subseteq \mathfrak{C}_{\mathfrak{B}}^{\sharp}(\mathfrak{C}_{\mathfrak{B}}^{\flat}(Y)) \subseteq Y.$$

∎

Theorem 2 ([77], Theorem 4.22). *Let $\langle 2^U, \mathfrak{D}_{\mathfrak{B}}, \mathfrak{C}_{\mathfrak{B}}^{\flat}, \mathfrak{C}_{\mathfrak{B}}^{\sharp} \rangle$ be a lower semi–strong $\mathfrak{B}$-approximation space.*

The pair of maps $(2^U, \partial\mathfrak{C}_{\mathfrak{B}}^{\sharp}, \mathfrak{C}_{\mathfrak{B}}^{\flat}, 2^U)$ forms a partial Galois connection on $(2^U, \subseteq)$ if and only if the $\mathfrak{B}$-sets are pairwise disjoint.

Proof. Clearly, $\partial\mathfrak{C}_{\mathfrak{B}}^{\sharp}$ is a monotone partial map, and $\mathfrak{C}_{\mathfrak{B}}^{\flat}$ is a monotone total map. Thus the conditions (1–2) in Def. 2 are trivially satisfied. Proposition 28 implies condition (3) in Def. 2, Propositions 29 and 30 implies condition (4) in Def. 2. ∎

7   Applications

To demonstrate the effectiveness of our approach let us see three real–life applications of it.

The first application will demonstrate the relationship of our approach with natural computing [94] via a biological example.

> Natural computing is the field of research that investigates models and computational techniques inspired by nature and, dually, *attempts to understand the world around us in terms of information processing.* ([94], p. 72, *The italics are mine.*)

In particular, we will show how our approach helps us *to understand* some behavioral features of the natural vegetation heritage of Hungary. This presentation is based on the so–called MÉTA program which is a recognition and evaluation system of the state of the natural and semi–natural vegetation heritage of Hungary [95,96].

The second example models Intrusion Detection Systems (IDS) in computer security in such a way that two separated approximation spaces are defined for anomalies and misuses at the same time. In this framework anomalies and misuses can be detected simultaneously.

The third application presents a general tool–based approximation framework which is a generalization of the previous example. The starting point is that we observe a class of objects and, as usual, we suppose that there are some well–defined features with which an object possesses or not. In practice, two relevant groups of objects can be separated. A group whose elements really possess some features in question and another group whose elements do not substantially possess the same features.

In general, the features of objects cannot directly be observed. We need tools to be at our disposal with which we are able to judge easily and unambiguously whether an object possesses a feature in question or not. However, as a rule, a property ascertained by a tool never coincides with the feature observed by the tool completely.

In the framework, the class of objects is modeled as an abstract set called the universe of discourse. The two separate groups of objects correspond two crisp subsets of the universe. They are disjoint and, in general, their union does not add up to the whole universe. For obvious reasons, the former can be marked with the adjective *positive*, whereas the latter with *negative*.

The objects classified by a tool are simply called tools as well. Notice that the complement of a tool is not necessarily a tool at the same time. We also distinguish two types of tools: the *positive* and *negative* ones. Positive (resp., negative) tools provide the opportunity to locate the positive (resp,. negative) subset. It is a natural assumption that the union of positive tools and the union of negative tools are disjoint and their union does not add up to the whole universe.

In the proposed tool–based approximation framework, two approximation spaces are defined on the same universe, a positive and a negative one with the base systems consisting of positive tools and negative tools, respectively. Any proportion of the observed objects can *simultaneously* be approximated in the two approximation spaces.

7.1   Natural Computing—A Biological Example

A Brief Outline of the MÉTA Program. The biological example is relying on the MÉTA program which is a grid-based, landscape-ecology-oriented, satellite-image supported, field vegetation mapping method of Hungarian habitats [95,97,98,96,99] (MÉTA stands for Magyarországi Élőhelyek Térképi Adatbázisa: GIS Database of the Hungarian Habitats). Its main goals include a nationwide survey of the actual state of (semi-)natural vegetation heritage of Hungary and the evaluation of the present state of Hungarian landscapes from a vegetation point of view.

The survey in MÉTA program was carried out on three spatial levels which are nested units of the survey: 1. quadrant, 2. hexagon, 3. habitat type inside the hexagon.

The basic units of the survey are the hexagons. A *hexagon grid* consists of cells of 35 hectares covering the territory of Hungary comprehensively. 267,813 hexagons cover the whole country.

For organizational reasons around 100 hexagons form a *quadrant*. Quadrants are also used for collecting certain vegetation data. The quadrants are the quarters of the base units of the European Flora Survey. Their territory is approximately 35 km^2 and there are 2,834 quadrants in Hungary.

In 1996 a new habitat classification system was developed in Hungary, called Á-NÉR (the Hungarian abbreviation stands for General National Habitat Classification System). This system has 112 *habitat types*, all with detailed and standardized descriptions [100]. For the MÉTA method the Á-NÉR system was partly extended and thoroughly revised [101,102]. These Á-NÉR habitat types are recorded as a list for each hexagon.

The data is mainly collected by a single field survey of the hexagons. The mapper estimates the actual status on the spot. Hexagons with more than 25% natural or semi-natural vegetation are compulsory to survey and to be thoroughly documented. In most cases satellite images and maps help to decide whether a hexagon is compulsory or not. During the field mapping each compulsory hexagon has to be examined by thematically travelling through the area that it covers. Its most dominant habitat type is recorded, as well as those types covering at least 25% of the hexagon. Moreover, the vegetation patches found "on the way" should also be recorded. Vegetation data of noncompulsory hexagons should be documented if these hexagons are crossed by the mapping route or the data can be derived from the satellite image. Collected data are stored in an MS-SQL 2000 database and are mainly recorded as codes.

The data for each habitat type collected by the MÉTA method at the hexagon level are as follows [96]:

- The *areal cover* of each recorded habitat type has to be given as a proportion of the hexagon using the categories $< 1, 1, 10, 50, 100\%$. Satellite images help the observers to make the estimation.
- *Spatial pattern* of each type should be documented so that it forms only 1-2, 3 or several distinct patches, or it has a diffuse spatial pattern in the hexagon.
- In order to establish the *naturalness-based habitat quality* of each vegetation type in the hexagon, the following standardized naturalness-based habitat evaluation was used: (1) totally degraded state; (2) heavily degraded state; (3) moderately degraded state; (4) semi-natural state; (5) natural state.
- In each hexagon for each occurring habitat the most characteristic ones from the 28 *threat types* (Th_1-Th_{28}) had to be selected that actually threaten the survival and maintenance of the habitat type in the MÉTA hexagon in the next 10-15 years [103]. The strength of the threats is not recorded. The presence of the discernible threats in each case has been documented. Maximum four threats could be given, others were to be written in notes column.

The threatening factors are as follows [96]: improper water management, improper pasturing or mowing, drainage, encroachment of shrubs and trees, burning, afforestation with improper species, woodland patches managed homogeneously, improper selection of trees for timber extraction, logging trees at low age, inappropriate plantation, keeping high densities of game, colonization by invasive plant species, tillage, building and construction, gardening, mining, establishment of a pond, trampling, pollution, rubbish, commercial collection of plants.

- Prediction of future changes of vegetation patches can be supported by the evaluation of the direct effect of the *neighborhood* (< 200 m) on the mapped stands. This evaluation defines whether the neighboring patches will aid or hinder the survival of the particular patch in the next few (10-15) years [103]. The categories are: (1) definitely positive (sustaining neighborhood), (2) slightly positive, (3) indifferent, (4) slightly negative, (5) definitely negative (destructive neighborhood).

    The neighborhood is negative, *e.g.*, if there is an intensively used arable field (chemicals, infiltration of fertilizer), expanding settlement, or spreading populations of invasive species surrounding the patch. Neighborhood is positive, if it serves as a source of species, provides proper micro-climate, buffers against degrading factors.

- The *connectedness* is the potential of dispersal of the species of one vegetation stand compared to the surrounding areas. It is documented at two spatial scales: within the distance of several hundred meters (hexagon), and several kilometers (quadrant). It is recorded whether the patches are (1) isolated (typical species of the habitat are not present in the surroundings), (2) connected (species are abundant) or (3) the connectedness is intermediate.

    Connectedness is also documented at the quadrant level. Categories indicate whether stands are properly connected, moderately connected or isolated. The first two categories denote any possibilities for dispersal through quadrant whereas the third category shows whether the dispersal is hindered.

Additional data for each habitat type at quadrant level (invasive species, connectedness, regeneration potential), and data for the landscape at hexagon level (potential natural vegetation, area of invasive species and old fields, land-use type, landscape health status) are collected, as well (see [96] for details).

Model of Behavioral Features of Natural Vegetation. In this section we present a model for the behavioral features of natural vegetation of Hungary. The model is relying on the results of MÉTA program and the set approximations including both Pawlakian rough set theory and partial approximation of sets.

Let $\mathfrak{H}$ denote the set of all hexagons of Hungarian landscapes. The hexagons are disjoint and cover the whole country, *i.e.*, they form a partition. Let π denote the equivalence relation corresponding to this partition. If A denotes an arbitrary area of the country, one can approximate A in the Pawlakian framework. So, Pawlakian lower $\underline{\pi}(A)$ and upper $\overline{\pi}(A)$ π-approximations can be determined in the universe $\mathfrak{H}$.

Now, we want to investigate the area A in relation to the threat types. First of all, we need the following classification of threat types which is applied by the MÉTA program [103]: $Th_1 =$ water shortage, $Th_2 =$ access water, $Th_3 =$ improper water dynamics, $Th_4 =$ overgrazing, $Th_5 =$ undergrazing, $Th_6 =$ improper grazing regime, Th_7 = abandonment from grazing, Th_8 = improper mowing, Th_9 = abandonment from mowing, Th_{10} = melioration, Th_{11} = encroachment of shrubs and trees, Th_{12} = non-natural burning, Th_{13} = afforestation with improper species, Th_{14} = woodland patches managed homogeneously, Th_{15} = improper selection of trees for timber extraction, Th_{16} = logging trees at low age, Th_{17} = new plantations on grasslands, Th_{18} = overpopulated game, Th_{19} = colonization by invasive plant species, Th_{20} = tillage, Th_{21} = building and construction, Th_{22} = spread of gardens threatens vegetation, Th_{23} = mines destroying vegetation, Th_{24} = establishment of a pond destroying vegetation, Th_{25} = trampling, Th_{26} = pollution, Th_{27} = rubbish, Th_{28} = commercial collection of plants.

According to the MÉTA method, any threat type determine a well-defined subset of hexagons in $\mathfrak{H}$. Let $\mathfrak{h}_{Th_1}, \mathfrak{h}_{Th_2}, \ldots, \mathfrak{h}_{Th_{28}} \in 2^{\mathfrak{H}}$ denote the sets of all hexagons in which threat types $Th_1, Th_2, \ldots, Th_{28}$ are found, respectively. For instance, $\mathfrak{h}_{Th_1}$ contains *all hexagons* which are threatened with the threat type Th_1. Of course, $\mathfrak{h}_{Th_1}, \mathfrak{h}_{Th_2}, \ldots, \mathfrak{h}_{Th_{28}}$ do not necessarily disjoint and their union do not necessarily cover the whole are of the country.

Let $\mathfrak{B} = \{\mathfrak{H}_{Th1}, \mathfrak{H}_{Th2}, \ldots, \mathfrak{H}_{Th28}\} \subseteq 2^{\mathfrak{H}}$ be the base system in the universe $2^{\mathfrak{H}}$. The base system $\mathfrak{B}$ can directly be applied only in the case when A is exactly built up by hexagons, which is rather an extreme case. It is self–explanatory that for the first time we apply Pawlakian π-approximations to A in the universe $\mathfrak{H}$. All the lower and upper π-approximations and π-boundary of A are already made up of hexagons. Thus, in the next step, we can apply the partial approximation of sets to the three sets in the universe $2^{\mathfrak{H}}$.

Now, let us consider all possible cases one by one.

Case (1) The lower $\mathfrak{B}$-approximation of the lower π-approximation of A is:

$$\mathfrak{C}^{\flat}_{\mathfrak{B}}(\underline{\pi}(A)) = \bigcup\{\mathfrak{h} \mid \mathfrak{h} \in \mathfrak{B}, \mathfrak{h} \subseteq \underline{\pi}(A)\}.$$

If $\mathfrak{C}^{\flat}_{\mathfrak{B}}(\underline{\pi}(A)) \neq \emptyset$, then $\{\mathfrak{h} \mid \mathfrak{h} \in \mathfrak{B}, \mathfrak{h} \subseteq \underline{\pi}(A)\}$ contains the threat types (to be more exact, the hexagons threatened with these threat types) which *certainly* and *exclusively* appear in A.

Case (2) The upper $\mathfrak{B}$-approximation of the lower π-approximation of A is:

$$\mathfrak{C}^{\sharp}_{\mathfrak{B}}(\underline{\pi}(A)) = \bigcup\{\mathfrak{h} \mid \mathfrak{h} \in \mathfrak{B}, \mathfrak{h} \cap \underline{\pi}(A) \neq \emptyset\}.$$

If $\mathfrak{C}^{\sharp}_{\mathfrak{B}}(\underline{\pi}(A)) \neq \emptyset$, then $\{\mathfrak{h} \mid \mathfrak{h} \in \mathfrak{B}, \mathfrak{h} \cap \underline{\pi}(A) \neq \emptyset\}$ contains the threat types which *certainly* but *not exclusively* appear in A.

Case (3) The lower $\mathfrak{B}$-approximation of the upper π-approximation of A is:

$$\mathfrak{C}^{\flat}_{\mathfrak{B}}(\overline{\pi}(A)) = \bigcup\{\mathfrak{h} \mid \mathfrak{h} \in \mathfrak{B}, \mathfrak{h} \subseteq \overline{\pi}(A)\}.$$

If $\mathfrak{C}^{\flat}_{\mathfrak{B}}(\overline{\pi}(A)) \neq \emptyset$, then $\{\mathfrak{h} \mid \mathfrak{h} \in \mathfrak{B}, \mathfrak{h} \subseteq \overline{\pi}(A)\}$ contains the threat types which *perhaps exclusively* but *not certainly* appear in A.

Case (4) The upper $\mathfrak{B}$-approximation of the upper π-approximation of A is:

$$\mathfrak{C}^{\sharp}_{\mathfrak{B}}(\overline{\pi}(A)) = \bigcup\{\mathfrak{h} \mid \mathfrak{h} \in \mathfrak{B}, \mathfrak{h} \cap \overline{\pi}(A) \neq \emptyset\}.$$

If $\mathfrak{C}_{\mathfrak{B}}^{\sharp}(\overline{\pi}(A)) \neq \emptyset$, then $\{\mathfrak{h} \mid \mathfrak{h} \in \mathfrak{B}, \mathfrak{h} \cap \overline{\pi}(A) \neq \emptyset\}$ contains all the threat types which may appear in A at all.

The last two cases, the $\mathfrak{B}$-approximations of the π-boundary of the area A, provide information about the *spread of the thread types* around A.

<u>Case (5)</u> The lower $\mathfrak{B}$-approximation of the π-boundary of A is:

$$\mathfrak{C}_{\mathfrak{B}}^{\flat}(\overline{\pi}(A) \setminus \underline{\pi}(A)) = \bigcup \{\mathfrak{h} \mid \mathfrak{h} \in \mathfrak{B}, \mathfrak{h} \subseteq \overline{\pi}(A) \setminus \underline{\pi}(A)\}.$$

If $\mathfrak{C}_{\mathfrak{B}}^{\flat}(\overline{\pi}(A) \setminus \underline{\pi}(A)) \neq \emptyset$, then $\{\mathfrak{h} \mid \mathfrak{h} \in \mathfrak{B}, \mathfrak{h} \subseteq \overline{\pi}(A) \setminus \underline{\pi}(A)\}$ contains the threat types which partly belong to A and partly not. These are the threat types which *spread from inside of A to outwards*.

<u>Case (6)</u> The upper $\mathfrak{B}$-approximation of the π-boundary of A is:

$$\mathfrak{C}_{\mathfrak{B}}^{\sharp}(\overline{\pi}(A) \setminus \underline{\pi}(A)) = \bigcup \{\mathfrak{h} \mid \mathfrak{h} \in \mathfrak{B}, \mathfrak{h} \cap (\overline{\pi}(A) \setminus \underline{\pi}(A)) \neq \emptyset\}.$$

If $\mathfrak{C}_{\mathfrak{B}}^{\sharp}(\overline{\pi}(A) \setminus \underline{\pi}(A)) \neq \emptyset$, then $\{\mathfrak{h} \mid \mathfrak{h} \in \mathfrak{B}, \mathfrak{h} \cap (\overline{\pi}(A) \setminus \underline{\pi}(A)) \neq \emptyset\}$ contains the threat types which may partly belong to A. These are the threat types which *spread from outside of A to inwards*—"*Hannibal ante portas*".[6]

7.2   Simultaneous Anomaly and Misuse Intrusion Detections

Introduction. Nowadays, people run their applications in a complex open computing environment including all sorts of interconnected devices. While this environment permanently changes, people watch their applications, work with one of them, and, in general, also follow details of other applications with attention. Many applications, at the same time, work unnoticeably in the background, and some of them, even by stealth. In order to meet the computer security challenge in human environments, Intrusion Detection Systems (IDS) have to be designed.

To a large extent, acceptable and/or unacceptable patterns in the behaviors of the observed system cannot be designed and/or forecast in advance. This strange situation is smartly described by B. Schneier:

> You have to imagine an intelligent and malicious adversary inside your system (the "Satan" of Satan's computer), constantly trying new ways to subvert it. You have to consider all the ways your system can fail, most of them having nothing to do with the design itself. You have to look at everything backwards, upside down, and sideways. You have to think like an alien. ([104], from the Foreword by B. Schneier)

Computer security has definitely different challenges in corporate information systems and nonprofessional human computing environments. In the former one there are many approaches for security policy specification. Traditionally, security policies are formulated along the so-called CIA taxonomy which sees security as the combination of three attributes—confidentiality, integrity, and availability [105].

---

[6] This characterization of the situation is due to T. Mihálydeák.

People in nonprofessional human computing environments are flooded by recommendations how they operate their system and use their applications. In headwords only: strong passwords creation tips and maintenance, virus protection, software downloading and installation, removable media risks, encryption and cryptographic means, system backups, incident handling, e-mail and internet use best practises, etc.

Nonprofessionals, of course, cannot convert these good pieces of advice into security policies, especially into formal ones. Meanwhile, arising from the human thinking, all nonprofessional users have *anticipated hypotheses* how an application or the whole system should or should not work [84]. These presupposes may range from informal expected behaviors, their constituents might call expected "milestones", to more formal ones described in user manuals and other development artifacts.

To built up a formal security model for computer systems, first, one has to understand what has to be protected and why. The answers determine the *security strategy* which is, in turn, expressed by *security policies* [106,107]. Security policies as a general rule *prescribe* and *proscribe* behaviors of software systems *in advance*, only with more or less knowledge about future applications.

We model a computer system as a semantic system model, a so–called *traced-based model*. A traced-based model describes the behaviors of a computer system as sets of execution traces. We focus *solely* on externally observable execution traces sent out by the observed computing system.

An important note. An information system, among others, consists of different software components of finite number. Each component has an individual behavior, and the global behavior of the whole system is the collection of the individual ones. The components can operate with each other. Their interconnections may be deliberate or *ad hoc*. Notice, however, that in both cases, the mechanism of these interconnections mostly remains concealed from the external observers. In particular, based on only external observations we cannot model these synchronization mechanisms.

According to the trace–based model, it is assumed that security policies specify the prescribed and proscribed behaviors of a computer system via the patterns of *acceptable* and *unacceptable* execution traces, respectively [104,108]. We also take into account the partial nature of security policies. Typically some policies may only apply to specific hardware appliances, software applications or type of information. For instance, possibly it is enough to enforce the information flow policy on such software processes which handle confidential information.

In order to meet the computer security challenge outlined above, a sort of sophisticated *Intrusion Detection System* (IDS) has to be developed. Intrusion detection techniques can be categorized in different ways. For a survey of intrusion detection methods, see, *e.g.*, [109,110].

Intrusion detection techniques are categorized into *anomaly* and *misuse* detections. Both techniques use patterns based on different types of data [110,111]. Anomaly detection, originally proposed by Denning [112], profiles *expected* behaviors to identify abnormal behaviors as anomalies which deviate from the defined profile. Misuse detection profiles patterns of known attacks, *i.e.*, *unexpected* behaviors, to identify abnormal behavior directly.

In our IDS model, the patterns of expected and unexpected execution traces provide positive and negative *reference sets*, while the patterns of acceptable and unacceptable execution traces determined by the security policies serve as positive and negative *tools*.

The Intrusion Detection Model. Let us assume that A denotes a nonempty finite set of *symbols*. A *string* is a finite or infinite sequence of symbols chosen from A. String containing no symbols is called the *empty string* and is denoted by λ. Let A^* and A^ω denote the set of all finite and infinite strings made up of symbols chosen from A, respectively. We also use the following notations: $A^+ = A^* \setminus \{\lambda\}$, $A^\infty = A^* \cup A^\omega$.

An *execution sequence* or *trace* consists of linearly ordered observable atomic actions concerning the observed computer system [113]. Types of atomic actions are the following:

- Let A_{req} be a finite nonempty set of externally observable required atomic actions. It is called the *required action set*.
- Let A_{uns} be a finite nonempty set of insecure atomic actions which may happen during the running time of the observed system. It is called the *unsafe action set*.
- Let A_{neu} be a finite nonempty set of additional atomic actions which in themselves may not influence the safety of the observed system. It is called the *neutral action set*.

Let us assume that A_{req}, A_{uns} and A_{neu} are pairwise disjoint. Let $\mathcal{A} = A_{req} \cup A_{uns} \cup A_{neu}$ which is called the *system action set*.

An *execution trace* $\sigma \in \mathcal{A}^\infty$ is a finite or infinite sequence of not necessarily different system atomic actions.

Definition 21. *By a computer system we mean an $(\mathcal{A}, \Sigma)$ pair, where $\Sigma(\neq \emptyset) \subseteq \mathcal{A}^\omega$.*

If the computer system terminates, we as usual model it as an infinite execution trace by infinitely stuttering the empty action λ.

$\mathcal{A}^*$ is the set of all possible finite observable execution traces generated by the computer system.

Definition 22. *By a system observation we mean an $(\mathcal{A}, \Sigma^{obs})$ pair, where $\Sigma^{obs}(\neq \emptyset) \subseteq \mathcal{A}^*$. A subset $S \subseteq \Sigma^{obs}$ is called the snapshot.*

Let $\mathcal{A}^*$ be the universe of discourse in our IDS model.

Let $P^+ \subseteq (A_{req} \cup A_{neu})^* \subsetneq \mathcal{A}^*$ denote the set of *expected* execution traces which describes the expected behavior of the running system (see Fig. 11).

$\mathcal{A}^* \setminus P^+$ can be seen as the abnormal behavior of the system which deviates from the previously defined expected profile. Its elements are called *anomalies* [112] (see Fig. 12).

According to our current available knowledge, however, only a subset $P^- \subseteq$ $\mathcal{A}^* \setminus P^+$ can really be modeled as the *unexpected* behavior of the system. Its elements are usually called *misuses*. Of course, the unexpected behavior P^- has its own right to be profiled (see Fig. 13).

Expected and unexpected behaviors serve as positive and negative reference sets in our IDS model.

Security strategy is expressed by security policies of finite number and modeled as a family of sets $\mathfrak{S} \subseteq 2^{\mathcal{A}^*}$. Prescriptions and proscriptions of security policies can also be represented by families of sets of execution traces denoted by $\mathfrak{S}^+$ and $\mathfrak{S}^-$ respectively.

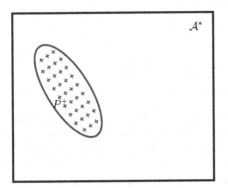

$\mathcal{A}^*$ is the universe of the IDS model: the set of all possible finite observable execution traces.

$P^+ \subseteq (A_{req} \cup A_{neu})^* \subsetneq \mathcal{A}^*$ profiles the expected behavior of the computer system.

Fig. 11. Initialization of the IDS model

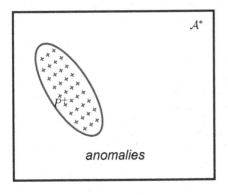

$\mathcal{A}^* \setminus P^+$ can be seen as the abnormal behavior of the system which deviates from the previously defined expected profile P^+.

Its elements are called *anomalies*.

Fig. 12. Anomalies

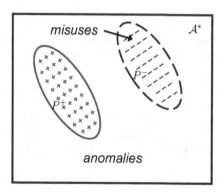

$P^- \subseteq \mathcal{A}^* \setminus P^+$ is the unexpected behavior of the system.

Its elements are called *misuses*.

P^+ and P^- are considered as positive and negative reference sets in our IDS model.

Fig. 13. Misuses

Let

$$\mathfrak{S}^+ = \{S_i^+ \mid S_i^+ \subseteq \mathcal{A}^*, i = 1, \dots, n^+\} \subseteq 2^{\mathcal{A}^*},$$
$$\mathfrak{S}^- = \{S_i^- \mid S_i^- \subseteq \mathcal{A}^*, i = 1, \dots, n^-\} \subseteq 2^{\mathcal{A}^*},$$

where $\mathfrak{S} = \mathfrak{S}^+ \cup \mathfrak{S}^-$. It is assumed that $\bigcup \mathfrak{S}^+ \cap \bigcup \mathfrak{S}^- = \emptyset$, *i.e.*, an execution trace cannot model a prescribed and proscribed behavior at the same time. Note that $\bigcup \mathfrak{S}^+$ (resp., $\bigcup \mathfrak{S}^-$) may contains execution traces with unsafe (resp., required and neutral) atomic actions.

Members of $\mathfrak{S}^+$ are called *acceptable* behaviors, whereas members of $\mathfrak{S}^-$ are called *unacceptable* behaviors. They serve as positive and negative tools in our IDS model. Acceptable/unacceptable behaviors and expected/unexpected behaviors *mutually justify each other*.

The sets in $\mathfrak{S}^+$ (resp., in $\mathfrak{S}^-$) are not necessarily pairwise disjoint and they do not cover $\mathcal{A}^*$ in general. That is, $\mathfrak{S}^+$ and $\mathfrak{S}^-$ are base systems over the universe $\mathcal{A}^*$. In other words, $\langle \mathcal{A}^*, \mathfrak{D}_{\mathfrak{S}^+}, \mathfrak{C}_{\mathfrak{S}^+}^\flat, \mathfrak{C}_{\mathfrak{S}^+}^\sharp \rangle$ and $\langle \mathcal{A}^*, \mathfrak{D}_{\mathfrak{S}^-}, \mathfrak{C}_{\mathfrak{S}^-}^\flat, \mathfrak{C}_{\mathfrak{S}^-}^\sharp \rangle$ form lower semi–strong approximation spaces (see Fig. 14 and 15).

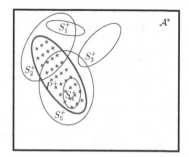

$\mathfrak{S}^+ = \{S_1^+, S_2^+, S_3^+, S_4^+, S_5^+\}$ is the prescriptions of the security policies.

$\mathfrak{S}^+$ is considered as positive tools in our IDS model.

Fig. 14. Approximation space $\langle \mathcal{A}^*, \mathfrak{D}_{\mathfrak{S}^+}, \mathfrak{C}_{\mathfrak{S}^+}^\flat, \mathfrak{C}_{\mathfrak{S}^+}^\sharp \rangle$ with a positive reference set P^+

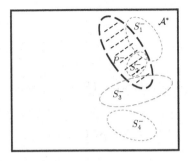

$\mathfrak{S}^- = \{S_1^-, S_2^-, S_3^-, S_4^-\}$
is the proscriptions of
the security policies.

$\mathfrak{S}^-$ is considered as
negative tools in our IDS model.

Fig. 15. Approximation space $\langle \mathcal{A}^*, \mathfrak{D}_{\mathfrak{S}^-}, \mathfrak{C}_{\mathfrak{S}^-}^\flat, \mathfrak{C}_{\mathfrak{S}^-}^\sharp \rangle$ with a negative reference set P^-

The complete IDS model is depicted in Fig. 16.

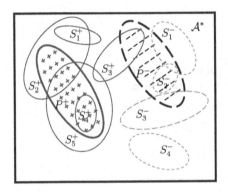

Formally, approximation spaces:
$\langle \mathcal{A}^*, \mathfrak{D}_{\mathfrak{S}^+}, \mathfrak{C}_{\mathfrak{S}^+}^\flat, \mathfrak{C}_{\mathfrak{S}^+}^\sharp \rangle$,
$\langle \mathcal{A}^*, \mathfrak{D}_{\mathfrak{S}^-}, \mathfrak{C}_{\mathfrak{S}^-}^\flat, \mathfrak{C}_{\mathfrak{S}^-}^\sharp \rangle.$;
positive/negative tools: $\mathfrak{S}^+, \mathfrak{S}^-$;
positive/negative reference sets: P^+, P^-.

Informally, acceptable/unacceptable
behaviors: $\mathfrak{S}^+, \mathfrak{S}^-$;
expected/unexpected behaviors: P^+, P^-.

Fig. 16. The complete IDS model

Using the running example, we show how this model can be applied.

<u>Step 1</u>. *Mutual justifying the reference sets and tools*

(1) P^+ is $\mathfrak{S}^-$-consistent and $\mathfrak{S}^+$-complete (see Fig. 17).

(2) P^- is $\mathfrak{S}^+$-inconsistent (see Fig. 18), P^- is $\mathfrak{S}^-$-incomplete (see Fig. 19).

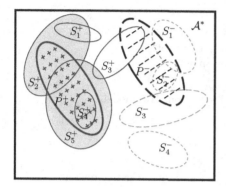

P^+ is $\mathfrak{S}^-$-consistent: $\mathfrak{C}^\sharp_{\mathfrak{S}^-}(P^+) = \emptyset$.

P^+ is $\mathfrak{S}^+$-complete:
$P^+ \subseteq \mathfrak{C}^\sharp_{\mathfrak{S}^+}(P^+) = S_2^+ \cup S_4^+ \cup S_5^+$.

Fig. 17. P^+ is $\mathfrak{S}^-$-consistent and $\mathfrak{S}^+$-complete

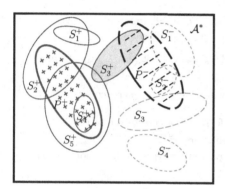

P^- is $\mathfrak{S}^+$-inconsistent:
$\mathfrak{C}^\sharp_{\mathfrak{S}^+}(A^-) = S_3^+ \neq \emptyset$.

Fig. 18. P^- is $\mathfrak{S}^+$-inconsistent

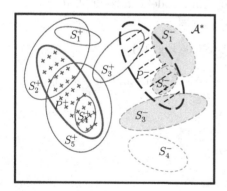

P^- is $\mathfrak{S}^-$-incomplete:
$P^- \not\subseteq \mathfrak{C}^\sharp_{\mathfrak{S}^-}(P^-) = S_1^- \cup S_2^- \cup S_3^-$.

Fig. 19. P^- is $\mathfrak{S}^-$-incomplete

<u>Step 2.</u> *Rebuilding positive and negative tools*

(1) Since P^+ was $\mathfrak{S}^-$-consistent, there was nothing to be done.

(2) Since P^+ was $\mathfrak{S}^+$-complete, P^+ was removed from the framework (see Fig. 20).

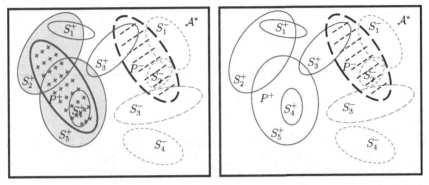

Fig. 20. P^+ is removed from the framework

(3) P^- was $\mathfrak{S}^+$-inconsistent, we *decided* that the positive tool S_3^+ was reasonable (see Fig. 21).

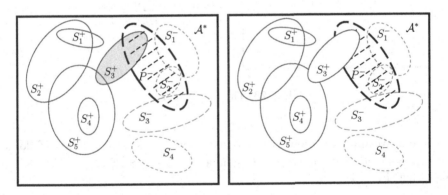

Fig. 21. We *decide* that the positive tool S_3^+ is reasonable

(4) P^- was $\mathfrak{S}^-$-incomplete, we *decided* that we augmented negative tools with S_5^-, S_6^- patterned upon one or more elements of the uncovered subset of P^-. Then P^- was removed from the framework (see Fig. 22).

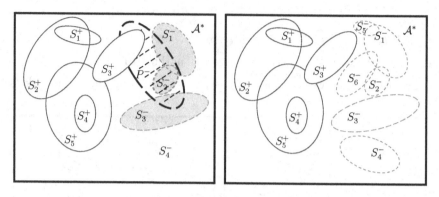

Fig. 22. New negative tools: S_5^-, S_6^-

By the end of Steps 1 and 2, we obtained the rebuilt positive tools $\mathfrak{S}_r^+ = \mathfrak{S}^+$, and the rebuilt negative tools $\mathfrak{S}_r^- = \mathfrak{S}^- \cup \{S_5^-, S_6^-\}$ (see Fig. 23).

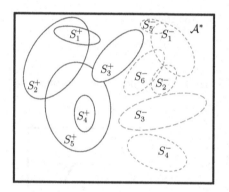

The rebuilt positive tools:
$\mathfrak{S}_r^+ = \mathfrak{S}^+$

The rebuilt negative tools:
$\mathfrak{S}_r^- = \mathfrak{S}^- \cup \{S_5^-, S_6^-\}$.

Fig. 23. The rebuilt tools

Step 3. We apply the rebuilt tools to justify snapshots of the system as follows.

A possible analysis based on the sample snapshots $S_1, S_2, S_3 \subseteq \Sigma^{obs}$ is the following (Fig. 24):

- $\mathfrak{C}_{\mathfrak{S}_r^+}^{\flat}(S_2)$ contains all prescriptions of the security policies being actually in force which in full pertain to the snapshot S_2.
  Since $\mathfrak{C}_{\mathfrak{S}_r^+}^{\flat}(S_2) = S_4^+$, thus S_4^+ is the only prescription which in full belongs to the snapshot S_2.
- $\mathfrak{C}_{\mathfrak{S}_r^+}^{\sharp}(S_2)$ contains all prescriptions of the security policies being actually in force which *possibly* pertain to the snapshot S_2.
  Since $\mathfrak{C}_{\mathfrak{S}_r^+}^{\sharp}(S_2) = S_4^+ \cup S_5^+$, thus only S_4^+, S_5^+ are the prescriptions which on the whole or in part belong to the snapshot S_2.

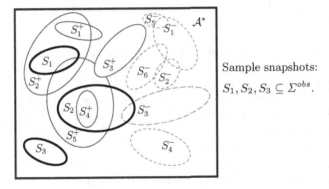

Fig. 24. Sample snapshots

- Acceptable execution traces in $\mathfrak{C}^{\sharp}_{\mathfrak{S}^+_r}(S_2) \setminus \mathfrak{C}^{\flat}_{\mathfrak{S}^+_r}(S_2) = S_5^+ \setminus S_4^+$ are *abstained* because they cannot be uniquely classified either as belonging to S_2 or as not belonging to S_2 with respect to the prescriptions of the security policies.
- $S_2 \not\subseteq \mathfrak{C}^{\sharp}_{\mathfrak{S}^+_r}(S_2)$ and so the execution traces in the subset $S_2 \setminus \mathfrak{C}^{\sharp}_{\mathfrak{S}^+_r}(S_2)$ of S_2 are anomalous. Moreover, since $\mathfrak{C}^{\sharp}_{\mathfrak{S}^-_r}(S_2 \setminus \mathfrak{C}^{\sharp}_{\mathfrak{S}^+_r}(S_2)) = S_3^-$, the execution traces in $S_3^- \cap (S_2 \setminus \mathfrak{C}^{\sharp}_{\mathfrak{S}^+_r}(S_2))$ are actually unacceptable.

A similar analysis can be made in the case of the snapshot S_1.

Notice that the snapshot S_3 cannot be justified at all with the prescriptions and proscriptions of the security policies being actually in force.

7.3   A General Tool–Based Approximation Framework

This section generalize the method presented in Subsection 7.2.

Let U be a nonempty set. Let $A^+, A^- \in 2^U$ be nonempty subsets of U in such a way that $A^+ \cap A^- = \emptyset$. A^+ and A^- are called the *positive reference set* and *negative reference set*, respectively. In general, $A^+ \cap A^- = \emptyset$ is the only requirement for A^+ and A^-. Of course, additional relations between them may be prescribed.

Furthermore, let $\mathfrak{T}^+, \mathfrak{T}^- \subseteq 2^U$ be two nonempty families of nonemty subsets of U such that $\bigcup \mathfrak{T}^+ \cap \bigcup \mathfrak{T}^- = \emptyset$. $\mathfrak{T}^+$ is called *positive* or $\mathfrak{T}^+$-*tools*, $\mathfrak{T}^-$ is called *negative* or $\mathfrak{T}^-$-*tools*. For each subset $T^+ \in \mathfrak{T}^+$ (resp., $T^- \in \mathfrak{T}^-$) it is *easy* to decide whether an element of U belongs to T^+ (resp., T^-) or not.

The sets in $\mathfrak{T}^+$ are not necessarily pairwise disjoint, so they are not in $\mathfrak{T}^-$. Neither $\bigcup \mathfrak{T}^+$ nor $\bigcup \mathfrak{T}^-$ covers U.

Note that, the adjectives *positive* and *negative* claim nothing else but that the sets A^+ (resp., $\mathfrak{T}^+$) and A^- (resp., $\mathfrak{T}^-$) are well separated.

Remark 16. Such pair of sets as above A^+ and A^- and/or $\mathfrak{T}^+$ and $\mathfrak{T}^-$ have been studied by several authors and in several environments. For instance, as orthopairs [114] or equivalently nested pairs [115], interval sets [116], bipolarity [117], etc.

According to the general set theoretic approximation framework, $\mathfrak{T}^+$ and $\mathfrak{T}^-$ are *primary* tools and let $\mathfrak{D}_{\mathfrak{T}^+}$ and $\mathfrak{D}_{\mathfrak{T}^-}$ denote their *derived* tools as usual. Then the quadruples $\langle U, \mathfrak{D}_{\mathfrak{T}^+}, \mathfrak{C}^\flat_{\mathfrak{T}^+}, \mathfrak{C}^\sharp_{\mathfrak{T}^+} \rangle$ and $\langle U, \mathfrak{D}_{\mathfrak{T}^-}, \mathfrak{C}^\flat_{\mathfrak{T}^-}, \mathfrak{C}^\sharp_{\mathfrak{T}^-} \rangle$ form lower semi–strong $\mathfrak{T}^+$-approximation and $\mathfrak{T}^-$-approximation spaces, respectively.

Borrowing the terminology from the inductive logic programming [118], the mutual relationships between A^+ and A^- can be characterized by the available $\mathfrak{T}^+$-tools and $\mathfrak{T}^-$-tools as follows.

It is said that
- A^+ is $\mathfrak{T}^+$-*complete* if $A^+ \subseteq \mathfrak{C}^\sharp_{\mathfrak{T}^+}(A^+)$, otherwise A^+ is $\mathfrak{T}^+$-*incomplete*;
- A^+ and $\mathfrak{T}^-$ are consistent, or A^+ is $\mathfrak{T}^-$-*consistent* for short, if $\mathfrak{C}^\sharp_{\mathfrak{T}^-}(A^+) = \emptyset$, otherwise A^+ and $\mathfrak{T}^-$ are inconsistent, or A^+ is $\mathfrak{T}^-$-*inconsistent* for short.

It is said that
- A^- is $\mathfrak{T}^-$-*complete* if $A^- \subseteq \mathfrak{C}^\sharp_{\mathfrak{T}^-}(A^-)$, otherwise A^- is $\mathfrak{T}^-$-*incomplete*;
- A^- and $\mathfrak{T}^+$ are consistent, or A^- is $\mathfrak{T}^+$-*consistent* for short, if $\mathfrak{C}^\sharp_{\mathfrak{T}^+}(A^-) = \emptyset$, otherwise A^- and $\mathfrak{T}^+$ are inconsistent, or A^- is $\mathfrak{T}^+$-*inconsistent* for short.

From a pure combinatorial point of view, according to the previous terminology, a positive reference set A^+ may be
- $\mathfrak{T}^+$-*complete* and $\mathfrak{T}^-$-*consistent*,
- $\mathfrak{T}^+$-*complete* and $\mathfrak{T}^-$-*inconsistent*,
- $\mathfrak{T}^+$-*incomplete* and $\mathfrak{T}^-$-*consistent*,
- $\mathfrak{T}^+$-*incomplete* and $\mathfrak{T}^-$-*inconsistent*;

a negative reference set A^- may be
- $\mathfrak{T}^-$-*complete* and $\mathfrak{T}^+$-*consistent*,
- $\mathfrak{T}^-$-*complete* and $\mathfrak{T}^+$-*inconsistent*,
- $\mathfrak{T}^-$-*incomplete* and $\mathfrak{T}^+$-*consistent*,
- $\mathfrak{T}^-$-*incomplete* and $\mathfrak{T}^+$-*inconsistent*.

There may be in sum $4 \cdot 4 = 16$ different compound situations. However, some of them are impossible by constraint $\bigcup \mathfrak{T}^+ \cap \bigcup \mathfrak{T}^- = \emptyset$. Now, let us consider all the possible and impossible cases one by one.

Case (1) A^+ is $\mathfrak{T}^+$-complete, $\mathfrak{T}^-$-consistent;
            A^- is $\mathfrak{T}^-$-complete, $\mathfrak{T}^+$-consistent.
            It is a possible case (see, *e.g.*, Fig. 25).

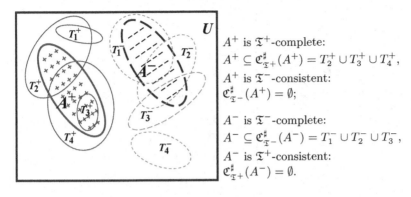

A^+ is $\mathfrak{T}^+$-complete:
$A^+ \subseteq \mathfrak{C}^\sharp_{\mathfrak{T}+}(A^+) = T_2^+ \cup T_3^+ \cup T_4^+,$
A^+ is $\mathfrak{T}^-$-consistent:
$\mathfrak{C}^\sharp_{\mathfrak{T}-}(A^+) = \emptyset;$

A^- is $\mathfrak{T}^-$-complete:
$A^- \subseteq \mathfrak{C}^\sharp_{\mathfrak{T}-}(A^-) = T_1^- \cup T_2^- \cup T_3^-,$
A^- is $\mathfrak{T}^+$-consistent:
$\mathfrak{C}^\sharp_{\mathfrak{T}+}(A^-) = \emptyset.$

Fig. 25. Case (1) It is a possible case because $\bigcup \mathfrak{T}^+ \cap \bigcup \mathfrak{T}^- = \emptyset$

<u>Case (2)</u> A^+ is $\mathfrak{T}^+$-complete, $\mathfrak{T}^-$-consistent;
   A^- is $\mathfrak{T}^-$-complete, $\mathfrak{T}^+$-inconsistent.
   It is an impossible case (see, *e.g.*, Fig. 26).

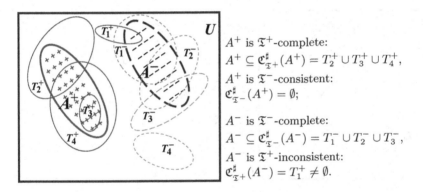

A^+ is $\mathfrak{T}^+$-complete:
$A^+ \subseteq \mathfrak{C}^\sharp_{\mathfrak{T}+}(A^+) = T_2^+ \cup T_3^+ \cup T_4^+,$
A^+ is $\mathfrak{T}^-$-consistent:
$\mathfrak{C}^\sharp_{\mathfrak{T}-}(A^+) = \emptyset;$

A^- is $\mathfrak{T}^-$-complete:
$A^- \subseteq \mathfrak{C}^\sharp_{\mathfrak{T}-}(A^-) = T_1^- \cup T_2^- \cup T_3^-,$
A^- is $\mathfrak{T}^+$-inconsistent:
$\mathfrak{C}^\sharp_{\mathfrak{T}+}(A^-) = T_1^+ \neq \emptyset.$

Fig. 26. Case (2) It is an impossible case because $\bigcup \mathfrak{T}^+ \cap \bigcup \mathfrak{T}^- \neq \emptyset$

<u>Case (3)</u> A^+ is $\mathfrak{T}^+$-complete, $\mathfrak{T}^-$-consistent;
   A^- is $\mathfrak{T}^-$-incomplete, $\mathfrak{T}^+$-consistent.
   It is a possible case (see, *e.g.*, Fig. 27).

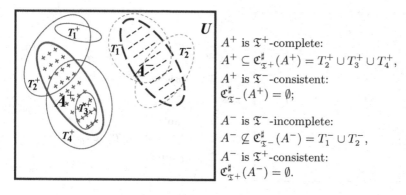

A^+ is $\mathfrak{T}^+$-complete:
$A^+ \subseteq \mathfrak{C}^\sharp_{\mathfrak{T}+}(A^+) = T_2^+ \cup T_3^+ \cup T_4^+$,
A^+ is $\mathfrak{T}^-$-consistent:
$\mathfrak{C}^\sharp_{\mathfrak{T}-}(A^+) = \emptyset$;

A^- is $\mathfrak{T}^-$-incomplete:
$A^- \not\subseteq \mathfrak{C}^\sharp_{\mathfrak{T}-}(A^-) = T_1^- \cup T_2^-$,
A^- is $\mathfrak{T}^+$-consistent:
$\mathfrak{C}^\sharp_{\mathfrak{T}+}(A^-) = \emptyset$.

Fig. 27. Case (3) It is a possible case because $\bigcup \mathfrak{T}^+ \cap \bigcup \mathfrak{T}^- = \emptyset$

Case (4) A^+ is $\mathfrak{T}^+$-complete, $\mathfrak{T}^-$-consistent;
A^- is $\mathfrak{T}^-$-incomplete, $\mathfrak{T}^+$-inconsistent.
It is a possible case (see, *e.g.*, Fig. 28).

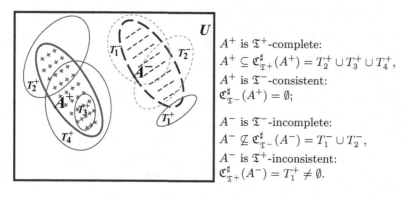

A^+ is $\mathfrak{T}^+$-complete:
$A^+ \subseteq \mathfrak{C}^\sharp_{\mathfrak{T}+}(A^+) = T_2^+ \cup T_3^+ \cup T_4^+$,
A^+ is $\mathfrak{T}^-$-consistent:
$\mathfrak{C}^\sharp_{\mathfrak{T}-}(A^+) = \emptyset$;

A^- is $\mathfrak{T}^-$-incomplete:
$A^- \not\subseteq \mathfrak{C}^\sharp_{\mathfrak{T}-}(A^-) = T_1^- \cup T_2^-$,
A^- is $\mathfrak{T}^+$-inconsistent:
$\mathfrak{C}^\sharp_{\mathfrak{T}+}(A^-) = T_1^+ \neq \emptyset$.

Fig. 28. Case (4) It is a possible case because $\bigcup \mathfrak{T}^+ \cap \bigcup \mathfrak{T}^- = \emptyset$

Case (5) A^+ is $\mathfrak{T}^+$-complete, $\mathfrak{T}^-$-inconsistent;
A^- is $\mathfrak{T}^-$-complete, $\mathfrak{T}^+$-consistent.
It is an impossible case (see, *e.g.*, Fig. 29).

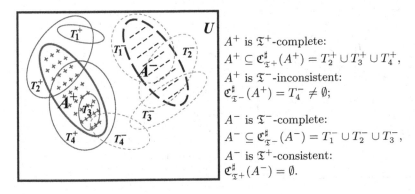

A^+ is $\mathfrak{T}^+$-complete:
$$A^+ \subseteq \mathfrak{C}^\sharp_{\mathfrak{T}+}(A^+) = T_2^+ \cup T_3^+ \cup T_4^+,$$
A^+ is $\mathfrak{T}^-$-inconsistent:
$$\mathfrak{C}^\sharp_{\mathfrak{T}-}(A^+) = T_4^- \neq \emptyset;$$

A^- is $\mathfrak{T}^-$-complete:
$$A^- \subseteq \mathfrak{C}^\sharp_{\mathfrak{T}-}(A^-) = T_1^- \cup T_2^- \cup T_3^-,$$
A^- is $\mathfrak{T}^+$-consistent:
$$\mathfrak{C}^\sharp_{\mathfrak{T}+}(A^-) = \emptyset.$$

Fig. 29. Case (5) It is an impossible case because $\bigcup \mathfrak{T}^+ \cap \bigcup \mathfrak{T}^- \neq \emptyset$

Case (6) A^+ is $\mathfrak{T}^+$-complete, $\mathfrak{T}^-$-inconsistent;
_____ A^- is $\mathfrak{T}^-$-complete, $\mathfrak{T}^+$-inconsistent.
   It is an impossible case (see, *e.g.*, Fig. 30).

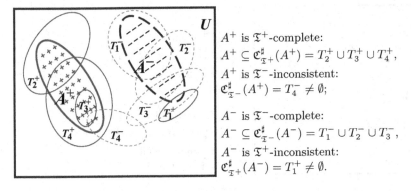

A^+ is $\mathfrak{T}^+$-complete:
$$A^+ \subseteq \mathfrak{C}^\sharp_{\mathfrak{T}+}(A^+) = T_2^+ \cup T_3^+ \cup T_4^+,$$
A^+ is $\mathfrak{T}^-$-inconsistent:
$$\mathfrak{C}^\sharp_{\mathfrak{T}-}(A^+) = T_4^- \neq \emptyset;$$

A^- is $\mathfrak{T}^-$-complete:
$$A^- \subseteq \mathfrak{C}^\sharp_{\mathfrak{T}-}(A^-) = T_1^- \cup T_2^- \cup T_3^-,$$
A^- is $\mathfrak{T}^+$-inconsistent:
$$\mathfrak{C}^\sharp_{\mathfrak{T}+}(A^-) = T_1^+ \neq \emptyset.$$

Fig. 30. Case (6) It is an impossible case because $\bigcup \mathfrak{T}^+ \cap \bigcup \mathfrak{T}^- \neq \emptyset$

Case (7) A^+ is $\mathfrak{T}^+$-complete, $\mathfrak{T}^-$-inconsistent;
_____ A^- is $\mathfrak{T}^-$-incomplete, $\mathfrak{T}^+$-consistent.
   It is an impossible case (see, *e.g.*, Fig. 31).

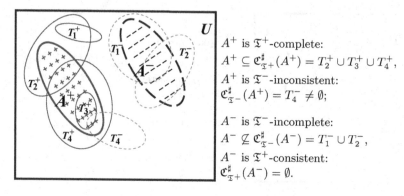

A^+ is $\mathfrak{T}^+$-complete:
$$A^+ \subseteq \mathfrak{C}^\sharp_{\mathfrak{T}+}(A^+) = T_2^+ \cup T_3^+ \cup T_4^+,$$
A^+ is $\mathfrak{T}^-$-inconsistent:
$$\mathfrak{C}^\sharp_{\mathfrak{T}-}(A^+) = T_4^- \neq \emptyset;$$

A^- is $\mathfrak{T}^-$-incomplete:
$$A^- \not\subseteq \mathfrak{C}^\sharp_{\mathfrak{T}-}(A^-) = T_1^- \cup T_2^-,$$
A^- is $\mathfrak{T}^+$-consistent:
$$\mathfrak{C}^\sharp_{\mathfrak{T}+}(A^-) = \emptyset.$$

Fig. 31. Case (7) It is an impossible case because $\bigcup \mathfrak{T}^+ \cap \bigcup \mathfrak{T}^- \neq \emptyset$

Case (8) A^+ is $\mathfrak{T}^+$-complete, $\mathfrak{T}^-$-inconsistent;
 A^- is $\mathfrak{T}^-$-incomplete, $\mathfrak{T}^+$-inconsistent.
 It is an impossible case (see, *e.g.*, Fig. 32).

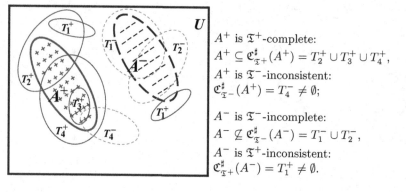

A^+ is $\mathfrak{T}^+$-complete:
$$A^+ \subseteq \mathfrak{C}^\sharp_{\mathfrak{T}+}(A^+) = T_2^+ \cup T_3^+ \cup T_4^+,$$
A^+ is $\mathfrak{T}^-$-inconsistent:
$$\mathfrak{C}^\sharp_{\mathfrak{T}-}(A^+) = T_4^- \neq \emptyset;$$

A^- is $\mathfrak{T}^-$-incomplete:
$$A^- \not\subseteq \mathfrak{C}^\sharp_{\mathfrak{T}-}(A^-) = T_1^- \cup T_2^-,$$
A^- is $\mathfrak{T}^+$-inconsistent:
$$\mathfrak{C}^\sharp_{\mathfrak{T}+}(A^-) = T_1^+ \neq \emptyset.$$

Fig. 32. Case (8) It is an impossible case because $\bigcup \mathfrak{T}^+ \cap \bigcup \mathfrak{T}^- \neq \emptyset$

Case (9) A^+ is $\mathfrak{T}^+$-incomplete, $\mathfrak{T}^-$-consistent;
 A^- is $\mathfrak{T}^-$-complete, $\mathfrak{T}^+$-consistent.
 It is a possible case (see, *e.g.*, Fig. 33).

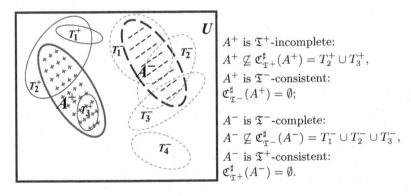

A^+ is $\mathfrak{T}^+$-incomplete:

$A^+ \not\subseteq \mathfrak{C}^\sharp_{\mathfrak{T}+}(A^+) = T_2^+ \cup T_3^+$,

A^+ is $\mathfrak{T}^-$-consistent:

$\mathfrak{C}^\sharp_{\mathfrak{T}-}(A^+) = \emptyset$;

A^- is $\mathfrak{T}^-$-complete:

$A^- \not\subseteq \mathfrak{C}^\sharp_{\mathfrak{T}-}(A^-) = T_1^- \cup T_2^- \cup T_3^-$,

A^- is $\mathfrak{T}^+$-consistent:

$\mathfrak{C}^\sharp_{\mathfrak{T}+}(A^-) = \emptyset$.

Fig. 33. Case (9) It is a possible case because $\bigcup \mathfrak{T}^+ \cap \bigcup \mathfrak{T}^- = \emptyset$

Case (10) A^+ is $\mathfrak{T}^+$-incomplete, $\mathfrak{T}^-$-consistent;
$\quad\quad\quad A^-$ is $\mathfrak{T}^-$-complete, $\mathfrak{T}^+$-inconsistent.
$\quad\quad\quad$ It is an impossible case (see, *e.g.*, Fig. 34).

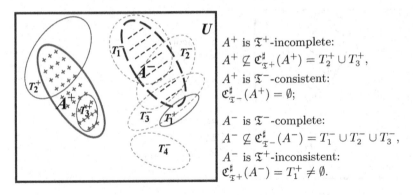

A^+ is $\mathfrak{T}^+$-incomplete:

$A^+ \not\subseteq \mathfrak{C}^\sharp_{\mathfrak{T}+}(A^+) = T_2^+ \cup T_3^+$,

A^+ is $\mathfrak{T}^-$-consistent:

$\mathfrak{C}^\sharp_{\mathfrak{T}-}(A^+) = \emptyset$;

A^- is $\mathfrak{T}^-$-complete:

$A^- \not\subseteq \mathfrak{C}^\sharp_{\mathfrak{T}-}(A^-) = T_1^- \cup T_2^- \cup T_3^-$,

A^- is $\mathfrak{T}^+$-inconsistent:

$\mathfrak{C}^\sharp_{\mathfrak{T}+}(A^-) = T_1^+ \neq \emptyset$.

Fig. 34. Case (10) It is an impossible case because $\bigcup \mathfrak{T}^+ \cap \bigcup \mathfrak{T}^- \neq \emptyset$

Case (11) A^+ is $\mathfrak{T}^+$-incomplete, $\mathfrak{T}^-$-consistent;
$\quad\quad\quad A^-$ is $\mathfrak{T}^-$-incomplete, $\mathfrak{T}^+$-consistent.
$\quad\quad\quad$ It is a possible case (see, *e.g.*, Fig. 35).

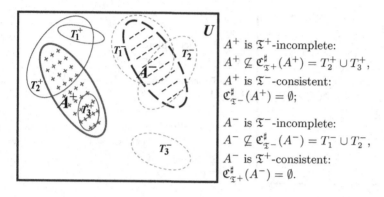

A^+ is $\mathfrak{T}^+$-incomplete:
$A^+ \not\subseteq \mathfrak{C}^\sharp_{\mathfrak{T}+}(A^+) = T_2^+ \cup T_3^+$,
A^+ is $\mathfrak{T}^-$-consistent:
$\mathfrak{C}^\sharp_{\mathfrak{T}-}(A^+) = \emptyset$;

A^- is $\mathfrak{T}^-$-incomplete:
$A^- \not\subseteq \mathfrak{C}^\sharp_{\mathfrak{T}-}(A^-) = T_1^- \cup T_2^-$,
A^- is $\mathfrak{T}^+$-consistent:
$\mathfrak{C}^\sharp_{\mathfrak{T}+}(A^-) = \emptyset$.

Fig. 35. Case (11) It is a possible case because $\bigcup \mathfrak{T}^+ \cap \bigcup \mathfrak{T}^- = \emptyset$

Case (12) A^+ is $\mathfrak{T}^+$-incomplete, $\mathfrak{T}^-$-consistent;
A^- is $\mathfrak{T}^-$-incomplete, $\mathfrak{T}^+$-inconsistent.
It is a possible case (see, *e.g.*, Fig. 36).

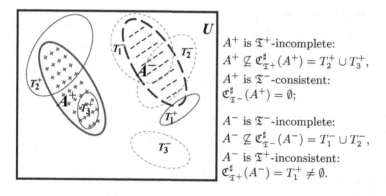

A^+ is $\mathfrak{T}^+$-incomplete:
$A^+ \not\subseteq \mathfrak{C}^\sharp_{\mathfrak{T}+}(A^+) = T_2^+ \cup T_3^+$,
A^+ is $\mathfrak{T}^-$-consistent:
$\mathfrak{C}^\sharp_{\mathfrak{T}-}(A^+) = \emptyset$;

A^- is $\mathfrak{T}^-$-incomplete:
$A^- \not\subseteq \mathfrak{C}^\sharp_{\mathfrak{T}-}(A^-) = T_1^- \cup T_2^-$,
A^- is $\mathfrak{T}^+$-inconsistent:
$\mathfrak{C}^\sharp_{\mathfrak{T}+}(A^-) = T_1^+ \neq \emptyset$.

Fig. 36. Case (12) It is a possible case because $\bigcup \mathfrak{T}^+ \cap \bigcup \mathfrak{T}^- = \emptyset$

Case (13) A^+ is $\mathfrak{T}^+$-incomplete, $\mathfrak{T}^-$-inconsistent;
A^- is $\mathfrak{T}^-$-complete, $\mathfrak{T}^+$-consistent.
It is a possible case (see, *e.g.*, Fig. 37).

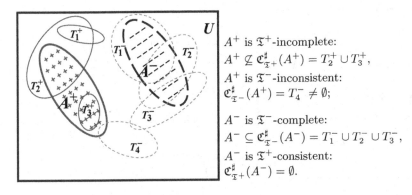

A^+ is $\mathfrak{T}^+$-incomplete:
$$A^+ \nsubseteq \mathfrak{C}^\sharp_{\mathfrak{T}+}(A^+) = T_2^+ \cup T_3^+,$$
A^+ is $\mathfrak{T}^-$-inconsistent:
$$\mathfrak{C}^\sharp_{\mathfrak{T}-}(A^+) = T_4^- \neq \emptyset;$$

A^- is $\mathfrak{T}^-$-complete:
$$A^- \subseteq \mathfrak{C}^\sharp_{\mathfrak{T}-}(A^-) = T_1^- \cup T_2^- \cup T_3^-,$$
A^- is $\mathfrak{T}^+$-consistent:
$$\mathfrak{C}^\sharp_{\mathfrak{T}+}(A^-) = \emptyset.$$

Fig. 37. Case (13) It is a possible case because $\bigcup \mathfrak{T}^+ \cap \bigcup \mathfrak{T}^- = \emptyset$

Case (14) A^+ is $\mathfrak{T}^+$-incomplete, $\mathfrak{T}^-$-inconsistent;
$\quad\quad\quad A^-$ is $\mathfrak{T}^-$-complete, $\mathfrak{T}^+$-inconsistent.
$\quad\quad\quad$ It is an impossible case (see, *e.g.*, Fig. 38).

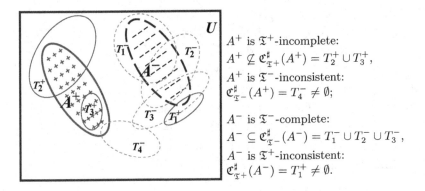

A^+ is $\mathfrak{T}^+$-incomplete:
$$A^+ \nsubseteq \mathfrak{C}^\sharp_{\mathfrak{T}+}(A^+) = T_2^+ \cup T_3^+,$$
A^+ is $\mathfrak{T}^-$-inconsistent:
$$\mathfrak{C}^\sharp_{\mathfrak{T}-}(A^+) = T_4^- \neq \emptyset;$$

A^- is $\mathfrak{T}^-$-complete:
$$A^- \subseteq \mathfrak{C}^\sharp_{\mathfrak{T}-}(A^-) = T_1^- \cup T_2^- \cup T_3^-,$$
A^- is $\mathfrak{T}^+$-inconsistent:
$$\mathfrak{C}^\sharp_{\mathfrak{T}+}(A^-) = T_1^+ \neq \emptyset.$$

Fig. 38. Case (14) It is an impossible case because $\bigcup \mathfrak{T}^+ \cap \bigcup \mathfrak{T}^- \neq \emptyset$

Case (15) A^+ is $\mathfrak{T}^+$-incomplete, $\mathfrak{T}^-$-inconsistent;
$\quad\quad\quad A^-$ is $\mathfrak{T}^-$-incomplete, $\mathfrak{T}^+$-consistent.
$\quad\quad\quad$ It is a possible case (see, *e.g.*, Fig. 39).

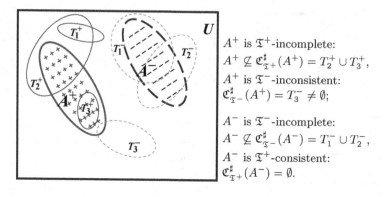

A^+ is $\mathfrak{T}^+$-incomplete:
$A^+ \not\subseteq \mathfrak{C}^\sharp_{\mathfrak{T}+}(A^+) = T_2^+ \cup T_3^+$,
A^+ is $\mathfrak{T}^-$-inconsistent:
$\mathfrak{C}^\sharp_{\mathfrak{T}-}(A^+) = T_3^- \neq \emptyset$;

A^- is $\mathfrak{T}^-$-incomplete:
$A^- \not\subseteq \mathfrak{C}^\sharp_{\mathfrak{T}-}(A^-) = T_1^- \cup T_2^-$,
A^- is $\mathfrak{T}^+$-consistent:
$\mathfrak{C}^\sharp_{\mathfrak{T}+}(A^-) = \emptyset$.

Fig. 39. Case (15) It is a possible case because $\bigcup \mathfrak{T}^+ \cap \bigcup \mathfrak{T}^- \neq \emptyset$

Case (16) A^+ is $\mathfrak{T}^+$-incomplete, $\mathfrak{T}^-$-inconsistent;
A^- is $\mathfrak{T}^-$-incomplete, $\mathfrak{T}^+$-inconsistent.
It is a possible case (see, *e.g.*, Fig. 40).

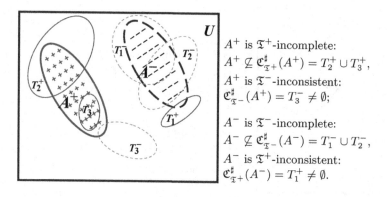

A^+ is $\mathfrak{T}^+$-incomplete:
$A^+ \not\subseteq \mathfrak{C}^\sharp_{\mathfrak{T}+}(A^+) = T_2^+ \cup T_3^+$,
A^+ is $\mathfrak{T}^-$-inconsistent:
$\mathfrak{C}^\sharp_{\mathfrak{T}-}(A^+) = T_3^- \neq \emptyset$;

A^- is $\mathfrak{T}^-$-incomplete:
$A^- \not\subseteq \mathfrak{C}^\sharp_{\mathfrak{T}-}(A^-) = T_1^- \cup T_2^-$,
A^- is $\mathfrak{T}^+$-inconsistent:
$\mathfrak{C}^\sharp_{\mathfrak{T}+}(A^-) = T_1^+ \neq \emptyset$.

Fig. 40. Case (16) It is a possible case because $\bigcup \mathfrak{T}^+ \cap \bigcup \mathfrak{T}^- \neq \emptyset$

In order to build up the general tool-based approximation framework, at the beginning, let us assume that a positive reference set A^+ and a negative reference set A^- are at our disposal together with the suitable positive tools $\mathfrak{T}^+$ and negative tools $\mathfrak{T}^-$. Initially, we only presuppose that $A^+ \cap A^- = \emptyset$ and $\bigcup \mathfrak{T}^+ \cap \bigcup \mathfrak{T}^- = \emptyset$. The framework can be built up and used in the following three consecutive steps.

1. **Mutual justifying the reference sets and tools**
   Step 1 is intended to reveal consistencies/inconsistencies between positive (resp., negative) reference sets and negative (resp., positive) tools and the completeness/incompleteness of reference sets in terms of $\mathfrak{T}^+$ and $\mathfrak{T}^-$-approximation spaces.

2. Rebuilding positive and negative tools

Step 2 is intended to resolve inconsistencies completely and eliminate incompleteness as far as possible.

The case of consistency. There is nothing to be done.

The case of inconsistency. We have to *decide* within the context of the observed field

- if A^+ or the concerned negative tools is reasonable or not, and/or
- whether A^- or the concerned positive tools is reasonable or not.

The case of completeness. We remove the covered positive and/or negative reference sets from the framework.

The case of incompleteness. We may *decide* within the context of the observed field either to remove the uncovered subset from A^+ (resp., A^-) completely or in part, or to augment the positive (resp., negative) tools with new subsets of which elements are patterned upon one or more elements of the uncovered subset of A^+ (resp., A^-).

These new subsets may contain any elements of the universe, provided that they can easily be determined.

For the new tools $\mathfrak{T}^+$ and/or $\mathfrak{T}^-$, $\bigcup \mathfrak{T}^+ \cap \bigcup \mathfrak{T}^- = \emptyset$ should also be fulfilled.

In this step, all decisions should be made by domain experts on professional criteria within the context of the observed field.

We obtain new rebuilt tools, denoted by $\mathfrak{T}_r^+$ and $\mathfrak{T}_r^-$, by the end of the steps 1 and 2.

3. Applying rebuilt tools

Step 3 is intended to justify any subset of the universe in terms of partial approximation of sets based on rebuilt positive and negative tools as usual.

8 Conclusion and Future Work

In this article, first, we have proposed a general set theoretic approximation framework. We have pointed out that the classic Pawlakian rough set theory can be treated in a natural way within it. Next, we have presented a special approximation scheme based on the partial covering of the universe which can also be treated naturally within the same framework. We have cleared up which conditions have to be satisfied by the novel upper and lower approximations so that they form a regular or partial Galois connection.

Further researches are planned in two directions. On the one hand, the granular property of definable sets is a strict requirement. It seems to be worth to investigate what would happen if we relaxed this requirement in diverse ways. Another research direction is to apply rough set and/or partial approximation idea to membrane computing. However, membrane computing works with multisets, and so first of all, an adequate approximation framework for multisets has to be worked out.

Acknowledgements. First of all, I want to thank my supervisors, Dr. Tamás Mihálydeák CSc for the stimulating discussions presented in my thesis, and Prof. Attila Pethő for his support when writing my thesis.

The author is grateful for the invitation from Prof. Andrzej Skowron to submit the article based on my thesis to the Transactions on Rough Sets, and also for the inspiration of Prof. James Peters. The author would like to thank the anonymous reviewers for their helpful comments and suggestions.

The author is also grateful to Bea Deák for improving the linguistic quality of the article.

References

1. Csajbók, Z.: Approximation of Sets Based on Partial Covering. PhD thesis, University of Debrecen, Debrecen (2011); Supervisors: Dr. Tamás Mihálydeák CSc and Prof. Attila Pethő
2. Pawlak, Z.: Information systems theoretical foundations. Information Systems 6(3), 205–218 (1981)
3. Pawlak, Z.: Rough sets. International Journal of Computer and Information Sciences 11(5), 341–356 (1982)
4. Pawlak, Z.: Rough Sets: Theoretical Aspects of Reasoning about Data. Kluwer Academic Publishers, Dordrecht (1991)
5. Pawlak, Z., Skowron, A.: Rough sets: Some extensions. Information Sciences 177, 28–40 (2007)
6. Skowron, A.: Vague concepts: A rough-set approach. In: De Baets, B., De Caluwe, R., De Tré, G., Fodor, J., Kacprzyk, J., Zadrożny, S. (eds.) Proceedings of EUROFUSE 2004, pp. 480–493. Akademicka Oficyna Wydawnicza EXIT, Warszawa (2004)
7. Skowron, A.: On topology in information system. Bulletin of the Polish Academy of Sciences, Mathematics 36, 477–479 (1988)
8. Xu, F., Yao, Y.Y., Miao, D.: Rough Set Approximations in Formal Concept Analysis and Knowledge Spaces. In: An, A., Matwin, S., Raś, Z.W., Ślęzak, D. (eds.) Foundations of Intelligent Systems. LNCS (LNAI), vol. 4994, pp. 319–328. Springer, Heidelberg (2008)
9. Yao, Y., Chen, Y.: Rough Set Approximations in Formal Concept Analysis. In: Peters, J.F., Skowron, A. (eds.) Transactions on Rough Sets V. LNCS, vol. 4100, pp. 285–305. Springer, Heidelberg (2006)
10. Yao, Y.Y.: On Generalizing Rough Set Theory. In: Wang, G., Liu, Q., Yao, Y., Skowron, A. (eds.) RSFDGrC 2003. LNCS (LNAI), vol. 2639, pp. 44–51. Springer, Heidelberg (2003)
11. Rasiowa, H., Skowron, A.: Rough Concept Logic. In: Skowron, A. (ed.) SCT 1984. LNCS, vol. 208, pp. 288–297. Springer, Heidelberg (1985)
12. Rasiowa, H., Skowron, A.: Approximation logic. In: Bibel, W., Jantke, K.P. (eds.) Mathematical Methods of Specification and Synthesis of Software Systems. Mathematical Research, vol. 31, pp. 123–139. Akademie Verlag, Berlin (1985)
13. Yao, Y.Y., Lin, T.Y.: Generalization of rough sets using modal logics. Intelligent Automation and Soft Computing, An International Journal 2, 103–120 (1996)
14. Orlowska, E.S.: Logic of vague concepts. Bulletin of the Section of Logic 11(3-4), 115–126 (1982)

15. Orlowska, E.: Algebraic Aspects of the Relational Knowledge Representation: Modal Relation Algebras. In: Pearce, D., Wansing, H. (eds.) All-Berlin 1990. LNCS, vol. 619, pp. 1–22. Springer, Heidelberg (1992)
16. Balbiani, P., Iliev, P., Vakarelov, D.: A modal logic for Pawlak's approximation spaces with rough cardinality n. Fundam. Inform. 83(4), 451–464 (2008)
17. Vakarelov, D.: A Modal Characterization of Indiscernibility and Similarity Relations in Pawlak's Information Systems. In: Ślęzak, D., Wang, G., Szczuka, M.S., Düntsch, I., Yao, Y. (eds.) RSFDGrC 2005. LNCS (LNAI), vol. 3641, pp. 12–22. Springer, Heidelberg (2005)
18. Nenov, Y., Vakarelov, D.: Modal logics for mereotopological relations. In: Areces, C., Goldblatt, R. (eds.) Advances in Modal Logic, College Publications, pp. 249–272 (2008)
19. Järvinen, J., Kondo, M., Kortelainen, J.: Modal-like operators in boolean lattices, Galois connections and fixed points. Fundamenta Informaticae 76(1-2), 129–145 (2007)
20. Zhu, W., Wang, F.Y.: On three types of covering-based rough sets. IEEE Trans. Knowl. Data Eng. 19(8), 1131–1144 (2007)
21. Lin, T.Y.: Update and illustration on granular computing: Practices, theory and future directions. In: Hu, X., Lin, T.Y., Raghavan, V.V., Grzymala-Busse, J.W., Liu, Q., Broder, A.Z. (eds.) 2010 IEEE International Conference on Granular Computing, GrC 2010, San Jose, California, USA, August 14-16, pp. 32–33. IEEE Computer Society (2010)
22. Skowron, A., Swiniarski, R., Synak, P.: Approximation Spaces and Information Granulation. In: Peters, J.F., Skowron, A. (eds.) Transactions on Rough Sets III. LNCS, vol. 3400, pp. 175–189. Springer, Heidelberg (2005)
23. Skowron, A., Stepaniuk, J.: Information granules: Towards foundations of granular computing. International Journal of Intelligent Systems 16(1), 57–85 (2001)
24. Yao, Y.Y.: Information granulation and rough set approximation. International Journal of Intelligent Systems 16(1), 87–104 (2001)
25. Zadeh, L.A.: Granular Computing and Rough Set Theory. In: Kryszkiewicz, M., Peters, J.F., Rybiński, H., Skowron, A. (eds.) RSEISP 2007. LNCS (LNAI), vol. 4585, pp. 1–4. Springer, Heidelberg (2007)
26. Polkowski, L., Skowron, A., Żytkow, J.: Rough foundations for rough sets. In: Lin, T.Y., Wildberger, A.M. (eds.) Soft Computing: Rough Sets, Fuzzy Logic, Neural Networks, Uncertainty Management, Knowledge Discovery, pp. 55–58. Simulation Councils, Inc., San Diego (1995)
27. Skowron, A., Stepaniuk, J.: Tolerance approximation spaces. Fundamenta Informaticae 27(2-3), 245–253 (1996)
28. Słowiński, R., Vanderpooten, D.: Similarity relation as a basis for rough approximations. In: Wang, P. (ed.) Advances in Machine Intelligence and Soft-Computing, vol. IV, pp. 17–33. Duke University Press, Durham (1997)
29. Nieminen, J.: Rough tolerance equality. Fundamenta Informaticae 11, 289–294 (1988)
30. Marcus, S.: Tolerance rough sets, Čech topology, learning processes. Bulletin of the Polish Academy of Sciences, Technical Sciences 42(3), 471–478 (1994)
31. Järvinen, J.: Lattice Theory for Rough Sets. In: Peters, J.F., Skowron, A., Düntsch, I., Grzymała-Busse, J.W., Orłowska, E., Polkowski, L. (eds.) Transactions on Rough Sets VI. LNCS, vol. 4374, pp. 400–498. Springer, Heidelberg (2007)
32. Zhu, W.: Relationship between generalized rough sets based on binary relation and covering. Information Sciences 179(3), 210–225 (2009)

33. Yao, Y.Y.: On Generalizing Pawlak Approximation Operators. In: Polkowski, L., Skowron, A. (eds.) RSCTC 1998. LNCS (LNAI), vol. 1424, pp. 298–307. Springer, Heidelberg (1998)

34. Skowron, A., Stepaniuk, J.: Generalized approximation spaces. In: Lin, T., Wildberger, A. (eds.) The Third International Workshop on Rough Sets and Soft Computing Proceedings (RSSC 1994), November 10-12. San Jose State University, San Jose (1994)

35. Skowron, A., Stepaniuk, J.: Generalized approximation spaces. ICS Research Report 41/94, Warsaw University of Technology (1994)

36. Skowron, A., Stepaniuk, J.: Generalized approximation spaces. In: Lin, T., Wildberger, A. (eds.) Soft Computing, pp. 18–21. Simulation Councils, Inc., San Diego (1995)

37. Pawlak, Z., Polkowski, L., Skowron, A.: Rough sets: An approach to vagueness. In: Rivero, L.C., Doorn, J., Ferraggine, V. (eds.) Encyclopedia of Database Technologies and Applications, pp. 575–580. Idea Group Inc., Hershey (2005)

38. Revett, K., Gorunescu, F., Salem, A.B.M.: Feature selection in parkinson's disease: A rough sets approach. In: [119], pp. 425–428

39. Salem, A.B.M., Revett, K., El-Dahshan, E.S.A.: Machine learning in electrocardiogram diagnosis. In: [119], pp. 429–433

40. Zhu, P.: Covering rough sets based on neighborhoods: An approach without using neighborhoods. International Journal of Approximate Reasoning 52(3), 461–472 (2011)

41. Pawlak, Z., Skowron, A.: Rudiments of rough sets. Information Sciences 177(1), 3–27 (2007)

42. Pawlak, Z., Skowron, A.: Rough sets and boolean reasoning. Information Sciences 177(1), 41–73 (2007)

43. Chikalov, I., Lozin, V., Lozina, I., Moshkov, M., Nguyen, H.S., Skowron, A., Zielosko, B.: Three Approaches to Data Analysis. ISRL, vol. 41. Springer, Heidelberg (2013)

44. Skowron, A., Suraj, Z. (eds.): Rough sets and intelligent systems – Professor Zdzisław Pawlak in memoriam. Volume 1. ISRL, vol. 42. Springer, Heidelberg (2013)

45. Skowron, A., Suraj, Z. (eds.): Rough Sets and Intelligent Systems - Professor Zdzisław Pawlak in Memoriam. ISRL, vol. 43. Springer, Heidelberg (2013)

46. Skowron, A., Chakraborty, M.K., Grzymała-Busse, J., Marek, V., Pal, S.K., Peters, J., Rozenberg, G.: Ślęzak, D., Słowiński, R., Tsumoto, S., Wakulicz-Deja, A., Wang, G., Ziarko, W.: Professor Zdzisław Pawlak (1926-2006); Founder of the polish school of artificial intelligence. In: [44], 1–56

47. Skowron, A.: List of works by professor Zdzisław Pawlak (1926-2006). In: [44], 57–74

48. Nguyen, H.S., Skowron, A.: From rudiments to challenges. In: [44] 75–173

49. Keefe, R.: Theories of Vagueness. Cambridge Studies in Philosophy. Cambridge University Press, Cambridge (2000)

50. Pawlak, Z.: Some Issues on Rough Sets. In: Peters, J.F., Skowron, A., Grzymała-Busse, J.W., Kostek, B., Swiniarski, R.W., Szczuka, M.S. (eds.) Transactions on Rough Sets I. LNCS, vol. 3100, pp. 1–58. Springer, Heidelberg (2004)

51. Russell, B.: Vagueness. Australasian Journal of Philosophy and Psychology 1, 84–92 (1923)

52. Sorensen, R.: Vagueness. In: Zalta, E. N. (ed.): The Stanford Encyclopedia of Philosophy (Fall 2008 Edition), http://plato.stanford.edu/archives/fall2008/entries/vagueness/ (updated on August 29, 2006), (last accesed on November 6, 2012)

Author Index

110. Wang, G., Long, C., Yu, W.: Rough set based solutions for network security. In: Dunin-Kęplicz, B., Jankowski, A., Skowron, A., Szczuka, M. (eds.) Monitoring, Security, and Rescue Techniques in Multiagent Systems. AISC, vol. 28, pp. 455–465. Springer, Berlin (2005)

111. Wang, W., Guan, X., Zhang, X., Yang, L.: Profiling program behavior for anomaly intrusion detection based on the transition and frequency property of computer audit data. Computers & Security 25(7), 539–550 (2006)

112. Denning, D.E.: An intrusion-detection model. IEEE Transactions on Software Engineering SE-13(2), 222–232 (1987)

113. Bauer, L., Ligatti, J., Walker, D.: More enforceable security policies. In: Foundations of Computer Security, Copenhagen, Denmark (July 2002)

114. Ciucci, D.: Orthopairs: A simple and widely usedway to model uncertainty. Fundamenta Informaticae 108(3-4), 287–304 (2011)

115. Iwinski, T.: Algebras for rough sets. Bulletin of the Polish Academy of Sciences, Series: Mathematics 35, 673–683 (1987)

116. Yao, Y., Li, X.: Comparison of rough-set and interval-set models for uncertain reasoning. Fundamenta Informaticae 27(2, 3), 289–298 (1996)

117. Dubois, D., Prade, H.: An introduction to bipolar representations of information and preference. International Journal of Intelligent Systems 23(8), 866–877 (2008)

118. Lavrač, N., Dzeroski, S.: Inductive Logic Programming: Techniques and Applications. Ellis Horwood, New York (1994)

119. Proceedings of the International Multiconference on Computer Science and Information Technology. In: IMCSIT 2009, Mrągowo, Poland, October 12-14. Polskie Towarzystwo Informatyczne - IEEE Computer Society Press (2009)

92. Järvinen, J.: Properties of rough approximations. Journal of Advanced Computational Intelligence and Intelligent Informatics 9(5), 502–505 (2005)
93. Järvinen, J.: Pawlak's information systems in terms of Galois connections and functional dependencies. Fundamenta Informaticae 75(1-4), 315–330 (2007)
94. Kari, L., Rozenberg, G.: The many facets of natural computing. Communications of the ACM 51(10), 72–83 (2008)
95. Krasser, D., Horváth, F., Illyés, E., Molnár, Z., Bíró, M., Botta-Dukát, Z., Bölöni, J., Oláh, K.: MÉTA programme - Vegetation Heritage of Hungary, http://www.novenyzetiterkep.hu/?q=en/english/node/70 (updated on September 9, 2012), (last accessed on November 7, 2012)
96. Molnár, Z., Bartha, S., Seregélyes, T., Illyés, E., Botta-Dukát, Z., Tímár, G., Horváth, F., Révész, A., Kun, A., Bölöni, J., Bíró, M., Bodonczi, L., Deák, A.J., Fogarasi, P., Horváth, A., Isépy, I., Karas, L., Kecskés, F., Molnár, C., Ortmann-né Ajkai, A., Rév, S.: A grid-based, satellite-image supported multi-attributed vegetation mapping method (MÉTA). Folia Geobotanica 42, 225–247 (2007)
97. Bölöni, J., Molnár, Z., Illyés, E., Kun, A.: A new habitat classification and manual for standardized habitat mapping. Annali di Botanica Nuova Serie 7, 105–126 (2007)
98. Horváth, F., Molnár, Z., Bölöni, J., Pataki, Z., Polgár, L., Révész, A., Oláh, K., Krasser, D., Illyés, E.: Fact sheet of the MÉTA database 1.2. Acta Botanica Hungarica 50(suppl.), 11–34 (2008)
99. Takács, G., Molnár, Z. (eds.): National Biodiversity Monitoring System XI. Habitat mapping. MTA Ökológiai és Botanikai Kutatóintézete (Institute of Ecology and Botany of the Hungarian Academy of Sciences) (Vácrátót) and Környezetvédelmi és Vízügyi Minisztérium (Ministry of Environment and Water) (Budapest), Vácrátót (2009)
100. Fekete, G., Molnár, Z., Horváth, F. (eds.): A magyarországi élőhelyek leírása és határozókönyve. A Nemzeti Élőhely-osztályozási Rendszer (Guide and description of the Hungarian habitats. The National Habitat Classification System). Természettudományi Múzeum, Budapest (1997) (in Hungarian)
101. Bölöni, J., Kun, A., Molnár, Z.: Élőhely-ismereti Útmutató (Habitat guide). MTA Ökológiai és Botanikai Kutatóintézete (Institute of Ecology and Botany of the Hungarian Academy of Sciences), Vácrátót (2003) (in Hungarian)
102. Molnár, Z., Bíró, M., Bölöni, J.: Appendix - English names of the Á-NÉR habitat types. Acta Botanica Hungarica 50(supl.), 249–255 (2008)
103. Molnár, Z., Bölöni, J., Horváth, F.: Threatening factors encountered: Actual endangerment of the hungarian (semi-)natural habitats. Acta Botanica Hungarica 50(supl.), 199–217 (2008)
104. Anderson, R.J.: Security Engineering: A Guide to Building Dependable Distributed Systems, 2nd edn. Wiley (2008)
105. Senior Officials Group-Information Systems Security: Information Technology Security Evaluation Criteria (ITSEC). Department of Trade and Industry (1991)
106. Caelli, W., Longley, D., Shain, M.: Information security handbook. Stockton Press, New York (1991)
107. Goguen, J.A., Meseguer, J.: Security policies and security models. In: Proceedings of the IEEE Symposium on Research in Security and Privacy, pp. 11–20. IEEE Computer Society Press, Oakland (1982)
108. Bishop, M.: Computer Security: Art and Science. Addison Wesley (2002)
109. Beghdad, R.: Modelling and solving the intrusion detection problem in computer networks. Computers & Security 23(8), 687–696 (2004)

72. Pomykała, J.A.: Some remarks on approximation. Demonstratio Mathematica 24(1-2), 95–104 (1991)
73. Zakowski, W.: Approximations in the space (U, II). Demonstratio Mathematica 16(3), 761–769 (1983)
74. Ciucci, D.: Approximation algebra and framework. Fundamenta Informaticae 94, 147–161 (2009)
75. Davey, B.A., Priestley, H.A.: Introduction to Lattices and Order, 2nd edn. Cambridge University Press, Cambridge (2002)
76. Denecke, K., Erné, M., Wismath, S. (eds.): Galois Connections and Applications. Kluwer Academic Publishers, Dordrecht (2004)
77. Csajbók, Z.: Approximation of sets based on partial covering. Theoretical Computer Science 412(42), 5820–5833 (2011); Rough Sets and Fuzzy Sets in Natural Computing
78. Csajbók, Z.: Simultaneous anomaly and misuse intrusion detections based on partial approximative set theory. In: Cotronis, Y., Danelutto, M., Papadopoulos, G.A. (eds.) Proceedings of PDP 2011, February 9-11, pp. 651–655. IEEE Computer Society Press, Los Alamitos (2011)
79. Csajbók, Z., Mihálydeák, T.: General Tool-Based Approximation Framework Based on Partial Approximation of Sets. In: Kuznetsov, S.O., Ślęzak, D., Hepting, D.H., Mirkin, B.G. (eds.) RSFDGrC 2011. LNCS, vol. 6743, pp. 52–59. Springer, Heidelberg (2011)
80. Birkhoff, G.: Lattice theory, 3rd edn. vol. 25. Colloquium Publications. American Mathematical Society, Providence, Rhode Island (1967)
81. Gierz, G., Hofmann, K., Keimel, K., Lawson, J., Mislove, M., Scott, D.: Continuous Lattices and Domains. Encyclopedia of Mathematics and its Applications, vol. 93. Cambridge University Press (2003)
82. Grätzer, G.: General Lattice Theory. Birkhäuser Verlag, Basel und Stuttgart (1978)
83. Miné, A.: Weakly Relational Numerical Abstract Domains. PhD thesis, École Polytechnique, Palaiseau, France (December 2004)
84. Pagliani, P., Chakraborty, M.: A Geometry of Approximation: Rough Set Theory Logic, Algebra and Topology of Conceptual Patterns (Trends in Logic). Springer Publishing Company, Incorporated (2008)
85. Järvinen, J.: Lattice Theory for Rough Sets. In: Peters, J.F., Skowron, A., Düntsch, I., Grzymała-Busse, J.W., Orłowska, E., Polkowski, L. (eds.) Transactions on Rough Sets VI. LNCS, vol. 4374, pp. 400–498. Springer, Heidelberg (2007)
86. Csajbók, Z.: On the partial approximation of sets. Acta Medicinae et Sociologica 2(2), 143–152 (2011)
87. Salomaa, A.: Computation and automata. In: Encyclopedia of Mathematics and its Applications, vol. 25. Cambridge University Press, New York (1985)
88. Stadler, P.F., Stadler, B.M.R.: Genotype phenotype maps. Biological Theory 3, 268–279 (2006)
89. Zhu, W., Wang, F.Y.: Reduction and axiomatization of covering generalized rough sets. Information Sciences 152(1), 217–230 (2003)
90. Yang, T., Li, Q.: Reduction about approximation spaces of covering generalized rough sets. International Journal of Approximate Reasoning 51(3), 335–345 (2010)
91. Mihálydeák, T.: On tarskian models of general type-theoretical languages. In: Drossos, C., Peppas, P., Tsinakis, C. (eds.) Proceedings of the 7th Panhellenic Logic Symposium, pp. 127–131. Patras University Press, Patras (2009)

53. Keefe, R., Smith, P.: Introduction: Theories of vagueness. In: [44] 1–57
54. Priest, G.: A site for sorites. In: Beall, J.C. (ed.) Liars and Heaps: New Essays on Paradox, pp. 9–23. Oxford University Press Inc., New York (2003)
55. Varzi, A.C.: Cut-offs and their neighbors. In: Beall, J.C. (ed.) Liars and Heaps: New Essays on Paradox, pp. 24–38. Oxford University Press Inc., New York (2003)
56. Keefe, R., Smith, P. (eds.): Vagueness: A Reader. MIT Press, Cambridge (1996)
57. American Association for Clinical Chemistry: Lab tests online. Glucose, http://labtestsonline.org/understanding/analytes/glucose/tab/glance (updated on March 23, 2012), (last accessed on November 7, 2012)
58. Frege, G.: Grundgesetzen der Arithmetik, begriffsschriftlich abgeleitet, vol. 2. Verlag von Hermann Pohle, Jena (1903)
59. Peirce, C.S.: Vague. In: Baldwin, J.M. (ed.) Dictionary of Philosophy and Psychology, vol. 748. MacMillan, New York (1902)
60. Pawlak, Z.: Vagueness - A Rough Set View. In: Mycielski, J., Rozenberg, G., Salomaa, A. (eds.) Structures in Logic and Computer Science. LNCS, vol. 1261, pp. 106–117. Springer, Heidelberg (1997)
61. Skowron, A.: Rough sets and vague concepts. Fundamenta Informaticae 64(1–4), 417–431 (2005)
62. Zadeh, L.A.: Fuzzy sets. Information and Control 8(3), 338–353 (1965)
63. Dubois, D., Esteva, F., Godo, L., Prade, H.: An information-based discussion of vagueness. In: Lefebvre, C., Cohen, H. (eds.) Handbook of Categorization in Cognitive Science, pp. 892–913. Elsevier (2005); Part 8. Machine Category learning
64. Lin, T.Y.: Approximation Theories: Granular Computing vs Rough Sets. In: Chan, C.-C., Grzymala-Busse, J.W., Ziarko, W.P. (eds.) RSCTC 2008. LNCS (LNAI), vol. 5306, pp. 520–529. Springer, Heidelberg (2008)
65. Lin, T.Y.: Granular computing: Practices, theories, and future directions. In: Meyers, R.A. (ed.) Encyclopedia of Complexity and Systems Science, pp. 4339–4355. Springer, Heidelberg (2009)
66. Odifreddi, P.: Classical Recursion Theory. The Theory of Functions and Sets of Natural Numbers. Studies in Logic and the Foundations of Mathematics, vol. 125. Elsevier, Amsterdam (1989)
67. Yao, Y., Yao, B.: Covering based rough set approximations. Information Sciences 200, 91–107 (2012)
68. Düntsch, I., Gediga, G.: Approximation Operators in Qualitative Data Analysis. In: de Swart, H., Orłowska, E., Schmidt, G., Roubens, M. (eds.) TARSKI 2003. LNCS, vol. 2929, pp. 214–230. Springer, Heidelberg (2003)
69. Ciucci, D.: A Unifying Abstract Approach for Rough Models. In: Wang, G., Li, T., Grzymala-Busse, J.W., Miao, D., Skowron, A., Yao, Y. (eds.) RSKT 2008. LNCS (LNAI), vol. 5009, pp. 371–378. Springer, Heidelberg (2008)
70. Csajbók, Z.: Partial approximative set theory: A generalization of the rough set theory. In: Martin, T., Muda, A.K., Abraham, A., Prade, H., Laurent, A., Laurent, D., Sans, V. (eds.) Proceedings of SoCPaR 2010, December 7–10, pp. 51–56. IEEE, Cergy Pontoise (2010)
71. Csajbók, Z.: A security model for personal information security management based on partial approximative set theory. In: Ganzha, M., Paprzycki, M. (eds.) Proceedings of the International Multiconference on Computer Science and Information Technology (IMCSIT 2010), Wisła, Poland, October 18–20, pp. 839–845. Polskie Towarzystwo Informatyczne – IEEE Computer Society Press, Los Alamitos (2010)